P9-EEL-500

ABOUT ISLAND PRESS

Island Press is the only nonprofit organization in the United States whose principal purpose is the publication of books on environmental issues and natural resource management. We provide solutions-oriented information to professionals, public officials, business and community leaders, and concerned citizens who are shaping responses to environmental problems.

In 2006, Island Press celebrates its twenty-second anniversary as the leading provider of timely and practical books that take a multidisciplinary approach to critical environmental concerns. Our growing list of titles reflects our commitment to bringing the best of an expanding body of literature to the environmental community throughout North America and the world.

Support for Island Press is provided by the Agua Fund, The Geraldine R. Dodge Foundation, Doris Duke Charitable Foundation, The William and Flora Hewlett Foundation, Kendeda Sustainability Fund of the Tides Foundation, Forrest C. Lattner Foundation, The Henry Luce Foundation, The John D. and Catherine T. MacArthur Foundation, The Marisla Foundation, The Andrew W. Mellon Foundation, Gordon and Betty Moore Foundation, The Curtis and Edith Munson Foundation, Oak Foundation, The Overbrook Foundation, The David and Lucile Packard Foundation, The Winslow Foundation, and other generous donors.

The opinions expressed in this book are those of the author(s) and do not necessarily reflect the views of these foundations.

BRIDGING SCALES
A · N · D
KNOWLEDGE SYSTEMS

Concepts and Applications in Ecosystem Assessment

A contribution to the

MILLENNIUM ECOSYSTEM

ASSESSMENT

BRIDGING SCALES AND KNOWLEDGE SYSTEMS

Concepts and Applications in Ecosystem Assessment

EDITED BY

WALTER V. REID

FIKRET BERKES

THOMAS WILBANKS

DORIS CAPISTRANO

WASHINGTON • COVELO • LONDON

Island Press, 1718 Connecticut Ave., NW, Suite 300, Washington, D.C. 20009.

ISLAND PRESS is a trademark of The Center for Resource Economics.

Library of Congress Cataloging-in-Publication data.

Bridging scales and knowledge systems : concepts and applications in ecosystem assessment / Millennium Ecosystem Assessment ; edited by Walter V. Reid ... [et al.].
p. cm.
ISBN 1-59726-037-1 (cloth : alk. paper) — ISBN 1-59726-038-X (pbk. : alk. paper)
1. Ecosystem management. 2. Human ecology. I. Reid, Walter V., 1956– II. Millennium Ecosystem Assessment (Program)
QH75.B695 2006
333.95—dc22
2006010082

British Cataloguing-in-Publication data available.

Printed on recycled, acid-free paper

Design by Joan Wolbier

Manufactured in the United States of America

10 9 8 7 6 5 4 3 2 1

CONTENTS

PREFACE

The Millennium Ecosystem Assessment (MA) was carried out between 2001 and 2005 to assess the consequences of ecosystem change for human well-being and to establish the basis for actions needed to enhance the conservation and sustainable use of ecosystems and their contributions to human well-being. The MA was originally conceived as a global scientific assessment that would be modeled on two intergovernmental processes that have contributed significantly to policy development in relation to the problems of climate change and stratospheric ozone depletion: the Intergovernmental Panel on Climate Change and the Ozone Assessment.

The very first meeting of the group tasked with exploring whether the MA should be launched, however, set the design of the assessment on a very different course. While many aspects of the MA process did still draw heavily on the experience of other international assessments, that first meeting and subsequent design team meetings introduced three novel dimensions. First, the group concluded that the assessment could not be done at a single global scale and would need to examine processes of ecosystem change and human impacts at other scales, including in particular the scale of individual communities. Second, it was evident that the audience for the findings of an assessment of these issues was much broader than the traditional audience of global assessments (national governments) and must include other stakeholders from business, nongovernmental organizations, indigenous people, and other civil society groups. Finally, it was clear that the knowledge base for an assessment of this nature could not be limited to the scientific literature but must draw on other "informal" sources of knowledge, including local, traditional, and practitioner's knowledge.

The MA was the largest assessment effort ever to attempt to incorporate all of these dimensions in its design, and in that regard it can be seen as an experiment or pilot in applying multiple scales and knowledge systems in an assessment. But, in fact, a tremendous depth of research and experience exists in relation to each of these dimensions of scale, stakeholders, and knowledge systems. Recognizing that this existing experience could significantly aid the MA process, and also recognizing that the MA itself provided an experiment that could further advance understanding of issues of scale and epistemology, the MA Sub-Global Working Group organized an international conference on these issues called Bridging Scales and Epistemologies: Linking Local Knowledge and Global Science in Multi-scale Assessments. More than two hundred people from fifty countries participated in that conference, which was held in March 2004 and hosted by the Bibliotheca Alexandrina in Alexandria, Egypt.

This book—*Bridging Scales and Knowledge Systems: Concepts and Applications in Ecosystem Assessment*—is one product of that conference. While the MA provides the motivation for this book, and while several chapters present experiences from the MA, this book, like the conference, reaches far beyond the MA process to explore the challenges, costs, and benefits of bridging scales and knowledge systems in assessment processes and in resource management. The issues explored in this book push the limits of science, politics, and social processes. Although a number of general lessons emerge, many questions remain unanswered about how to make such processes work, how to address issues of power and empowerment, and how to address technical issues of information scaling and knowledge validation. In this respect, the volume does not attempt to provide a blueprint, but it does illustrate the multiple dimensions of the challenges inherent in bridging scales and knowledge systems.

ACKNOWLEDGMENTS

We would like to thank the MA Sub-Global Working Group, for its initiative in organizing the March 2004 conference that led to this book, and Ismail Serageldin and the Bibliotheca Alexandrina, for hosting the conference. One of this book's editors (Doris Capistrano) was one cochair of the MA Sub-Global Working Group, and we would like to recognize the central role that the other Working Group cochair, Cristián Samper, played in designing the conference. We thank the conference international advisory committee, composed of Janis Alcorn, Alejandro Argumedo, Fikret Berkes, Marie Byström, Esther Camac, Doris Capistrano, William Clark, Angela Cropper, Elaine Elisabetsky, Carl Folke, Madhav Gadgil, Sandy Gauntlett, C. S. Holling, Louis Lebel, Liu Jiyuan, Akin Mabogunje, Jane Mogina, Harold Mooney, M. Granger Morgan, Douglas Nakashima, Thomas Rosswall, and Cristián Samper. We also thank the Conference Organizing Committee, which consisted of Carolina Katz Reid, Walter V. Reid, Chan Wai Leng, John Ehrmann, Marcus Lee, Nicolas Lucas, Ciara Raudsepp-Hearne, and Sara Suriani. Special thanks are due to Carolina Katz Reid for her tireless work as the conference organizer. We also thank the MA Board and Assessment Panel listed elsewhere in this volume.

We thank the sponsors of the conference and this publication: the Swedish International Biodiversity Programme, The Christensen Fund, the International Council for Science, the Canadian International Development Agency, Bibliotheca Alexandrina, and the MA. The MA, in turn, received significant financial support from the Global Environment Facility, the United Nations Foundation, The David and Lucile Packard Foundation, the World Bank, the United Nations Environment Programme, the Government of Norway, the Kingdom of Saudi Arabia, and other donors listed on the MA Web site at http://www.MAweb.org.

Each of the contributed chapters in this volume underwent peer review. We thank the reviewers for their significant contribution to this volume: Neil Adger, Katrina Brown, David Cash, Donna Craig, Chimere Diaw, Polly Ericksen, Christo Fabricius, Cathy Fogel, Keith Forbes, Tim Forsyth, Peter Frost, Cole Genge, Clark C. Gibson, Madhav Karki, Don Kash, Ann Kinzig, Rene Kuppe, Murari Lal, Micheline Manseau, Peter H. May, Ronald Mitchell, P. K. Muraleedharan, Timothy O'Riordan, P. Ramakrishnan, Maureen Reed, Benjamin Samson, Marja Spierenburg, Angelica Toniolo, Ellen Woodley, and Fernanda Zermoglio.

Walter V. Reid
Fikret Berkes
Thomas J. Wilbanks
Doris Capistrano

CHAPTER 1

Introduction

Walter V. Reid, Fikret Berkes,
Thomas J. Wilbanks, and Doris Capistrano

Local communities, national governments, and international institutions all face difficult choices concerning goals, priorities, investments, policies, and institutions needed to effectively address interlinked challenges concerning development and the environment (Millennium Ecosystem Assessment 2005a). They must make these choices in the face of substantial uncertainty about current conditions and the potential future consequences of actions taken, or not taken, today. One way to improve those decisions is to ensure that the best knowledge concerning the problem and potential solutions is available to decision makers and the public. Better knowledge does not guarantee that better choices will be made, but it does provide a sound basis for making better decisions and for holding decision makers accountable.

But how can knowledge concerning environment and development be best mobilized in support of decision making? Over the past thirty to forty years, many different mechanisms have been developed to assemble, assess, and synthesize information for use in decision processes, including environmental impact assessments, technology assessments, scientific advisory boards, national environmental reports, global environmental (or development or economic) reports, and global environmental assessments. Both the processes and scientific methods used for these types of "knowledge assessments" have evolved considerably during this time. Modern global assessments, for example, commonly make use of such tools as scenarios and integrated assessment models used infrequently in earlier assessments. And while the "product" (that is, the assessment report) was all that mattered in earlier assessments, more recent assessments increasingly generate a range of products to better respond to specific needs of diverse stakeholders and are often as heavily focused on the process of stakeholder engagement as they are on the product itself.

This book explores two issues at the cutting edge of the further development and evolution of knowledge assessments: how to address issues of scale and how to embrace different knowledge systems in assessments. More specifically, in the case of scale, there are many reasons to think that both the findings of an assessment and the use of those findings could be enhanced if the assessment incorporates information from multiple spatial and temporal scales and if "cross-scale" effects are examined. But what are the real costs and benefits of such multiscale assessments and, from a pragmatic standpoint, just how can they be implemented? In the case of knowledge systems, assessments traditionally have relied almost exclusively on scientific information, yet considerable knowledge relevant to decisions concerning the environment and development can be found outside of formal scientific disciplines. This includes knowledge held within businesses, knowledge held by local resource managers, and traditional knowledge passed down from one generation to the next. But how can a *science assessment* be transformed into a *knowledge assessment*? Scientific disciplines have well-developed means of validating information through peer review that would rule out incorporating many other forms of knowledge. How can multiple types of knowledge be incorporated in an assessment when each type of knowledge has its own mechanisms for determining validity and utility?

Although these issues of scale and knowledge systems could be dealt with separately and although the literature on the two issues tends to be distinct, in this book we expressly seek to examine the intersection of these issues for both pragmatic and heuristic reasons. From a pragmatic standpoint, while scientific knowledge dominates the considerations of global and long-term processes (such as climate change), local, traditional, and practitioner's knowledge often dominates the considerations of site-specific resource management issues, where detailed scientific studies may not exist. Thus, in order to deal with "multiple scales," an assessment cannot help but confront the need to deal with multiple types of knowledge, reflecting not only different paradigms but also, in some cases, different processes and phenomena. From a heuristic standpoint, the intersection of the issues of scale, knowledge systems, and assessment provides a rich opportunity for obtaining insights into not just how best to assess knowledge for the purposes of decision making but also how to further our understanding of basic socioecological processes.

The Millennium Ecosystem Assessment

This book was catalyzed by the Millennium Ecosystem Assessment (MA), a multiscale assessment of the consequences of ecosystem change for human well-being that was carried out between 2001 and 2005 (MA 2003, MA 2005a). The MA was one of the first global assessments to attempt to incorporate multiple scales and multiple knowledge systems. Recognizing that the base of experience on which to develop these dimensions of the assessment was quite limited, the MA organized an international conference—Bridging Scales and Epistemologies: Linking Local Knowledge and Global Science in Multi-scale Assessments—at the Bibliotheca Alexandrina in Alexandria, Egypt, in March 2004. The conference provided an opportunity for assessment practitioners, academic researchers, indigenous peoples, and individuals directly involved in the MA process to discuss theory, learn from case studies and practical experiences, and debate the strengths and weaknesses of various approaches. The following chapters are drawn from papers presented at that conference. We briefly describe the MA here to provide context and to help introduce the themes of the book, but most of the chapters address the issues of scale and knowledge systems more broadly.

The Millennium Ecosystem Assessment was called for by United Nations (UN) secretary-general Kofi Annan in 2000 in his report to the UN General Assembly *We the Peoples: The Role of the United Nations in the 21st Century* (Annan 2000). Governments subsequently supported establishing the assessment through decisions taken by three international conventions, and the MA was initiated in 2001. The MA was conducted under the auspices of the United Nations, with the secretariat coordinated by the United Nations Environment Programme. It was governed by a multistakeholder board that included representatives of international institutions, governments, business, nongovernmental organizations (NGOs), and indigenous peoples.

The MA was established in response to demands from both policy makers and scientists for an authoritative assessment of the state of the world's ecosystems and of the consequences of ecosystem change for human well-being. By the mid-1990s, many individuals involved in the work of international conventions, such as the Convention on Biological Diversity (CBD) and the Convention to Combat Desertification (CCD), had come to realize that the extensive needs for scientific assessments within the conventions were not being met through the mechanisms then in place. In contrast, such other international environmental conventions as the Framework Convention on Climate Change and the Vienna Convention for the Protection of the Ozone Layer did have effective assessment mechanisms—the Intergovernmental Panel on Climate Change (IPCC) and the Ozone Assessment, respectively—that were proving valuable to these treaties.

The scientific community was also encouraging the establishment of an IPCC-like process to establish scientific consensus on issues related to biodiversity and ecosystems in the belief that the urgency of the problem of ecosystem degradation demanded such an assessment. The major advances that had been made in ecological sciences, resource economics, and other fields during the 1980s and 1990s were poorly reflected in policy discussions concerning ecosystems (Reid 2000; Ayensu et al. 2000; J. C. Clark et al. 2002). Moreover, the scientific community was concerned that existing sectoral assessments (focused on climate, ozone, forests, agriculture, and so forth) were insufficient to address the interlinkages among different environmental problems and among their solutions (Watson et al. 1998).

The design of the MA sought to meet three criteria identified by the Harvard Global Environmental Assessment Project that generally underlie successful global scientific assessments (Clark and Dickson 1999):

- *First, they are scientifically credible.* To meet this criterion, the MA followed the basic procedures used in the IPCC. A team of highly regarded social and natural scientists cochaired the four MA working groups, and prominent scientists from around the world served as coordinating lead authors and lead authors. An independent Peer Review Board oversaw the review process. In the end, more than two thousand authors and expert reviewers were involved in preparing and reviewing the MA.
- *Second, they are politically legitimate.* An assessment is far more likely to be used by its intended audience if that audience has fully "bought in" to the process. In other words, if the intended users request the assessment, have a role in governing the assessment, are involved in its design, and are able to review and comment on draft findings, then they will be far more likely

to use the results. To ensure the legitimacy of the process, the decision to establish the MA was not taken until formal requests for the assessment had been made by international conventions. And, like the IPCC, all of the MA working groups were cochaired by developed and developing country experts and involved a geographically balanced group of authors.

- *Finally, successful assessments respond to decision makers' needs.* This is not to say that scientists do not have an opportunity to introduce new issues and findings that decision makers need to be aware of—they do. But the priority for the assessment is to inform decisions that are being faced or soon will be faced by decision makers. To meet the standard of utility, extensive consultations were made with intended MA users in governments, the private sector, and civil society.

When the idea for the MA first arose in early 1988, it could have been accurately described to be an "IPCC for ecosystems and human well-being." The assessment that was finally launched in 2001, however, differed in several important ways from the IPCC, in particular in relation to scale and knowledge systems. First, the MA was a multiscale assessment—that is, it included analyses at various levels of organization from local to national to international. By contrast, the IPCC was a global assessment, although it increasingly included regional analyses. In addition to the global component, the MA included thirty-three subglobal assessments carried out at the scale of individual communities, watersheds, countries, and regions. The subglobal assessments were not intended to serve as representative samples of all ecosystems; rather, they were designed to meet the needs of decision makers at the scales at which they were undertaken. At the same time, it was anticipated that the global assessment could be informed by findings of the subglobal assessments and vice versa.

Second, the MA included a mechanism allowing use of both published scientific information and traditional, indigenous, and practitioner's knowledge, while the IPCC uses only published scientific information. Much local and traditional knowledge was incorporated into many of the local MA subglobal assessments using this mechanism. While the mechanism allowed, in principle, for local, traditional, and practitioner's knowledge to also be incorporated into the global assessment products, this was quite rare in practice and only occurred to any significant extent in the global report prepared by the MA Sub-Global Working Group.

The primary reasons the MA adopted this multiscale approach and sought to incorporate multiple types of knowledge relate to the nature of ecological process and to the locus of authority for decisions affecting ecosystems. Compare the issues addressed in the MA, for example, with those addressed by the IPCC. Climate change is the classic example of a global environmental change. Although considerable local specificity exists as to the causes of emissions of greenhouse gases, once those gases are emitted they quickly mix in the atmosphere. The increased greenhouse gas concentrations in the atmosphere will have a global impact in that all countries are affected by this change (although, again, the local impacts differ from region to region). Also, decisions taken to address the problem must have a strong global component, although many decisions for emission reduction and—in particular—adaptation will be local (Kates and Wilbanks 2003; Wilbanks et al. 2003).

While ecosystem change and biodiversity loss are of global environmental concern, and although the problem and its solutions have global dimensions, the subglobal dimensions are often much more significant. Factors affecting ecosystems include drivers with global impacts such as climate change and species introductions, regional impacts such as regional trade or agricultural policies, and local impacts such as land use practices and the construction of irrigation systems. Changes to ecosystems can have global consequences, such as the contribution of deforestation to climate change; regional consequences, such as the impact of nutrient loading in agricultural ecosystems on coastal fisheries production; and local consequences, such as the impact of overharvesting or land degradation on local food security. Policy, institutional, technological, and behavioral responses to ecosystem-related issues can involve global actions, such as the creation of global financial mechanisms; regional actions, such as regional agreements for wetlands conservation for migratory bird protection; and local responses, such as a decision by a farmer to alter land management practices to conserve topsoil.

In light of this multiscale nature of both the issues involved and the decisions being made, it was clear that a strictly global assessment would be insufficient. Assessments at subglobal scales are needed because ecosystems are highly differentiated in space and time and because sound management requires careful local planning and action. Local assessments alone are insufficient, however, because some processes are global and because local goods, services, matter, and energy are often transferred across regions (Ayensu et al. 2000). These same considerations also caused the MA organizers to rethink the question of what type of knowledge should "count" in an ecosystem assessment.

For example, at the scale of an individual village, much of the knowledge concerning trends in ecosystems, impacts of ecosystem change on people, and potential responses to ecosystem change will often be held by the members of that community. Such information is unlikely to have been published in a scientific journal. The IPCC relies primarily on peer-reviewed information in order to ensure its credibility. But if a local assessment is to have any credibility at all for local decision makers, then clearly it would make little sense to use only the limited published information bearing on the conditions in a particular village when much better knowledge existed within the community itself.

Moreover, considerations of the legitimacy of the process also forced the reconsideration of policies for what sources of knowledge should be included in the assessment. Legitimacy can be conferred on a process in part through formal mechanisms (e.g., the involvement of particular stakeholders in governance roles), but many other less tangible elements are also involved in any particular stakeholder's decision about whether a process is legitimate and sufficiently trusted to be of use in the person's own decision making. The IPCC arrangements, as well as its reliance on scientific knowledge, were appropriate to ensure that the process was seen as legitimate by governments. But it was unlikely that the MA would be viewed as legitimate by other decision makers such as the business community and indigenous people if it expressly excluded their knowledge from the process.

The experience of the MA in using multiple scale and multiple knowledge systems was somewhat mixed (MA 2005b). Overall, it appears that both the assessment findings and the use of those findings were strengthened by incorporating these two dimensions. However, the mechanisms used by the assessment to address these issues fell short of the initial goals. Lessons from the MA experience are summarized in MA 2005b, and in particular in MA chapters by Ericksen et al. (2005) and Zermoglio et al. (2005).

Scale

We define the term *scale* to be the physical dimensions, in either space or time, of phenomena or observations (MA 2003). *Level,* in contrast, is a characterization of perceived influence; not a physical measure, it is what people accept it to be. A network of cooperating irrigation farmers can contain dozens or thousands of farmers, operating at different scales but on the same level, while state-run irrigation systems at both scales of dozens or thousands of farmers may be perceived to be operating at a "higher" level (Zermoglio et al. 2005). The term *cross-scale interactions* refers to situations where events or phenomena at one scale influence phenomena at another scale. The process of wetlands drainage, for example, takes place at local scales but can in turn influence regional hydrology (by lessening water storage capacity and thereby exacerbating floods) and global climate (by affecting rates of carbon emissions).

The meaning of scale in the context of an assessment is somewhat ambiguous. Environmental assessments are typically characterized by their geographic scale, such as a global, national, river basin, or local community assessment. That characterization means not that the assessment ignores factors operating at other scales but, rather, that the scale defines the primary area of interest (in terms of impacts, potential actions by decision makers, and so forth). Thus a "national"-scale assessment might include both considerations of global climate change and subnational problems of water pollution, but its focus would be on the national implications and the potential decisions that might be taken nationally.

The choice of scale for an assessment is not politically neutral, because that selection may intentionally or unintentionally privilege certain groups (MA 2003). Adopting a particular scale of assessment limits the types of problems that can be addressed, the modes of explanation, and the generalizations that are likely to be used in analysis. For example, users of a global assessment of ecosystem services would be interested in some issues, such as carbon sequestration, that may be of relatively little interest to users of a local assessment. In contrast, the users of a local assessment might be more interested in questions related to, for example, sanitation or local commodity prices that would not necessarily be the focus of a global assessment. Similarly, a global assessment is likely to implicitly devalue local knowledge (and the interests and concerns of the holders of that knowledge) since it is not in a form that can be readily aggregated to provide useful global information, while a local assessment would reinforce the importance of local knowledge and the perspectives of holders of that knowledge.

A large body of literature emphasizes the importance of considering temporal and spatial scale for understanding and assessing processes of social and ecological change (Clark 1985; Wilbanks and Kates 1999; Gunderson and Holling 2002; Giampietro 2003; Rotmans and Rothman 2003; Wilbanks 2003; MA 2003; Zermoglio et al. 2005). There are several ways in which an assessment can be conducted to better consider multiple scales. First, the assessment could simply include analyses undertaken at other space and time scales. Thus a national assessment could include a set of case studies undertaken at the scale of individual river basins within the country. Alternatively, the assessment could be composed of multiple semi-independent subassessments, each with its own user audience and own scale of analysis. The MA defines the former category to be "single scale assessments with multi-scale analyses" and the latter to be a "multi-scale assessment." (The MA, for example, is a multiscale assessment since each of the subglobal assessments included in the process was a semi-independent process with its own user group and assessment team.)

The potential benefits of a process that includes multiple scales differ somewhat depending on which of these two arrangements is used, but they fall into two basic categories: *information benefits* that might improve the accuracy, validity, or applicability of the assessment findings, and *impact benefits* that would improve the relevance, utility, ownership, and legitimacy of the assessment with decision makers.

Potential information benefits gained through considering multiple scales include the following (see Zermoglio et al. 2005).

- *Better problem definition.* A single-scale assessment tends to focus narrowly on the issues, theories, and information most relevant to that scale. Perspectives gained from other scales would contribute to a fuller understanding of the issues.
- *Improved analysis of scale-dependent processes.* Many ecological and social processes exhibit a characteristic scale. If a process is observed at a scale significantly smaller or larger than its characteristic scale, drawing the wrong conclusions would be likely (MA 2003).
- *Improved analysis of cross-scale effects.* For example, the direct cause of a change in an ecosystem is often intrinsically localized (a farmer cutting a patch of forest), while the indirect drivers of that change (for example, a subsidy to farmers for forest clearing) may operate at a regional or national scale.

- *Better understanding of causality.* The relationships among environmental, social, and economic processes are often too complex to fully understand when viewed at any single scale. Studies at additional scales are often needed to fully understand the implications of changes at any given scale.
- *Improved accuracy and reliability of findings.* Subglobal assessment activities can help to ground-truth the global findings.

Potential impact benefits gained through multiscale processes, particularly those that include separate user groups at different scales, include the following.

- *Improved relevance of the problem definition and assessment findings for users and decision makers.* An assessment focused on the specific needs of the users at a particular scale will be more relevant than an assessment in which those users have little input.
- *Improved scenarios.* Although commonly used in environmental assessments, scenarios are most useful in decision making if the decision makers play a direct role in their development.
- *Increased ownership by the intended users.* For example, the legitimacy of the global assessment could be enhanced for governments by the presence of subglobal assessments in individual countries. Similarly, the legitimacy of subglobal assessments for the users of those assessments could be enhanced by virtue of the inclusion of the assessment in a globally authorized assessment mechanism.

But as significant as these potential benefits may be, the challenges associated with designing and implementing a multiscale assessment are also significant. How should scales of analysis be selected? Is there an inherent tradeoff between a design based on scientific sampling and a design based on relevance to users at smaller scales? How can the information and findings from nested assessments be incorporated effectively in larger scale assessments (upscaled) and vice versa (downscaled)? Can common indicators or variables be measured at multiple scales? Can a common conceptual framework be used at multiple scales? Does the added cost and time of a multiscale assessment justify the benefits gained? And, as will be explored in the next section, how can the different types of knowledge present at different scales of analysis be incorporated effectively into a single assessment process?

Knowledge Systems

We define a knowledge system as a body of propositions actually adhered to (whether formal or otherwise) that are routinely used to claim truth (Feyerabend 1987). As described by Zermoglio et al. (2005): "Knowledge is a construction of a group's perceived reality, which the group members use to guide behavior toward each other and the world around them." Science is defined as systematized knowledge that can be replicated and that is validated through a process of academic peer review by an established community of recognized experts in formal research institutions (Zermoglio et al. 2005). Traditional ecological knowledge is a "cumulative body of knowledge, practice and beliefs, evolving by adaptive processes and handed down through generations by cultural transmission" about local ecology (Berkes 1999, 8). Traditional ecological knowledge may or may not be indigenous but has roots firmly in the past. Local knowledge refers to place-based experiential knowledge, knowledge that is largely oral and practice based, in contrast to that acquired by formal education or book learning (Gadgil et al. 2003; Zermoglio et al. 2005).

The norms and procedures of scientific research have evolved and persisted because they have provided a successful mechanism to advance understanding of social and natural systems. Given that background, it makes good sense to ground an assessment of the state of knowledge concerning a particular issue on formal scientific procedures of peer review and publication. Yet scientific knowledge is not the only source of knowledge and, in the case of issues concerning the management of ecosystems in particular locales, may not be the most valuable source of knowledge that can be brought to bear on a problem. In that context, how could an assessment of the state of knowledge *not* include local and traditional knowledge?

There are a number of reasons why incorporating multiple knowledge systems into integrated assessments of environmental and development issues should be beneficial (Warren, Slikkerveer, and Brokensha, 1995; MA 2003; Pahl-Wostl 2003). First, the incorporation of multiple systems of knowledge should increase the amount and quality of information available about a particular environmental or development issue. The experiential knowledge of a local farmer or resource manager, for example, may not meet the criteria of formal science, but it certainly could aid in the understanding and assessment of a local environmental issue. Incorporating multiple systems of knowledge can also potentially bring benefits similar to those obtained through interdisciplinary processes. Assessments are usually enhanced when they are informed by a variety of research disciplines and scientific perspectives. Scientists in different disciplines tend to frame issues in different ways, question assumptions that other disciplines may treat as facts, and broaden the nature of the evidence brought to bear on particular problems. The incorporation of different systems of knowledge in an assessment could produce similar benefits. People using different systems of knowledge, for example, will frame questions and define problems in different ways and have different perspectives on issues.

Second, the findings of an assessment for those individuals using different systems of knowledge should be more useful if multiple systems of knowledge are incorporated in the assessment. If an assessment is to be used by a local community, for example, then it should respond to problems and issues identified by those communities; thus the "local problem" definition is more important than a "scientific" definition of the problem. Similarly, if a business or local community is to use the findings of an assessment, then they must perceive the findings to be credible and the process to be legitimate. That perception will not exist if their knowledge and information are excluded from the assessment. They will not see the assessment as a credible source of information because they know that they may have better information, and they will not perceive the process to be legitimate because their holders of knowledge were excluded from the process.

Finally, the use of multiple knowledge systems can help empower groups that hold that knowledge (Agarwal 1995). For example, at one extreme an environmental or development assessment of a local community could be undertaken by external scientists, who gather data from the community, interview local people, categorize and interpret that information through their own knowledge system, and report their findings to local and regional decision makers. Such an assessment not only would tend to muffle that community's voice or influence in its own future but also could miss or misinterpret vital local information and lead to inappropriate decisions. In contrast, an assessment of that same community that involved both external experts and local experts, was guided by the needs of the community, and involved mechanisms to validate both the scientific and local knowledge of the problems and their solutions would both enhance the utility of the findings for the community and strengthen the ability of that community to influence change, in part through the recognition given to the utility and validity of the knowledge and perspectives of the community (Holt 2005).

The challenges to incorporating multiple knowledge systems in an assessment are significant. First, who establishes what appropriate "validation" of information is? The MA adopted a scientific mechanism of validation (triangulation of information, review by other communities, review at other scales, and so forth). Yet different people and different cultures use different systems for validating the "truth" of information. (Indeed, any individual may use his or her own different standards for examining the truth of information; for example, the process an individual uses to validate information about whether or not it is raining outside might use different standards from the process of validating information related to religious beliefs.) Thus, while an assessment like the MA might indeed obtain better information through the incorporation of local or indigenous knowledge (because it in essence transforms that knowledge into formal scientific knowledge through an implicit peer review or validation mechanism), do the findings of that assessment in fact have any greater value for the original holders of that information (Moller et al. 2004)? They may not, if the standards by which those communities are judging the truth or legitimacy of information are very different from the standards used by the assessment process.

Second, can an assessment like the MA, which is grounded in a formal Western scientific tradition, ever hope to be seen as being "legitimate, credible, and useful" to indigenous communities or other individuals who hold very different worldviews and use different standards for evaluating the utility of information? And, conversely, how can it be ensured that a knowledge assessment that utilizes local and traditional knowledge is also seen as credible within the scientific community?

Theory and Experiences in Bridging Scales and Epistemologies

The chapters in this book explore theoretical issues related to bridging scales and knowledge systems as well as practical experiences and case studies involving issues of scale and knowledge in assessments. The volume begins with a set of chapters that focus primarily on issues of scale. Chapter 2, "How Scale Matters: Some Concepts and Findings," by Thomas Wilbanks, provides an overview of concepts related to how geographic scale matters in conducting integrative nature-society assessments and examines in particular how phenomena and processes differ between scales and how phenomena and processes at different scales affect each other. While chapter 2 focuses on the questions of how understanding and knowledge can be enhanced through considerations of scale, chapter 3, "The Politics of Scale in Environmental Assessments," by Louis Lebel, explores the political questions of who gains and who loses from the choice of scales in scientific assessments. This chapter argues that political considerations often define the choice of scales and that this choice, in turn, tends to further privilege the favored or more powerful resource users.

Chapter 4, "Assessing Ecosystem Services at Different Scales in the Portugal Millennium Ecosystem Assessment," by Henrique Pereira, Tiago Domingos and Luís Vicente, and chapter 5, "A Synthesis of Data and Methods across Scales to Connect Local Policy Decisions to Regional Environmental Conditions: The Case of the Cascadia Scorecard," by Chris Davis, then provide case study examples that explore the practical issues involved in bridging scales in knowledge assessments.

Chapter 6, "Scales of Governance in Carbon Sinks: Global Priorities and Local Realities," by Emily Boyd, also serves as a case study of the issue of scale in assessments. It focuses in particular on the disconnect that often exists between the framing and findings of global assessments and the on-the-ground realities of the actors who may be called on to take action in response to those assessments. The chapter demonstrates the benefits that assessments conducted at different scales can provide in understanding problems while also underlining the tremendous challenges that exist in developing institutions that can serve to bridge global and local institutions in the context of assessments.

Chapter 7, "What Counts as Local Knowledge in Global Environmental Assessments and Conventions?" (by Peter Brosius) turns more specifically to issues of knowledge systems, examining how local knowledge is constituted in global environmental assessments. Local and traditional knowledge is often seen as inseparable from its social context. While individual "facts" held in local knowledge systems (e.g., the timing of migration of a particular species) might be readily integrated in global scientific assessments, local communities generally hold a much broader set of knowledge that could also be of value in global assessments. But just how that knowledge is used depends on the "politics of translation" and the receptivity of global institutions to the more expansive definition of knowledge held by these communities. In chapter 8, "Bridging the Gap or Crossing a Bridge? Indigenous Knowledge and the Language of Law and Policy," Michael Davis explores the division that exists between indigenous knowledge and Western science, examines the basis on which local, traditional, and indigenous knowledge has been marginalized and assimilated by the dominant discourse of science, and then explores the extent to which legal instruments may be able to help restore the diversity of worldviews.

The next five chapters—chapters 9 through 13—present case studies of different attempts to bridge scales and knowledge systems in assessments and resource management. These chapters provide a rich set of lessons concerning methods that work or fail and the costs and benefits associated with these efforts. A number of themes recur in these chapters: the importance of "boundary organizations" that help to negotiate and facilitate the interactions across scales or knowledge systems; the tension that exists between mutually agreed use of local knowledge and the risk of knowledge being "extracted" for use in ways that do not return local benefits and may even result in local costs; and the challenge of knowledge validation. Yet, given the numerous problems that such bridging efforts face, overall this set of experiences is surprisingly positive. In the right context, with the right institutions, the potential for mechanisms to effectively bridge scales and knowledge systems in ways that benefit all stakeholders clearly exists.

But while positive examples do exist, chapter 14, "Barriers to Local-level Ecosystem Assessment and Participatory Management in Brazil," by Cristiana S. Seixas, uses four case studies of participatory fisheries management in Brazil to highlight some of the very real challenges involved in making such assessments a reality. Given the history of centralized decision making in most countries, the weak capacity of many local communities to engage in assessment or policy processes, and the continuing tendency to dismiss the value of local knowledge, it is clear that many barriers remain.

The last two chapters before the final, synthesis chapter examine how the structures and tools used in global environmental assessments might be modified to better address issues related to scale and knowledge systems. Chapter 15, "Integrating Epistemologies through Scenarios," by Elena Bennett and Monika Zurek, argues that a tool now commonly used in global assessments—scenario development—in fact provides a potentially valuable mechanism for bridging knowledge systems in assessment processes. Scenarios are most effective when they are developed jointly by experts and "users" in part because it is only in this way that they adequately represent the worldview of the potential decision makers and can thereby be relevant to those decision makers. This feature should lend itself to processes involving multiple knowledge systems, and Bennett and Zurek provide several case studies where this has been the case.

Chapter 16, "The Politics of Bridging Scales and Epistemologies: Science and Democracy in Global Environmental Governance," by Clark Miller and Paul Erickson, pulls together many of the threads of the earlier chapters to argue that the regionalization of "global" assessments can act to strengthen global civil society by fostering a deeper engagement of groups in the processes, strengthening regional voices in global governance, and providing a forum that respects the diversity of cultures.

The final chapter provides a short synthesis of lessons and conclusions from the previous chapters in the volume.

References

Agarwal, A. 1995. Dismantling the divide between indigenous and scientific knowledge. *Development and Change* 26:413–39.

Annan, K. A. 2000. *We the peoples: The role of the United Nations in the 21st century.* New York: United Nations.

Ayensu, E., D. R. Claasen, M. Collins, A. Dearing, L. Fresco, M. Gadgil, H. Gitay, et al. 2000. International ecosystem assessment. *Science* 286:685–86.

Berkes, F. 1999. *Sacred ecology: Traditional ecological knowledge and resource management.* Philadelphia: Taylor and Francis.

Clark, J. C., S. R. Carpenter, M. Barber, S. Collins, A. Dobson, J. A. Foley, D. M. Lodge, et al. 2002. Ecological forecasts: An emerging imperative. *Science* 293:657–60.

Clark, W. C. 1985. Scale of climate impacts. *Climatic Change* 7:5–27.

Clark, W. C., and N. M. Dickson. 1999. The global environmental assessment project: Learning from efforts to link science and policy in an interdependent world. *Acclimations* 8:6–7.

Ericksen, P., E. Woodley, G. Cundill, W. Reid, L. Vicente, C. Raudsepp-Hearne, J. Mogina, and P. Olsson. 2005. Using multiple knowledge systems in sub-global assessments: Benefits and challenges. In Millennium Ecosystem Assessment, *Multiscale assessments: Findings of the Sub-Global Assessments Working Group,* vol. 4, *Ecosystems and human well-being,* 85–117. Washington, DC: Island Press.

Feyerabend, P. 1987. *Farewell to reason.* London: Verso.

Gadgil, M., P. Olsson, F. Berkes, and C. Folke. 2003. Exploring the role of local ecological knowledge in ecosystem management: Three case studies. In *Navigating social-ecological systems,* ed. F. Berkes, J. Colding, and C. Folke, 189–209. Cambridge: Cambridge University Press.

Giampietro, M. 2003. *Multi-scale integrated analysis of ecosystems.* London: CRC Press.

Gunderson, L. H., and C. S. Holling. 2002. *Panarchy: Understanding transformations in human and natural systems.* Washington, DC: Island Press.

Holt, F. L. 2005. The catch-22 of conservation: Indigenous peoples, biologists and cultural change. *Human Ecology* 33:199–215.

Intergovernmental Panel on Climate Change. 2001. *Climate change 2001: Synthesis report.* Cambridge: Cambridge University Press.

Kates, R., and T. Wilbanks. 2003. Making the global local: Responding to climate change concerns from the bottom up. *Environment* 45 (3): 12–23.

Millennium Ecosystem Assessment (MA). 2003. *Ecosystems and human well-being: A framework for assessment.* Washington, DC: Island Press.

———. 2005a. *Millennium ecosystem assessment synthesis.* Washington, DC: Island Press.

———. 2005b. *Multiscale assessments: Findings of the Sub-Global Assessments Working Group.* Vol. 4, *Ecosystems and human well-being.* Washington, DC: Island Press.

Moller, H., F. Berkes, P. O. Lyver, and M. Kislalioglu. 2004. Combining science and traditional ecological knowledge: Monitoring populations for co-management. *Ecology and Society* 9 (3): 2. http://www.ecologyandsociety.org/vol9/iss3/art2 (accessed May 20, 2006).

Pahl-Wostl, C. 2003. Polycentric integrated assessment. In *Scaling issues in integrated assessment,* ed. J. Rotmans and D. S. Rothman, 237–61. Lisse, The Netherlands: Swets and Zeitlinger.

Reid, W. V. 2000. Ecosystem data to guide hard choices. *Issues in Science and Technology* 16 (3): 37–44.

Rotmans, J., and D. S. Rothman, eds. 2003. *Scaling in integrated assessment.* Lisse, The Netherlands: Swets and Zeitlinger.

Warren, D. M., L. J. Slikkerveer, and D. Brokensha, eds. 1995. *The cultural dimension of development: Indigenous knowledge systems.* London: Intermediate Technology Publications.

Watson, R. T., J. A. Dixon, S. P. Hamburg, A. C. Janetos, and R. H. Moss. 1998. *Protecting our planet—securing our future.* Washington, DC: United Nations Environment Programme, U.S. National Aeronautics and Space Administration, and World Bank.

Wilbanks, T. J. 2003. Geographic scaling issues in integrated assessments of climate change. In *Scaling in integrated assessment,* ed. J. Rotmans and D. S. Rothman, 5–34. Lisse, The Netherlands: Swets and Zeitlinger.

Wilbanks, T. J., S. M. Kane, P. N. Leiby, R. D. Perlack, C. Settle, J. F. Shogren, and J. B. Smith. 2003. Possible responses to global climate change: Integrating mitigation and adaptation. *Environment* 45 (5): 28–38.

Wilbanks, T. J., and R. W. Kates. 1999. Global change in local places: How scale matters. *Climatic Change* 43 (3): 601–28.

Zermoglio, M. F., A. van Jaarsveld, W. Reid, J. Romm, O. Biggs, T. X. Yue, and L. Vicente. 2005. The multiscale approach. In Millennium Ecosystem Assessment, *Multiscale assessments: Findings of the Sub-Global Assessments Working Group,* vol. 4, *Ecosystems and human well-being,* 61–83. Washington, DC: Island Press.

BRIDGING SCALES

CHAPTER 2

How Scale Matters: Some Concepts and Findings

THOMAS J. WILBANKS

This chapter summarizes a number of concepts related to how geographic scale matters in conducting large, integrative nature-society assessments, such as the Millennium Ecosystem Assessment (MA) and the reports of the Intergovernmental Panel on Climate Change (IPCC). Such concepts relate both to (a) how phenomena and processes differ between scales and (b) how phenomena and processes at different scales affect each other. The chapter also considers lessons learned about how geographic scale relates to knowledge bases. Although it notes that temporal and institutional scale are important, in line with the conceptual framework of the subglobal component of the MA it focuses on geographic scale.

These questions for nature-society assessments are, of course, related to one of the great overarching intellectual challenges across a wide range of sciences: understanding relationships between macroscale and microscale phenomena and processes (Wilbanks and Kates 1999). Examples include biologists and ecologists considering linkages between molecules and cells, on the one hand, and biomes and ecosystems, on the other, related to such issues as biocomplexity; economists considering relationships between individual consumers and firms, on the one hand, and national and global economies, on the other, related to such issues as efficiency and equity; and such other scientific fields as far afield as fluidics, which considers how the behavior of fluids changes with scale and how these differences interact. In the spirit of traditions such as general systems theory, it is not uncommon to explore applications of findings in one field about how scale matters as possible hypotheses for another (for an early example of explorations of how scale shapes interactions between form and function, see Thompson 1942).

Basic Concepts

Some basic concepts about how we consider geographic scale as an aspect of nature-society assessments are summarized in Wilbanks 2003, Millennium Ecosystem Assessment 2003, and Zermoglio et al. 2005. Millennium Ecosystem Assessment 2003 defines scale as the physical dimension of a phenomenon or process in space or time, expressed in physical units. According to this perspective, "a level of organization is not a scale, but it can have a scale" (MA 2003, 108; also see O'Neill and King 1998).

Arrayed along a geographic scale continuum from very small to very large, most processes of interest establish a number of dominant frequencies; they show a kind of lumpiness, organizing themselves more characteristically at some scales than others (see, for instance, Holling 1992). Recognizing this lumpiness, we can concentrate on the scales that are related to particular levels of system activity—for example, family, neighborhood, city, region, and country—and at any particular level subdivide space into a mosaic of "regions" to simplify the search for understanding. In many cases, smaller-scale mosaics are nested within larger-scale mosaics; therefore, we can often think in terms of spatial hierarchies.

Although some care is needed in extrapolating from one field of study to another, in some cases (e.g., in ecology) relationships exist between spatial and temporal scales. For instance, it appears that in many cases shorter-term phenomena are more dominant at local scales than at global scales, while long-term phenomena are the converse. On the other hand, in human systems infrastructure, decisions involving lifetimes of thirty years or more may be made at very local scales, while political perspectives at a national scale are often focused on very-near-term costs and benefits.

What we are discovering is that place is more than an intellectual and social construct; it is also a context for communication, exchange, and decision making. Place has meaning for local empowerment, directly related to equity. In fact, a sense of place is related to personal happiness in the face of global space-time compression (see, for example, Harvey 1989).

Based partly on such concepts, it has been suggested that geographic scale matters in seeking an integrated understanding of global change processes and that understanding linkages between scales is an important part of the search for knowledge (Wilbanks and Kates 1999; also see Kates and Wilbanks 2003 and Association of American Geographers 2003). Several of the reasons have to do with *how the world works.* The forces that drive environmental systems arise from different domains of nature and society—for example, Clark has shown that distinctive systems embedded in global change processes operate at different geographic and temporal scales (Clark 1985). Within this universe of different domains, local and regional domains relate to global ones in two general ways: systemic and cumulative (Turner et al. 1990). Systemic changes involve fundamental changes in the functioning of a global system, such as effects of emissions of ozone-depleting gases on the stratosphere, which may be triggered by local actions (and certainly may affect them) but that transcend simple additive relationships at a global scale. Cumulative changes result from an accumulation of localized changes, such as groundwater depletion or species extinction; the resulting systemic changes are not global, although their effects may have global significance.

A second reason that scale can matter is that the scale of *agency*—the direct causation of actions—is often intrinsically localized, while at the same time such agency takes place in the context of *structure*: a set of institutions and other regularized, often formal relationships whose scale is regional, national, or global. Land use decisions are a familiar example.

A third reason that scale can matter is that the driving forces behind environmental change involve interactions of processes at different locations and areal extents and different time scales, with varying effects related to geographic and temporal proximity and structure. Looking only at a local scale can miss some of these interactions, as can looking only at a global scale.

Several additional reasons why scale matters have to do with *how we learn about the world.* One of the strongest is the argument that complex relations among environmental, economic, and social processes that underlie environmental systems are too complex to unravel at any scale beyond the relatively local (National Academy of Sciences/National Research Council 1999). A second reason is that a portfolio of observations at a detailed scale is almost certain to contain more variance than observations at a very general scale; the greater variety of observed processes and relationships at a more local scale can provide for greater learning about the substantive questions being asked. In other words, variance often contains information rather than "noise." A third reason is that research experience in a variety of fields tells us that researchers looking at a particular issue from the top down can reach conclusions that differ dramatically from those of researchers looking at that very same issue from the bottom up. The scale embodied in the perspective can frame the investigation and shape the results, which suggests that full learning requires attention at a variety of scales.

These reasons, of course, do not mean that global-local linkages are salient for every question being asked about nature-society systems. What they suggest is more modest: that examinations of such changes should normally take time to consider linkages among different scales, geographic and temporal, and whether those linkages might be important to the questions at hand (Wilbanks, forthcoming).

In any case, they also suggest that integrated assessments of nature-society relationships should be sensitive to multiple scales rather than focused on a single scale (Wilbanks 2003; AAG 2003). One reason is that selection of a single scale can frame an investigation too narrowly because questions and research approaches characteristic of that scale tend to dominate and because upscaling or downscaling information from other scales requires compromises that often lose information or introduce biases. Another reason is that phenomena, processes, structures, technologies, and stresses operate differently at different scales and thus the implications for action can depend on the scale of observation. Figure 2.1 is an example from recent research.

Yet another reason is that a particular scale may be more or less important at different points in a single cause-consequence continuum and therefore less appropriate for exploring some of the points. Figure 2.2 is an example.

Finally, institutions important for decision making about the processes being examined operate at different scales. For these reasons, no single scale is ideal for broad-based investigation, although comparative studies at a single scale can contribute important insights (e.g., Schellnhuber and Wenzel 1998; Schellnhuber, Lüdeke, and Petschel-Held 2003; AAG 2003).

Findings about Scale Differences

A number of recent nature-society assessments, in addition to the Millennium Ecosystem Assessment, have helped to illuminate issues related to how scale matters in such assessments (AAG 2003; National Assessment of Climate

Figure 2.1

Scale matters in comparing net benefits from mitigation and adaptation responses to concerns about climate change impacts. For instance, at a global scale, mitigation ("avoidance") tends to appear preferable because many potentially dangerous impacts could be beyond capacities to adapt, whereas at a regional scale in an industrialized country, adaptation can appear preferable because many of the benefits of mitigation actions are external to the region. *(From Wilbanks et al., forthcoming.)*

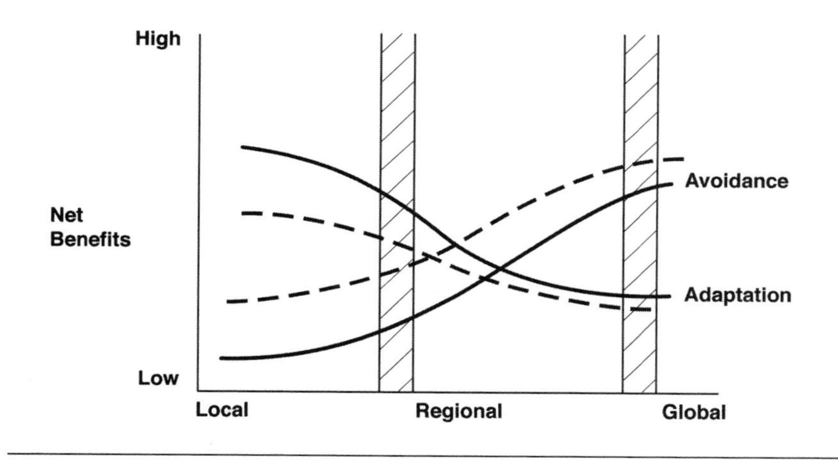

Figure 2.2

Climate change and its consequences include a number of different processes, which often differ in the scale domains where consequences are focused. *(From Kates and Wilbanks, 2003.)*

Changes and Consequences**

Scale Domains		Driving Forces			Emissions/Sink Changes				Radiative Forcing			Climate Change			Impacts				Responses		
																			Mitigation		
		Population	Affluence	Technological Change	Fossil Fuels	Agriculture	Wastes	Deforestation	Trace Gases	Aerosols	Reflectivity	Temperature	Precipitation	Extreme Events	Ecosystems	Agriculture	Coasts	Health	Sequestration	Prevention	Adaptation
Global																					
Regional	Continental																				
	Sub-continental																				
	Economic/Political/Unions																				
	Large Nations																				
Large Area	Small Nations, States, Provinces																				
	Large Basins																				
	5-10° Grids																				
Local	1° Grids																				
	Small Basins																				
	Cities																				
	Firms																				
	Households																				

*Depicts the scale of actions, not necessarily the locus of decision making.
**Dashed lines indicate occasional consequences or a lower level of confidence.

Figure 2.3

The first assessment of consequences of climate change in Canada found that the variance in net effects was considerably greater at local scales than at larger scales, as illustrated here. The solid lines depict net benefits without adaptive response; dotted lines indicate net effects of adaptation. *(From Environment Canada 1997.)*

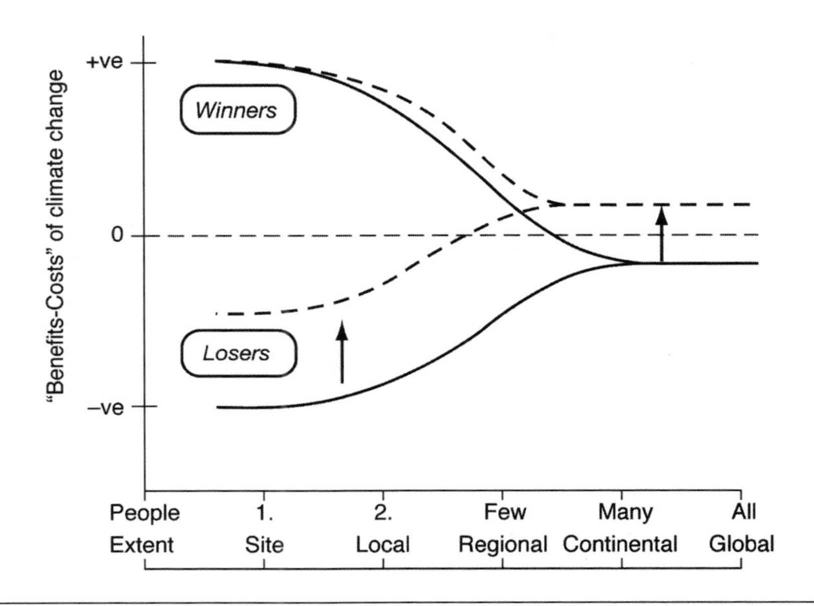

Change 2000; NAS/NRC, 1999). More recent findings have emerged from regional and local studies by the Assessments of Impacts and Adaptations to Climate Change (AIACC) project (http://www.aiaccproject.org) and the subglobal component of the Millennium Ecosystem Assessment (MA 2005). Also see listings of integrated studies of nature-society systems (http://sustsci.harvard.edu/integstudies.htm), local analyses of climate change adaptation experiences and potentials (http://www.sei.se/oxford/), and studies of environmental change vulnerabilities at various scales (http://www.vulnerabilitynet.org).

These investigations indicate that, as expected, observations of many variables at a more localized scale show greater variance and volatility. In other words, larger scales lose valuable information. Figure 2.3, from the Canadian national climate change impact assessment (Environment Canada 1997), was one of the earliest empirical findings of this nature in nature-society studies, supporting the theoretical expectations mentioned above.

Figure 2.4

Illustrations of several possible hypotheses about how scale matters. "L" indicates "Local"; "G" indicates "Global." There is room for considerable insight and innovativeness in suggesting other such hypotheses.

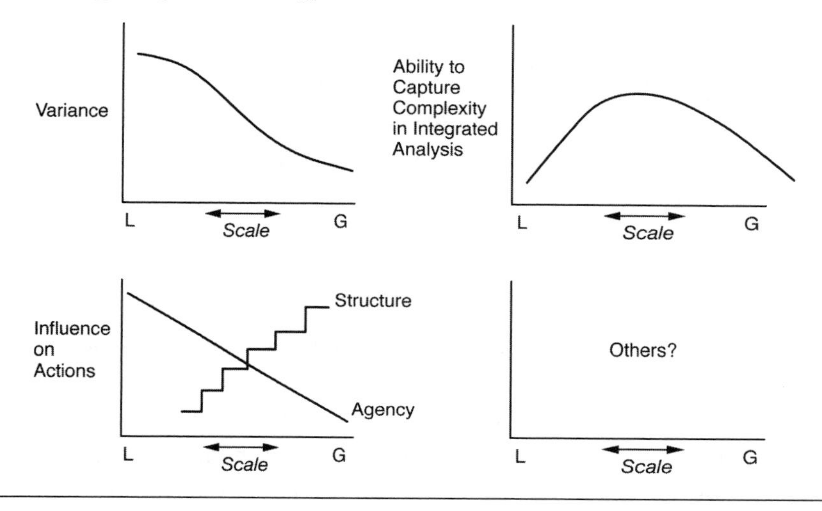

The literature also finds that analyses and assessments at different scales tend to be associated with different research paradigms and styles. As one example, in analyses of climate change responses, work at a global or national scale tends to be characterized by quantitative analysis, using net present value metrics, while work at a small-regional or local scale tends to involve integrated assessments, including significant stakeholder involvement (Wilbanks et al., forthcoming).

Downscaling and upscaling, in fact, are likely to contribute different insights; for instance, bottom-up investigations often provide different understandings compared with top-down investigations. As one illustration, the Global Change in Local Places (GCLP) project, by the American Association of Geographers, found that top-down assessments of potentials of technologies to reduce greenhouse gas emissions in local places tended to overestimate those potentials because they were not sensitive to local obstacles and constraints, whereas bottom-up assessments tended to underestimate the potentials because they were not fully informed about directions of technological and policy changes (Kates and Wilbanks 2003; AAG 2003).

Other findings include (a) that different scales are related to different institutional roles, and that the scale of decisions is often poorly matched with the scale of the processes being decided upon, and (b) that the choice of a scale and a set of boundaries is not politically neutral, even if the choice is not based on political considerations (MA 2003).

Even though proposing a theoretical structure at this stage in our knowledge development would seem premature, it is possible to imagine moving in that direction by considering and testing a number of hypotheses that seem reasonable based on what we know so far. Figure 2.4 illustrates just a few of the relationships that might be explored.

Findings about Scale Relationships

Similarly, recent assessments have suggested findings about how phenomena and processes at different scales are linked with each other, although the knowledge base about cross-scale relationships is not as well developed as it is about scale differences. Most significantly, perhaps, GCLP indicates that in many cases cross-scale *interactions* are more significant than aggregate *differences* between scales (e.g., Kates and Wilbanks 2003; AAG 2003). For instance, local actions shape cumulative environmental conditions and democratic policy making at larger scales, while local actions are affected in turn by market signals, institutional structures, and technology portfolios arising at larger scales (figure 2.5). It is in the intertwining of local activity with larger structures that most nature-society phenomena and processes play out.

Cross-scale interactions can be considered in terms of certain basic dimensions they demonstrate:

- *Strength: powerful or weak.* Consider, for example, top-down regulatory controls versus bottom-up messages through representative democracies.
- *Constancy: constant or intermittent; periodic or irregular.* Consider, for instance, gradual climate change versus technology breakthroughs.
- *Directionality: mainly in one direction or the other, or mutual.* Most often, directionality distinguishes top-down interactions, such as through corporate management frameworks, from feedbacks in both directions through democratic government processes supported by an active free press.
- *Resolution: focused or broadcast.* An example is specific location problem solving versus general information provision.

Figure 2.5

Macroscale and microscale processes and phenomena interact across scales in ways such as shown here. For instance, local actions shaped by larger driving forces add up to impacts on large-scale processes. Institutional responses at larger scales, shaped by democratic support or opposition from smaller scales, lead to large-scale structures that provide enablement (or constraints) for local-scale adaptive behavior.

(From Association of American Geographers 2003.)

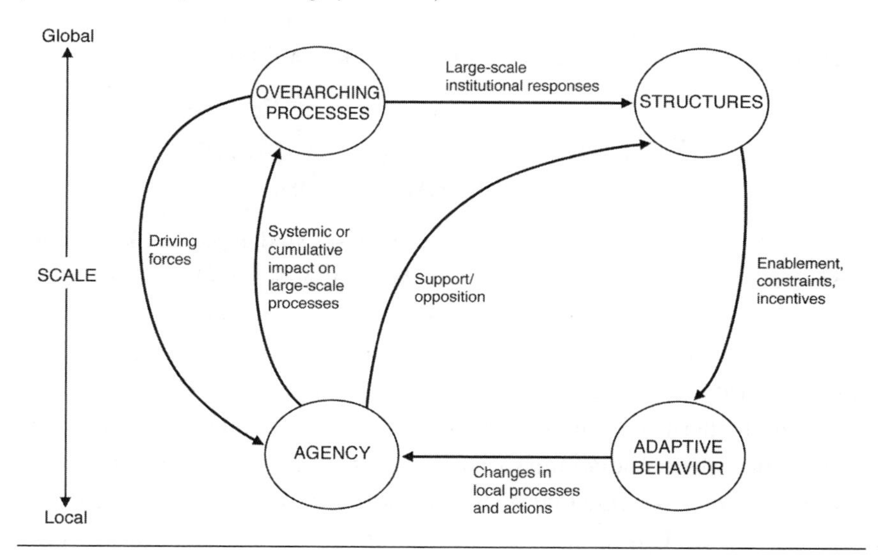

- *Context: additive or contradictory, in connection with other processes operating.* For instance, government policies that reinforce market signals have a different effect than do policies that differ from market signals.
- *Effect: stabilizing or destabilizing; controlling or enabling.* Among the many examples, terrorism arising from relatively local grievances can destabilize larger-scale units, while actions that provide conflict resolution can be stabilizing.
- *Intent: explicit and/or implicit.* Determining the intent of actions, for instance by large and small government, is not always easy, but intent is a fundamental aspect of cross-scale interactions, their effects, and their sustainability.

It is clear that cross-scale interactions are often associated with distinctive bridging-type institutional roles (Cash 2001); but in many cases involving human systems, relationships are too complicated to be incorporated into the kinds of hierarchy theory characteristic of ecological research (personal relationships, information flow, emission dispersal, etc.), and in many cases important kinds of data about the interactions are elusive (e.g., relationships between local phenomena and national or international corporate decision making).

In some cases, increasing understanding—at least at the current state of knowledge—seems to call for laying out rich narrative "story lines" and then exploring the connections from multiple base points (e.g., Root and Schneider's call [1995] for "strategic cyclical scaling"). Figure 2.5 illustrates such a story line. (For another example, see Kok et al. 2004.)

Findings about Scale Aspects of Knowledge Bases

Several of these findings speak directly to scale-related aspects of knowledge bases, especially the value of knowledge bases at a local scale and complexities in relating these knowledge bases to extra-local structures.

One set of findings addresses the potential value of local-scale studies and what that value may depend on. One finding shows that it is only in relatively focused place-based research that complex relationships among environmental, economic, and social processes can be traced, especially when the researchers are armed with specific local knowledge (NAS/NRC 1999). Of particular value is local knowledge about phenomena and processes not captured by data available to larger-scale analysts and modelers. For instance, data might not exist to document a relationship between temperature differences and health indicators in a local area, but a focus group discussion among local health care providers might provide a rich base of knowledge on the subject (e.g., Oak Ridge National Laboratory and Cochin University of Science and Technology 2003).

Another finding is that involving information and education infrastructures with a demonstrated commitment to the local area is the best way to facilitate cross-scale dialogues (Cash and Moser 2000). For example, communications channeled from global experts to local experts and then communicated by the local experts in interactions with local decision makers work far better than communications channeled between global experts and global decision makers and then between global decision makers and local decision makers (Cash and Moser 2000). Local experts are uniquely suited to help relate general and local knowledge because they are repositories themselves of both kinds of knowledge and because their life experiences include extensive contact networks in both worlds (AAG 2003).

Yet another value is the greater variance in information and experience that exists among a sample of local areas compared with a sample of larger areas; this increased variance offers greater opportunities for detailed understandings of complex causes and consequences (AAG 2003). In fact, observations of local circumstances—and especially adaptive behavior—can challenge generalizations about the regions in which they are located—for instance, the sub-global assessment component of the MA found that, in some cases, localities within regions considered highly stressed were relatively stable while some localities within regions considered relatively stable were relatively stressed (Pereira, Reyers, and Watanabe 2005; also see Kok 2001, 115).

Moreover, a local scale can be especially valuable in identifying sometimes overlooked or understudied issues related to environmental change (AAG 2003). By involving more information exchange between subject-matter analysts and the wider citizenry, a local scale contributes to a social process of learning and response. It can also help clarify differences in the local consequences of national and large-regional actions.

On the other hand, isolated local case studies are often problematic as bases for broader generalizations, especially when they are selected in part because of interesting kinds of unique characteristics. Their value for larger-scale understandings is enhanced when they are carefully chosen for comparability, use a common study protocol, and are compared with control studies, but this kind of interarea coordination is often difficult to arrange and implement. Local studies, based on local knowledge bases and perspectives, may also be limited by a lack of local understanding of larger-scale driving forces and trends, such as technological change (AAG 2003).

A second set of findings concerns challenges in relating the local and the global. In studies of climate change issues, at least, it is clear that some of the driving forces operate at a global scale while many of the phenomena that underlie environmental processes operate at a local scale (AAG 2003). Understanding climate change processes and responses requires attention to multiple scales and how they relate to one another. Studies of climate change issues at the two scales, however, are often poorly linked. Those moving from global toward local scales typically use climate change scenarios derived from global models as starting points, despite the absence of regional or local specificity in such models. Studies moving from the local toward the global are typically less quantitative, more participative, and related to different research traditions and disciplinary paradigms (Wilbanks et al., forthcoming). Connecting the two has generally been a challenge for what has been termed "analytic-deliberative" approaches (NAS/NRC 1996), where structured group processes—informed by a variety of analyses and bodies of experience—confront not only issues regarding different research approaches but also the fact that researchers examining an issue from a local perspective may reach conclusions different from those reached by analysts who view the very same issue from a global vantage point (and where it is possible that neither is wrong).

As indicated above, what we know from both research literature and practice is that local actions occur within the context of externally determined structures, from government and corporate policies to demography and technological change. For instance, a study of potentials to reduce greenhouse gas emissions in several local areas of the United States through local action concluded that local actions could yield substantial results under four sets of conditions: growing evidence of climate change impacts, related to largely external climate change forecasts and impact assessments; policy interventions that directly or indirectly associate emission reductions with local benefits, also largely externally derived; market or governmental incentives and assistance for local innovation; and technological change and improvement to enlarge the options available for local choice and application (AAG 2003). Unfortunately, few of these conditions exist at present.

A third set of lessons asks at what scale local studies should be performed. As reported above, lessons from assessments so far indicate that there is no one scale for every purpose. Because scale is related to function, and because different functions have different scales, a starting point in determining the appropriate scale is to clarify the functions of particular interest. In many cases, scales between assessments and the activities they consider are seriously mismatched (AAG 2003). Moreover, the scale selected affects the results by establishing boundaries between what is in and what is not, which can have social and political implications even if the selection is not politically motivated (MA 2003). In many cases, if the analysis is intended to inform decisions by particular institutions, it is worth considering whether to relate the scale to units for which or in which decisions are made.

Concluding Observations

Whereas a decade ago there might have been some debate about whether scale matters in far-reaching, integrative nature-society analyses and assessments, this issue seems to have been settled, with the debate shifting from the *importance* of multiscale assessments to (at least in the case of the U.S. Climate Change Research Program) the *practicality* of science-based assessments at the regional and local scales at the current state of data, tools, and knowledge (especially the forecasting of changes and impacts at a fine-grained scale).
The challenges are as follows:

- to show that regional and local assessments can be at least as sound scientifically as global assessments, where such initiatives as AIACC are very encouraging
- to show that qualitative deliberations and stakeholder participation, which are usually more important at a local scale, can contribute to the science of nature-society assessments as well as to their political acceptability
- to develop more effective approaches for facilitating open two-way interaction between experts, institutions, and interests across scales (for instance, developing guidelines for local assessments that are widely acceptable and useful, and replacing or supplementing quantitative large-scale scenarios with rich, informative narratives of different pathways for conceivable change).

In the longer run, of course, we will need to develop conceptual and methodological frameworks that incorporate both scale *differences* and scale *relationships.* But this development itself will need to include both top-down and bottom-up interactions, keeping its approaches consistent with its understandings of its subject.

References

Association of American Geographers (AAG). 2003. *Global change and local places: Estimating, understanding, and reducing greenhouse gases.* Cambridge: Association of American Geographers GCLP Research Team, Cambridge University Press.

Cash, D. W. 2001. In order to aid in diffusing useful and practical information . . . : Agricultural extension and boundary organizations. *Science, Technology, and Human Values* 26:431–53.

Cash, D. W., and S. C. Moser. 2000. Linking global and local scales: Designing dynamic assessment and management processes. *Global Environmental Change* 10:109–20.

Clark, W. C. 1985. Scales of climate impacts. *Climatic Change* 7:5–27.

Environment Canada. 1997. *The Canada Country Study: Climate impacts and adaptation.* Downsview, Ontario: Adaptation and Impacts Research Group.

Harvey, D. 1989. *The growth of postmodernity.* Baltimore: Johns Hopkins University Press.

Holling, C. S. 1992. Cross-scale morphology, geometry, and dynamics of ecosystems. *Ecological Monographs* 62:447–502.

Kates, R., and T. Wilbanks. 2003. Making the global local: Responding to climate change concerns from the bottom up. *Environment* 45 (3): 12–23.

Kok, K. 2001. Scaling the land use system: A modelling approach with case studies for Central America. PhD diss., Wageningen University, The Netherlands.

Kok, K., D. S. Rothman, M. Patel, and D. de Groot. 2004. Multi-scale participatory local scenario development: Using Mediterranean scenarios as boundary conditions. Paper presented at Bridging Scales and Epistemologies workshop, Millennium Ecosystem Assessment, Alexandria, Egypt. http://www.millennium assessment.org/documents/bridging/papers/kok.kasper.pdf (accessed May 28, 2006).

Millennium Ecosystem Assessment (MA). 2003. Dealing with scale. In *Ecosystems and human well-being: A framework for assessment,* 107–26. Washington, DC: Island Press.

———. 2005. *Multiscale assessments: Findings of the Sub-Global Assessments Working Group.* Vol. 4, *Ecosystems and human well-being.* Washington, DC: Island Press.

National Academy of Sciences/National Research Council (NAS/NRC). 1996. *Understanding risk: Informing decisions in a democratic society.* Washington, DC: U.S. National Research Council of the National Academies, National Academies Press.

———. 1999. *Our common journey: A transition toward sustainability.* Washington, DC: National Academies Press.

National Assessment of Climate Change. 2000. *Climate change impacts on the United States: The potential consequences of climate variability and change: Overview.* Cambridge: Cambridge University Press.

Oak Ridge National Laboratory (ORNL) and Cochin University of Science and Technology (CUSAT). 2003. *Possible vulnerabilities of Cochin, India to climate change impacts and response strategies to increase resilience.* Oak Ridge, TN: ORNL; Cochin, India: CUSAT.

O'Neill, R. V., and A. W. King. 1998. Homage to St. Michael: Or why are there so many books on scale? In *Ecological scale: Theory and applications,* ed. D. L. Peterson and V. T. Parker, 3–15. New York: Columbia University Press.

Pereira, H., B. Reyers, and M. Watanabe. 2005. Conditions and trends of ecosystem services and biodiversity. Chap. 8 in Millennium Ecosystem Assessment, *Multiscale assessments: Findings of the Sub-Global Assessments Working Group,* vol. 4, *Ecosystems and human well-being.* Washington, DC: Island Press.

Root, T. L., and S. H. Schneider. 1995. Ecology and climate: Research strategies and implications. *Science* 269:334–41.

Schellnhuber, H.-J., M.K.B. Lüdeke, and G. Petschel-Held. 2003. The syndromes approach to scaling: Describing global change on an intermediate functional scale. In *Scaling issues in integrated assessment,* ed. J. Rotmans and D. Rothman, 205–36. Lisse, The Netherlands: Swets and Zeitlinger.

Schellnhuber, H.-J., and V. Wenzel, eds. 1998. *Earth system analysis.* Berlin: Springer.

Thompson, D. 1942. *On growth and form.* Cambridge: Cambridge University Press.

Turner, B. L., II, R. E. Kasperson, W. B. Meyer, K. M. Dow, D. Golding, J. X. Kasperson, R. C. Mitchell, and S. J. Ratick. 1990. Two types of global environmental change: Definitional and spatial scale issues in their human dimensions. *Global Environmental Change* 1:14–22.

Wilbanks, T. J. 2003. Geographic scaling issues in integrated assessments of climate change. In *Scaling issues in integrated assessment,* ed. J. Rotmans and D. Rothman, 5–34. Lisse, The Netherlands: Swets and Zeitlinger.

———. Forthcoming. Scale in integrated environmental assessment. In *Integrated environmental assessment,* ed. J. Rotmans, M. Hulme, and D. Stanners. European Forum on Integrated Environmental Assessment (EFIEA).

Wilbanks, T. J., S. M. Kane, P. N. Leiby, R. D. Perlack, C. Settle, J. F. Shogren, and J. B. Smith. 2003. Possible responses to global climate change: Integrating mitigation and adaptation. *Environment* 45 (5): 28–38.

Wilbanks, T. J., and R. W. Kates. 1999. Global change in local places. *Climatic Change* 43:601–28.

Wilbanks, T. J., P. Leiby, R. Perlack, J. T. Ensminger, and S. Wright. Forthcoming. Toward an integrated analysis of mitigation and adaptation: Some preliminary findings. In *Mitigation and Adaptation Strategies for Global Change,* special issue, "Challenges in Integrating Mitigation and Adaptation as Responses to Climate Change."

Zermoglio, M. F., A. van Jaarsveld, W. V. Reid, J. Romm, O. Biggs, T. Yue, and L. Vicente. 2005. The multiscale approach. Chap. 4 in Millennium Ecosystem Assessment, *Multiscale assessments: Findings of the Sub-Global Assessments Working Group,* vol. 4, *Ecosystems and human well-being.* Washington, DC: Island Press.

CHAPTER 3

The Politics of Scale in Environmental Assessments

Louis Lebel

This chapter argues that choice of scale in scientific assessments of environmental changes are not unambiguously defined by biophysical characteristics. Nor are they politically neutral. Actors contest spatial, temporal, and jurisdictional levels in the assessment process directly as well as through their influence on decisions about inclusion criteria for issues and sources of information, analytical methods, and rhetorical devices in communication. Moreover, choices of scale have political implications because they focus scrutiny on the activities of, and the impacts on, subsets of actors, who in turn may try to influence scale choices strategically. The power of one group of actors to alter the behavior of another group can be through directly influencing the decision making or, more subtly, through shifting the agendas or shaping the contexts in which knowledge is organized or decisions are made. On the other hand, the knowledge produced and shared through the social process of assessment may help build coalitions of interests that make collective action more likely.

This chapter examines some of the main pathways through which "politics of scale" (Brenner 2001; Cox 1998; Meadowcroft 2002; Swyngedouw 2000) are reproduced in environmental assessments. The term *environmental assessments* here means activities involving multiple actors gathering, reviewing, synthesizing, and communicating information about environmental conditions, trends, drivers, impacts, and plausible futures with the aim of either raising awareness of a particular issue or supporting environmental governance processes already under way. Such assessments are a social process of communication and interaction that involves much more than simply "producing reports." Most environmental assessments are about local impacts of individual infrastructure development projects. Increasingly, however, environmental assessments are also being made across multiple projects, at regional, national, and international levels. A key feature of the latter is that they consider drivers, changes, and consequences that are multilevel and transboundary. These are sometimes called "strategic environmental assessments" (Fischer and Seaton 2002). It is with this latter, diverse class of assessments that this chapter is mostly concerned.

The first three sections here explore the ways in which politics of scale are expressed in the framing, conduct, and use of environmental assessments. Next is a classification of strategies and mechanisms that produce scale politics. The chapter ends by discussing the implications for the design and conduct of assessments.

Assessments

Environmental assessments vary greatly in purpose, organization, and scope. International assessments of climate change, ozone, and acid rain have focused most on inventorying emissions, atmospheric concentration changes, and immediate biophysical impacts rather than on considering underlying drivers or socioeconomic and health impacts (Jager et al. 2001). Some of the most frequently shared characteristics of environmental assessment as a social process are summarized in figure 3.1.

This conceptual figure highlights two key points about environmental assessments. First, scientific understanding and political interests interact in various arenas (framing, assessing, and using) to produce an assessment not just at a single interface. Second, issues do not emerge and information does not flow unidirectionally through these arenas; rather, they engage in co-evolving interactions among scientists, policy makers, and the wider public that are continually reframing, reassessing, and reusing the assessment.

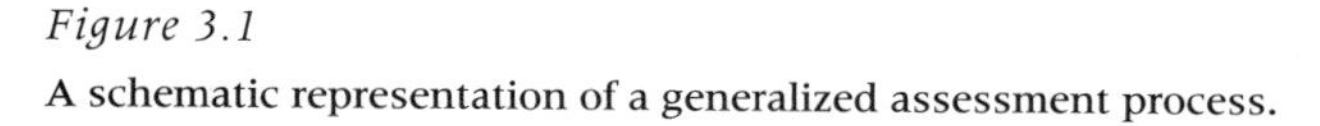

Figure 3.1
A schematic representation of a generalized assessment process.

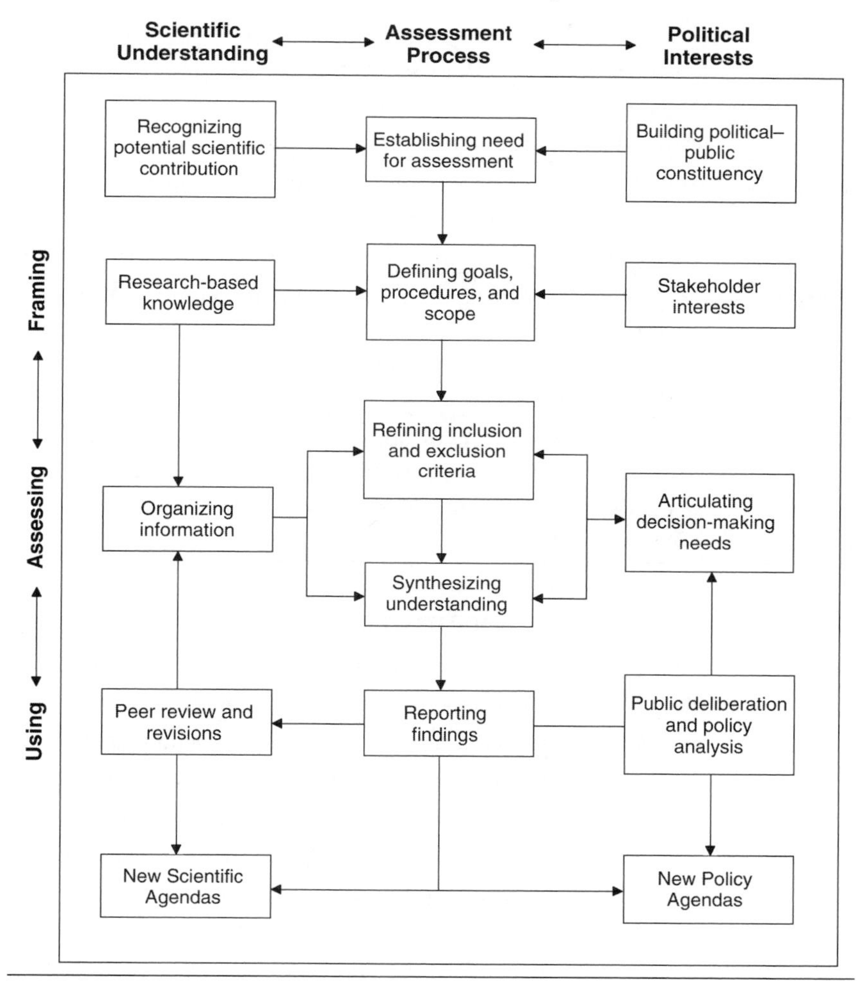

Thus, decision makers do not have to wait for the final assessment to test out new policy agendas, and nor do scientists.

The analysis in this paper was guided by an initial set of questions addressing politics of scale (table 3.1). The first three groups of questions deal with the more visible and direct arenas, whereas the fourth considers the impacts of institutions in shaping what we know and the context in which assessments are carried out.

Table 3.1
Analytical framework of initial questions for exploring the politics of scale in environmental assessment

Arena	Illustrative Analytical Questions
Framing	How were boundaries, resolutions, and levels decided and defined? What rationale is given for choice of scales?
Assessing	Which sources of scale-dependent knowledge were considered, and how were they combined? Was any effort made to address potential bias introduced by choices of scale?
Using	How were the findings of the assessment communicated and incorporated into decision-making processes at different levels? Did the assessment lead to new scale-dependent policy or science agendas?
Shaping	What are the most important scale-dependent institutions shaping the knowledge system and the social context in which assessments are framed, conducted, and used? Do these reproduce scale biases?

Framing

Politics are often most intense in the initial stages when the terms of reference for an assessment are set. Even labeling an assessment as local, regional, national, international, or transboundary can be considered a political act. The framing of an assessment, including the adoption of particular spatial, jurisdictional, and temporal boundaries, matters because it entrains the types of problems that are addressed, the kinds of data sought, the methods of analysis employed, and the scope of explanations allowed.

Choices of boundaries and levels are critical because they are used to decide who is a stakeholder. Scale-dependent interests are likely to be "articulated" only if they are represented. Tightly set boundaries can ensure that off-site, higher-level interests are only weakly represented, to the advantage of local interests, and vice versa.

States, for example, are often very keen to keep assessments about large infrastructure projects at or below the national level even when transboundary impacts are likely or certain. The framing and marketing of feasibility studies for the "Thai water grid" policy, which includes significant diversion of water from neighboring regions in Myanmar and Lao People's Democratic Republic, is a fine example of "nationalizing" a politically sensitive transboundary issue (Lebel, Garden, and Imamura 2005).

Assessments may be "internationalized" as a way of diffusing what would otherwise be perceived as direct criticism of individual nations. One consequence is to shift priorities upward to large-scale, shared changes. In assessments of health impacts from climate change in Africa, for example, participation by developing countries depends on funding. National agencies are often willing to accept the "scale" priorities of donors as they conduct their "country studies," even though these may not fit closely with needs (Ogunseitan 2003). Thus the focus of the Intergovernmental Panel on Climate Change (IPCC) on climate change–induced burdens from malaria, schistosomiasis, and dengue/dengue hemorrhagic fever get much higher priority at national levels than they would have otherwise relative to other disease burdens faced by a nation (Ogunseitan 2003). Scale choices matter because they may result in "loss of opportunities for articulating local solutions to global problems with serious local repercussions" (Ogunseitan 2003).

International assessments of air pollution issues also shift over time in the number of chemical species considered, progressively becoming more inclusive. Thus the early focus on carbon dioxide finally gave way to more comprehensive assessments and institutional responses for other greenhouse gases in climate protection long after their effects were known to science. Similarly, the early emphasis on sulfur dioxide meant that the response to other reactants important for acid deposition was delayed (Jager et al. 2001). Although these are examples of expanding causal pathways rather than simple changes in space and time, the different longevities and transport ranges of various atmospheric pollutants imply changes in scales of modeling, monitoring, and analysis.

The range of ecosystem services that are directly used and acknowledged as having important support functions depends on sociocultural contexts, which are restricted in space. As an assessment is conducted at progressively larger scales, the number of services that are fully shared among places, and thus that can be mapped "wall-to-wall," drops. The local services that would be visible in a local assessment may no longer be visible in a subglobal or global assessment. The same basic ecosystem processes (e.g., net primary production by trees) can be seen as providing different services at different scales: slope stabilization at a local scale and timber at a regional scale but carbon sequestration at the global scale.

Although space is the most easily recognized domain of scale politics, choice of time scales can also matter. If an assessment focuses on short-term concerns, then "important" goods and services are those that are already or are about to be threatened, such as freshwater resources for drinking or fuel wood supplies and food production. On the other hand, if the users are more concerned with decisions that may have consequences over time spans of several decades to centuries, then such issues as alterations to carbon balance or opportunity and resilience costs of biodiversity loss become much more important. Politicians and scientists make value judgments when setting the goals of assessments, and these often include decisions about boundaries and resolution.

Dimitrov (2003) notes that the international assessment on forest provided good information on rate of deforestation and cover as well as some understanding about causes but produced very little information about nontimber goods and services. The United Nations Food and Agriculture Organisation (1995) asserted that "it is highly unlikely that it will be possible, in the near future, to make comprehensive inventories of non-wood goods and services on a global basis."

The Millennium Ecosystem Assessment (MA) aims to assess the current status and future threats to the world's ecosystem goods and services (Millennium Ecosystem Assessment 2003, 2005). The MA is noteworthy among international assessments for the emphasis it placed, from early on, on the importance of scale—and hence on the value of attempting a multilevel assessment because the interests of many actors are scale dependent. However, the dominant logic, at least initially, was instrumental rather than normative. A multilevel assessment was desirable because such an assessment could help test the validity of up- and down-scaling exercises in global models rather than because, for ecosystem goods and services, it would be the right thing to do. The decision in favor of a multilevel analysis may itself be seen as an outcome of scale politics, where certain groups of researchers were really pushing for global-level analysis and had to concede that other levels were also crucial for an effective assessment. Having opened the door to interests and research at regional and community scales, the MA then found that issues of scale were often points of debate between subglobal assessments prioritizing issues important for their region and the global working group looking to make assessments at larger scales.

Assessing

Scale biases arise from the resolution of the instrumentation used, the density and spatial distribution of the observation networks, the scope of the mapping, the scales at which experimental manipulations are feasible and ethical, the choices of statistical methods, and the assumptions made in models. Expanding the scale of analysis to new levels (scaling up) or disaggregating vulnerabilities and impacts spatially (scaling down) also introduces scale biases.

How scale-dependent information is assembled is particularly important in assessing land use changes because these are already "value-laden" issues. A good illustration is the way various forest assessments treat swidden or rotational forest-agriculture systems. At the patch scale, a recently burned hillside being prepared for upland rice and other crops may look like a disaster site, but when moved up a spatial scale to a landscape and up a temporal scale to a decade, the practice can appear when glimpsed—and is when measured—much more benign than, for example, permanently converting forest to annual crop agriculture, at least in terms of maintaining biodiversity and ecosystem functions (such as time-averaged carbon stocks).

Remote sensing has become an important tool for regional and international environmental assessments. It has the advantage of providing repeatable large-scale coverage of variables that are often correlated with environmental states and, with complementary groundwork, ecosystem functions. On the other hand, the ease with which images can be obtained and processed means remote sensing is also frequently misused. Thus arguments about the data to be used for assessments often rested primarily on the data's capacity for "covering" the spatial area to be assessed wall-to-wall rather than on what indicators or measurements were really needed to determine the status of an environmental function.

Assessments regularly emphasize quantitative data strongly, a prejudice that works against insights—for example, related to gender or household security—that come from smaller-scale, in-depth case studies, rendering these insights invisible to analysis. Alternative "assessment technologies," such as rapid rural appraisal and its cousins, were in part created to empower local interests and to resist homogenizing analyses (Chambers 1997; Scott 1998). Traditional knowledge may complement and extend instrument-based observations allowing assessments to consider longer time frames necessary to capture rare disturbances (Berkes 1999). Unfortunately, however, strong prejudices within many branches of science persist against nonconventional sources of observations even before their utility has been adequately tested (Forsyth 1998).

The power of maps (Crampton 2001) in assessments and plans is rarely critically examined, but here scale choices, resolution, and classes undoubtedly influence the information actually communicated. Evans (2004), in a study of urban regeneration in the Vincent Drive "brownfield" area near Birmingham, England, describes how the different phases of the ecological assessment produced findings that depended very much on how different land uses were classified, mapped, and visually displayed. One of the most problematic aspects of assessing environmental changes is the norms societies place on "naturalness" or other baselines used for comparison to judge and value impacts. These matter for resolution, for example, on whether different patches of vegetation in an urbanized landscape are classified as the same "ecologically" (Evans 2004), and for temporal scales, and on whether vegetation in succession after disturbance is considered "natural." Finally, Ross (1998) noted that using larger areas in an assessment (of a smaller, fixed area) means that a "smaller" proportion of people would be counted as affected. The opportunities for molding findings to fit interests are ever present.

Models are important tools in assessment and frequently get rescaled. As the ozone regime was unfolding, initial detailed grid models of the atmosphere to predict exposures important for human health were unjustifiably scaled up to develop transport models that could help assess source-receptor relationships (Farrell, VanDeveer, and Jager 2001). Like maps, models and statistics can be used to both hide and reveal scale-dependent relations. Superficial rescaling (up or down) is frequent, because technically there is nothing to stop naive or strategic users from doing so with their desktop computers and the available models.

Decision makers repeatedly demand for climate change assessments to be downscaled to the national and subnational jurisdictions for which they have some responsibility and decision-making influence. This has proven challenging, though significant progress has been made, at least at larger regional scales (Intergovernmental Panel on Climate Change 1997). On the other hand, many dubious local assessments are made using climate model outputs inappropriately scaled down, for example, for highly uncertain rainfall. These system failures can be interpreted as researchers willing to carry out work they know is dubious for money and as decision makers seeking to be seen as doing something even when they know it does not really mean anything. Policy may demand levels of resolution that science cannot deliver. It takes honesty and humbleness to say, "We cannot do that (yet)."

Multilevel participation may be a practical way to reduce biases in assessment resulting from the failure to take into account key level-dependent knowledge and interests. The U.S. National Assessment of Possible Consequences of Climate Variability and Change was an intense $14 million, three-year assessment exercise that included sectoral, regional, and national assessments (Wolfe, Kerchner, and Wilbanks 2001). One key challenge it faced was how to elicit and make use of diverse public input effectively. Although it moderately improved stakeholder involvement, this did not necessarily mean the assessment was better or more successful as defined by conventional criteria, such as reducing technical uncertainties and influencing policy decisions. Unfortunately, the impacts of various forms of public participation on the quality of assessments and environmental decision making have rarely been studied (Rayner 2003).

Using

International environmental assessments vary greatly in how much influence they have on decision making (Social Learning Group 2001a). Part of this variation can be attributed to scale politics. For example, several international assessment about forests have not resulted in any progress toward an international forest regime, in part because cross-border consequences of changes in forest cover are poorly understood (Dimitrov 2003) and countries with substantial forest resources still left to exploit are keen to keep full control of rents flowing from harvests in their territories. On the other hand, where stakes are modest, states may be quite willing to participate in collective responses to assessments.

How assessments influence policy at different spatial scales may result more from the changes in institutional form and the capacities of authorities at the corresponding jurisdictional levels than from a direct result of the choice of geographical areas to which these apply. Systems of property right for same kinds of ecosystem goods and services may shift with jurisdictional level, reflecting practical limitations of monitoring and enforcement with expanding scales (Berkes 2002; Young 1994). Cooperation among individuals at small scales, and among collective entities like organizations or provinces at larger scales, is qualitatively different and may result in different institutional arrangements in support of collective actions (Ostrom 2003; Young 2002b). Institutional forms are in part scale dependent, and multilevel assessments of response options must account for this.

The focal level of an assessment may not be maintained once it enters public discourse. Actors may strategically shift the scale of findings. Mass media frequently "play up" events or risks to make these seem larger in magnitude, duration, or extent, especially when they involve people with which the media's target consumers empathize. Misreporting of environmental assessments is common and is used strategically by some actors. Even without shifting claims about the scale at which an issue is framed, actors may nevertheless strategically link with other powerful actors to expand the size of their network, a process aptly described as "constructing a scale of engagement" (Cox 1998).

Vertical interplay among institutions (Young 2002a), which clearly involves a politics of scale spanning jurisdictional levels, may be influenced by, and help shape, environmental assessments. This is a common way that international assessments and regimes interact, even in the face of substantial uncertainties in the assessments. The ozone regime, for example, was formed even though at the time substantial uncertainty existed about the extent of ozone depletion—but not about the serious consequences for plants and human health (Dimitrov 2003; Haas 1992). Likewise, the RAINS (Regional Air Pollution and Simulation) model strongly influenced the Convention on Long-Range Transboundary Air Pollution regime even though it was based on uncertain data inputs, simplistic assumptions of horizontal transport, and a crude resolution of 150- by 150-square-kilometer grid cells (Dimitrov 2003; Lidskog and Sundqvist 2002). The models also produced findings on emissions at policy-relevant national level. Perceptions of neutrality, however, were extremely important. Thus assessments both shape and are shaped by changes in scale use and understanding.

Shaping

The extent and persistence of scale-dependent interests, capacities, and beliefs help explain why the politics of scale emerges repeatedly in environmental assessments. Scale-dependent interests arise with respect to the benefits received from resource flows or ecosystem services and also with respect to the exposures to involuntary risks or apportioning blame.

Scale-dependent capacities include livelihood skills and access rights to environmental resources or other resources needed to exploit them. Critical links in social networks may also be largely constrained to particular levels; for most people, these are primarily local. Actors working at the jurisdictional levels of the nation-state often command greater resources, have better access to information, and therefore can perform better in contests over scale choice. Local government often does not have the expertise, the financial resources, or the access to do the preparatory work in making submissions or in commissioning level-relevant research. Control of the resources needed to carry out an assessment, such as funding of the secretariat, as well as access to information, avenues of political endorsement, media reporting, and participation in meetings can all influence what kind of scale-dependent information gets to the table (e.g., Goldman 2004).

Finally, scale-dependent beliefs arise both out of the networks of interaction and learning of the actors and out of the scales of direct experiences. Thus, while biophysical phenomena may often involve generalized knowledge, assessing impacts on health or livelihoods, for example, often suggests much more heterogeneous outcomes that are context specific and bound to local-level understanding.

The existence of "global" assessments on climate change, biodiversity, and now ecosystem goods and services reflects not just the realities of widespread, cumulative, and interactive changes in the "Earth System" but also the power of the earth system discourses (Adger et al. 2001; Dryzek 1997). The skills with which important processes of change have been identified have led to regular explicit and implicit calls for planetary or earth system management (Sachs 1993)—that is, that the "proper" scale (for management, or decision making) is global, beyond the nation-state. If misapplied, the global change discourse can be too strong and can displace policy attention about serious problems of environmental change at much smaller scales—such as securing clean drinking water and eliminating exposure to local air pollution in and outside the home, both common and hugely important problems for health in the developing world. On the other hand, an alternative discourse within global environmental change research recognizes the cross-scale and multiscale nature of these changes and the dangers of prioritizing those processes that a group of researchers or a discipline happens to study at the expense of others. Rather, global environmental change here is seen as an important confounder of what are already important processes at regional and local scales (Tyson et al. 2002).

Scale-dependent interests, capacities, and beliefs lay the foundations for actors to adopt different strategies—cooperative and resistive—in engaging scale politics (table 3.2). When combined, these two strategies produce additional strategies: bundling and merging scales.

An excellent example of shifting upward is the way the director of the World Health Organization's Roll Back Malaria initiative effectively used global warming impact assessments on the spread of malaria risk to the northern states as a way of securing financial support for essential malarial control programs in Africa (Ogunseitan 2003).

Opponents of large dam projects in Thailand and in the Mekong Region have been quick to latch onto larger-scale assessments and institutional processes. The World Commission on Dams report (2000) provided guidelines for approaches to negotiation at the appraisal stage. It could thus be argued that this report has had an impact on the Pak Mun dam politics in Thailand, as it was one of the eight highlighted case studies in the report.

Similarly, international-level climate change and ozone assessments have helped legitimize national-level assessments (Jager et al. 2001). Typically, new knowledge provided by an assessment can help interests to form by at least two pathways (Dimitrov 2003). First, it can provide information that allows actors to make improved strategic calculations about how to maximize their own benefits. Second, it can expose, across levels, shared interests that were not formally perceived.

International networks of scientists exchanging data and visiting one another's laboratories and field sites can make it very difficult to control or manipulate the scientific information that goes into assessments, should anyone wish to do so (Haas 1992). In the cases of both ozone and forests, powerful actors have not successfully suppressed or manipulated information counter to their interests (Dimitrov 2003). Scientific networks provide credibility to the political processes in environmental assessment by increasing the consensus at the international level.

This brings us to the last, and undoubtedly most uncomfortable, class of mechanisms by which politics of scale unfold: through shaping the knowledge systems and the contexts in which assessments are defined and conducted. Here the first question to ask is: what is not assessed at a particular level and why not?

Here is one example to illustrate the idea. Consumption growth, the ultimate driver of cumulative environmental changes at multiple levels, has not received

Table 3.2

Actors' strategies and their institutional contexts in engaging the politics of scale in environmental assessments

Strategies	
Rescaling to interest	Actors shift issues and analytical methods down/back or up/forward levels along a scale in ways that support or protect their own interests—for example, toward levels where they have greater capacities from access to or control of resources, or away from levels that would associate them with blame.
Rescaling to beliefs	Actors shift scales to fit their beliefs (about causes, changes, or consequences) that, in turn, have been shaped by formal and informal institutions with all their inherent scale biases, whether disciplinary, political, ideological, or cultural.
Rescaling to capacities	Actors shift scales where the issue is considered to fit the levels at which they have the greatest influence on negotiations—even if their interests are at another level.
Bundling to conceal	Actors bundle more difficult (e.g., controversial) issues from other levels with easier issues at their preferred level in the hope that the more difficult issues will then be "accepted" by others with less scrutiny or so that actors can be seen to negotiate on one level without having to "trade" at another.
Merging for consensus	Actors in assessment processes are frequently under pressure to reach consensus. Those with the most at stake, such as coordinators, may thus push for narrowing the scope to levels for which consensus can be reached by dropping controversial levels.

anything like the sustained attention given to population growth. Impacts of consumption growth conveniently stop at the edge of the farmer's field, and drivers of consumption growth travel no farther than the first market. But many of the key environmental changes are driven by consumption growth for products that are traded and consumed in distant locations (Lebel 2004; Princen,

Maniates, and Conca 2002). From the point of view of environmental assessments, this is a critical rescaling because it means that the stakeholders with interests in the resources, in the driving environmental changes, and in the response options have become multilevel. Developed countries have strong interests in keeping the analysis local with respect to commodity networks and distant with respect to *places with problems.* And they have pushed this strategy successfully: studies of deforestation, desertification, and other forms of land degradation in developing countries abound, but the key driver is invariably identified as excessive population growth. Environmental and poverty assessments in developing countries thus often end up targeting population policies. This targeting has been driven largely by an underlying reasoning whereby differences in competitive human fertility are seen as a threat to the long-term dominance of those developed economies (and, more pointedly, to the currently dominant or powerful ethnicities or races in these economies).

Funding agencies influence what scientists research. The scale specificity of this role for environmental issues varies over time and among players. This is most visible in the rises and falls for the support of international collaborative research on environmental risks, and in the difficulties faced by traditional science funders in handling studies into nonconventional knowledge, whether local, indigenous, or tacit. Nolin's comparison (1999) of national climate change research in four countries belonging to the Organisation for Economic Co-operation and Development is telling. He shows that once the reality of climate change was accepted and then placed on the policy agenda, funding for additional climate-oriented research, as per IPCC Working Group I, stabilized and maybe even fell. Overly successful communication can undermine a researcher's interests.

Much funding for global environmental change research—for example, from the Global Environmental Facility or the U.S. Country Studies Program—focused initially on emission mitigation, even in developing countries with relatively trivial cumulative contributions to the greenhouse gas emissions and where their needs were much more strongly related to vulnerability and adaptation measures. Mitigation issues are clearly global level, whereas those of adaptation and vulnerability are invariably constructed more usefully at the more local levels of states, provinces, and communities. Developing country partners play along with these irrelevant frameworks in return for financial support.

Formal and informal institutions shape what we know and what we think we need to know about scale in environmental assessments (Lebel et al. 2004).

Implications

Science and policy interact in framing, assessing, and using environmental assessments, and in each phase politics-of-scale issues arise with respect to the use and influence of knowledge. Consideration of scale-dependent interests, capacities, and beliefs—viewed through the lens of the politics of scale—support and extend several key generalizations that have emerged from comparative studies of global environmental assessments (Jasanoff and Wynne 1998; Social Learning Group 2001a, 2001b).

First, assessments are best conceived as social processes involving learning, coalition building, bargaining, and negotiation among researchers and decision makers across levels (Selin and Eckley 2003). Second, assessments handling cross-scale issues are most effective when they are perceived by participants as salient, credible, and legitimate (Social Learning Group 2001a). Third, assessment must engage with the scale-dependent understandings and capacities that lie at each level and must recognize the potential of boundary organizations to enhance the sharing of understanding across levels (Cash 2000; Cash and Moser 2000; Guston 2001; Young 1994).

The knowledge-governance interface is multilevel and multicentered. Environmental assessments are an increasingly important platform at this interface, which irregularly opens and closes to nonstate actors.

Multiple levels will often be highly desirable because, a priori, "the" appropriate spatial level is not known by ecologists, nor is the appropriate jurisdictional level known by social scientists. Key ecological processes may be level dependent or multilevel dependent, so obtaining information from multiple sources and analyzing it may provide a more precise understanding of a phenomenon at the focal level of interest. Because the spatial distribution of the social impacts may differ from that of the environmental impacts, a single-level or single-boundary definition may be misleading. A multilevel strategy also needs to be flexible; the choices of level may have to be modified and renegotiated over time as understanding and perceptions of causes, changes, and impacts change. Because governance arrangements at different levels do not function the same way (Young 1994), their requirements from assessments are also likely to differ. Thus multilevel assessment efforts are increasingly valuable to the realities of multilevel decision making.

Multilevel representation helps counteract the loss of legitimacy from single-level consideration of an issue, even if an assessment has a particular focal level of interest to which it must report. Effective representation will usually require providing special assistance or access channels to participating minorities and vulnerable people—to articulate needs, to contribute understanding, and to help with interpreting findings.

On the other hand, creating new levels for an assessment may be counterproductive or cause conflicts that prevent a needed assessment from proceeding. A well-framed assessment may have to exclude some levels of analysis to proceed. Actors, including assessors, however, should always be challenged to justify their scale positions and the scale choices made in assessments. Transparency in scale choices improves legitimacy because all actors start from a shared understanding of scope and assumptions and thus can challenge the choices if they are inappropriate.

Shifting to a multiple-level assessment does not remove scale politics. On the contrary, it can be expected to empower actors who can work effectively at multiple levels. Among these, we should not be surprised to find wealthier, better-educated, and more mobile international scientists and diplomats.

Centralized assessment processes, as in the early phases of IPCC, have served us well for understanding large-scale environmental changes but have been less help in supporting national and more local decision making to assess vulnerabilities and mitigation actions (Jasanoff and Wynne 1998). Ultimately, distributed systems of research, assessment, and management—such as the Pacific ENSO Applications Center, which partners the U.S. National Oceanic and Atmospheric Administration with Pacific Island climate agencies to develop and disseminate ENSO (El Niño/Southern Oscillation) forecasts (Cash 2000)—may better deal with multilevel issues (Cash et al., forthcoming).

International environmental assessments do not have to formally engage states. Indeed, avoiding processes of formal endorsement by government may allow assessments to address issues that would otherwise get bogged down in diplomatic battles of blame and counterblame. The World Commission on Dams, for example, set global standards without special reference to states, allowing highly sensitive national and transboundary issues of water infrastructure to at least be addressed (World Commission on Dams 2000).

In the future, citizen-led assessments calling on scientific experts to review and synthesize information according to their own terms of reference are likely to become more common. These assessments will often shift the focal level of interest downward to levels closer to those people experience in normal social interactions. A tough issue for localized assessments is how to downscale or place-transpose science done at lower resolution or elsewhere in a way that does not destroy the credibility of the assessment. Regardless of which levels are considered, the key is that legitimacy comes from the discursive interactions, not just from the formal endorsement of governments (Dryzek 1997).

The proliferation of national and international environmental assessments—perhaps relabeled as "integrated" or "sustainability" assessments—will likely continue. A knowledge system that encourages multiple competing research and development efforts and that allows alternative models and analyses to continually arise and compete for explanatory space is likely to be more robust. Similarly, assessments should encourage independent evaluations at different levels of the assessment process and use the output products in ways that facilitate social learning across levels and assessment exercises. Unfortunately, one of the likely side effects will be *assessment fatigue*, whereby more and more of the tired researcher's time is taken away from those observations and analyses needed to form the foundations of a credible environmental assessment.

Conclusions

Four messages arise from this chapter's consideration of the politics of scale in environmental assessments. First, scientists, policy makers, and citizens involved in environmental assessments should not be allowed to make scale choices secretively, because these choices matter too much. Rationale, criteria, and assumptions related to scale and level decisions need to be made transparent. Second, major uncertainties about scale dynamics and response options still exist for many pressing issues about the environment. Hence, it makes intuitive sense to start with the assumption that a multilevel assessment may be needed while also recognizing that this does not eliminate scale politics.

Third, many other important issues of governance that should be addressed in environmental assessments are not restricted to issues of scale—including politics of place and position as well as more general issues of transparency, accountability, representation, and responsibility. A politics of space and scale out in the open is usually a good sign for society, not a bad one.

Fourth, and more sinisterly, much of the politics of scale has been like a play of shadows, shaping the appearances of what is studied and assessed. Although scale negotiation clearly matters for the design, conduct, and outcome of environmental assessment processes, this does not mean that assessments can be rescaled with impunity. Biophysical processes have complex scale realities that society can misunderstand, or choose to ignore or distort for a while, but history tells us that ecosystems have a way of coming back to remind us of our past.

Acknowledgments

This chapter is based on research supported in part by a grant from the U.S. National Oceanic and Atmospheric Administration's Office of Global Programs for the Knowledge Systems for Sustainable Development Project, led by William C. Clark at the Kennedy School of Government, Harvard University. Thanks are also due to START (the global change SysTem for Analysis, Research and Training), the Asia-Pacific Network for Global Change Research, and the Millennium Ecosystem Assessment for their support for various science-policy and assessment activities in which I have been fortunate to participate.

References

Adger, W. N., T. A. Benjaminsen, K. Brown, and H. Svarstad. 2001. Advancing a political ecology of global environmental discourses. *Development and Change* 32:681–715.

Berkes, F. 1999. *Sacred ecology: Traditional ecological knowledge and management systems.* London: Taylor and Francis.

———. 2002. Cross-scale institutional linkages for commons management: Perspectives from the bottom up. In *The drama of the commons,* ed. E. Ostrom, T. Dietz, N. Dolsak, P. C. Stern, S. Stonich, and E. U. Weber, 293–321. Washington, DC: National Academy Press.

Blaikie, P. M., and J.S.S. Muldavin. 2004. Upstream, downstream, China, India: The politics of environment in the Himalayan region. *Annals of the Association of American Geographers* 94:520–48.

Brenner, N. 2001. The limits to scale? Methodological reflections on scalar structuration. *Progress in Human Geography* 25:591–614.

Cash, D. W. 2000. *Distributed assessment systems: An emerging paradigm of research, assessment and decision-making for environmental change.* Cambridge, MA: Global Environmental Assessment Project, Belfer Centre for Science and International Affairs, Harvard University.

Cash, D. W., N. W. Adger, F. Berkes, P. Garden, L. Lebel, P. Olsson, L. Pritchard, and O. R. Young. Forthcoming. Scale and cross-scale dynamics: Governance and information in a multi-level world. *Ecology and Society.*

Cash, D. W., and S. C. Moser. 2000. Linking global and local scales: Designing dynamic assessment and management processes. *Global Environmental Change* 10:109–20.

Chambers, R. 1997. *Whose reality counts? Putting the last first.* London: Intermediate Technology.

Conca, K. 2004. Ecology in the age of empire: A reply to (and extension of) Dalby's imperial thesis. *Global Environmental Politics* 4:12–19.

Cox, K. R. 1998. Spaces of dependence, spaces of engagement and the politics of scale, or: Looking for local politics. *Political Geography* 17:1–23.

Crampton, J. 2001. Maps as social constructions: Power, communication and visualisation. *Progress in Human Geography* 25:235–52.

CSIRO Sustainable Ecosystems. 2003. *Natural values: Exploring options for enhancing ecosystem services in the Goulburn Broken Catchment.* Canberra: CSIRO Sustainable Ecosystems.

Dimitrov, R. S. 2003. Knowledge, power and interests in environmental regime formation. *International Studies Quarterly* 47:123–50.

Dryzek, J. S. 1997. *The politics of the earth: Environmental discourses.* Oxford: Oxford University Press.

Dube, M. G., and K. R. Munkittrick. 2001. Integration of effects-based and stressor-based approaches into a holistic framework for cumulative effects assessment in aquatic ecosystems. *Human Ecology and Risk Assessment* 7:247–58.

Evans, J. 2004. Political ecology, scale and the reproduction of urban space: The case of Vincent drive. Working paper series, School of Geography, Earth and Environmental Sciences, University of Birmingham, England.

Farrell, A., S. D. VanDeveer, and J. Jager. 2001. Environmental assessments: Four under-appreciated elements of design. *Global Environmental Change* 11:311–33.

Fischer, T. B., and K. Seaton. 2002. Strategic environmental assessment: Effective planning instrument or lost concept? *Planning Practice and Research* 17:31–44.

Forsyth, T. 1998. Mountain myths revisited: Integrating natural and social environmental science. *Mountain Research and Development* 18:126–39.

———. 2003. *Critical political ecology: The politics of environmental science.* London: Routledge.

Goldman, M. 2004. Imperial science, imperial nature: Environmental knowledge for the World (Bank). In *Earthly politics: Local and global environmental governance,* ed. S. Jasanoff and M. L. Martello, 55–80. Cambridge, MA: MIT Press.

Guston, D. H. 2001. Boundary organizations in environmental policy and science: An introduction. *Science, Technology and Human Values* 26:399–408.

Haas, P. M. 1992. Banning chlorofluorocarbons: Epistemic community efforts to protect stratospheric ozone. *International Organisation* 46:187–224.

Hirsch, P. 2001. Globalisation, regionalisation and local voices: The Asian Development Bank and rescaled politics of environment in the Mekong region. *Singapore Journal of Tropical Geography* 22:237–51.

Intergovernmental Panel on Climate Change (IPCC). 1997. *The regional impacts of climate change: An assessment of vulnerability. Summary for policymakers.* Geneva, Switzerland: IPCC.

Jager, J., J. Cavender-Bares, N. M. Dickson, A. Fenech, E. A. Parson, V. Sokolov, F. L. Toth, C. Waterton, J. van der Sluijs, and J. van Eijndhoven. 2001. Risk assessment in the management of global environmental risks. In *Learning to manage global environmental risks: A functional analysis of social responses to climate change, ozone depletion and acid rain,* ed. S. L. Group. Cambridge, MA: MIT Press.

Jasanoff, S., and B. Wynne. 1998. Science and decisionmaking. In *Human choice and climate change: The societal framework,* ed. S. Rayner and E. L. Malone, 1–87. Columbus, OH: Batelle Press.

Joao, E. 2002. How scale affects environmental impact assessment. *Environmental Impact Assessment Review* 22:289–310.

Lebel, L. 2004. Transitions to sustainability in production-consumption systems. *Journal of Industrial Ecology* 9:1–3.

Lebel, L., A. Contreras, S. Pasong, and P. Garden. 2004. Nobody knows best: Alternative perspectives on forest management and governance in Southeast Asia: Politics, law and economics. *International Environment Agreements* 4:111–27.

Lebel, L., P. Garden, and M. Imamura. 2005. Politics of scale, position and place in the governance of water resources in the Mekong region. *Ecology and Society* 10 (2): 18. http://www.ecologyandsociety.org/vol10/iss2/art18/ (accessed May 25, 2006).

Lidskog, R., and G. Sundqvist. 2002. The role of science in environmental regimes: The case of LRTAP. *European Journal of International Relations* 8:77–101.

Meadowcroft, J. 2002. Politics and scale: Some implications for environmental governance. *Landscape and Urban Planning* 61:169–79.

Mekong River Commission. 2003. *State of the basin report.* Phnom Penh, Cambodia: Mekong River Commission.

Millennium Ecosystem Assessment. 2003. *Ecosystems and human well-being: A framework for assessment.* Washington, DC: Island Press.

———. 2005. *Ecosystems and human well-being: Synthesis.* Washington, DC: Island Press.

Nolin, J. 1999. Global policy and national research: The international shaping of climate research in four European Union countries. *Minerva* 37:125–40.

Ogunseitan, O. A. 2003. Framing environmental change in Africa: Cross-scale institutional constraints on progressing from rhetoric to action against vulnerability. *Global Environmental Change* 13:101–11.

Ostrom, E. 2003. How types of goods and property rights jointly affect collective action. *Journal of Theoretical Politics* 15:239–70.

Princen, T., M. Maniates, and K. Conca, eds. 2002. *Confronting consumption.* Cambridge, MA: MIT Press.

Rayner, S. 2003. Democracy in the age of assessment: Reflections on the roles of expertise and democracy in public-sector decision making. *Science and Public Policy* 30:163–70.

Ross, W. 1998. Cumulative effects assessment: Learning from Canadian case studies. Impact Assessment Project Appraisal 16:267–76.

Sachs, W., ed. 1993. *Global ecology: A new arena of political conflict.* London: Zed Books.

Scott, J. C. 1998. *Seeing like a state.* New Haven, CT: Yale University Press.

Selin, H., and N. Eckley. 2003. Science, politics, and persistent organic pollutants: The role of scientific assessments in international environmental co-operation. *International Environmental Agreements: Politics, Law and Economics* 3:17–42.

Social Learning Group. 2001a. *Learning to manage global environmental risks: A comparative history of social responses to climate change, ozone depletion and acid rain.* Cambridge, MA: MIT Press.

———. 2001b. *Learning to manage global environmental risks: A functional analysis of social responses to climate change, ozone depletion and acid rain.* Cambridge, MA: MIT Press.

Swyngedouw, E. 2000. Authoritarian governance, power and the politics of rescaling. *Environment and Planning D: Society and Space* 18:63–76.

Tyson, P., R. Fuchs, C. Fu, L. Lebel, A. P. Mitra, E. Odada, J. Perry, W. S. Steffen, and H. Virji, eds. 2002. *The earth system: Global-regional linkages.* Heidelberg, Germany: Springer-Verlag.

United Nations Food and Agriculture Organisation (FAO). 1995. *Forest resources assessment 1990: Global synthesis.* Rome: FAO.

Wolfe, A. K., N. Kerchner, and T. J. Wilbanks. 2001. Public involvement on a regional scale. *Environmental Impact Assessment Review* 21:431–48.

World Commission on Dams. 2000. *Dams and development: A new framework for decision-making.* London: Earthscan Publications.

Young, O. R. 1994. The problem of scale in human/environment relationships. *Journal of Theoretical Politics* 6:429–47.

———. 2002a. *The institutional dimensions of environmental change: Fit, interplay and scale.* Cambridge, MA: MIT Press.

———. 2002b. Institutional interplay: The environmental consequences of cross-scale interactions. In *The drama of the commons,* ed. E. Ostrom, T. Dietz, N. Dolsak, P. C. Stern, S. Stonich, and E. U. Weber, 263–92. Washington, DC: National Academy Press.

CHAPTER 4

Assessing Ecosystem Services at Different Scales in the Portugal Millennium Ecosystem Assessment

HENRIQUE M. PEREIRA, TIAGO DOMINGOS, AND LUÍS VICENTE

The Portugal Millennium Ecosystem Assessment (ptMA) (http://ecossistemas.org) is analyzing the condition of ecosystem services in Portugal, recent trends in those services, available policy responses, and scenarios for the next fifty years, following the conceptual framework of the Millennium Ecosystem Assessment (MA) (Millennium Ecosystem Assessment 2003). It is a multiscale assessment, carried out at the national, basin, and local scales. The assessment started in 2003, published a status report after two years (Pereira, Domingos, and Vicente 2004), and will publish the final results in 2006.

Of the thirty MA subglobal assessments, ptMA and the Southern African Millennium Ecosystem Assessment are the only two comprehensive multiscale assessments. As illustrated elsewhere in this volume, multiscale assessments allow the evaluation of the robustness and persistence of findings across scales and provide information benefits: more and better data, ground-truthing of data, and better analysis of the causes of change (Zermoglio et al. 2005).

Here we present the findings that resulted from conducting ptMA as a multiscale assessment and from ptMA itself being nested in the global MA. We start by describing how the different scales were analyzed in ptMA and how the MA framework was adapted. Next, we discuss how epistemologies were bridged within the assessment, including what barriers existed in the communication among experts in different fields, how the users interacted with the scientists, and how the scientists interacted with the local population in one of the case studies. We then present an overview of the findings of ptMA, including a discussion of the indirect drivers of ecosystem change, direct drivers and trends in ecosystems and their services, and future scenarios. The findings overview section provides a context for the next section, where we contrast the findings at multiple scales, from the local scale in ptMA to the global scale in the MA. The multiscale findings section also discusses how responses analyzed in ptMA at the local scale are related to changes in drivers and ecosystems at a larger scale.

Scales in the Portugal Assessment

At the national scale, ptMA is organized into ecosystems based on the global MA systems: marine, coastal, inland water, forest, *montado,* island, mountain, cultivated, and urban. *Montado,* which is not a global MA system, is an ecological and economically important ecosystem of Portugal and Spain (where it is called *dehesa*). *Montado* is an agroforestry system in which the main activities are cork, livestock, and cereal crop production. It is an evergreen oak woodland; the predominant tree species are the cork oak (*Quercus suber*) and the holm oak (*Quercus ilex*). *Montado* corresponds roughly to the intersection of the dryland and forest systems in Portugal.

The assessment has two case studies at the basin scale: the heavily human-influenced Mondego basin and the more pristine Mira basin. At the local level, the assessment has four case studies: Sistelo is a parish (about 340 inhabitants) located in a mountain range, where the main economic activity is agropastoralism; Quinta da França (QF) is a farm with an ongoing research program on agricultural sustainability; Herdade da Ribeira Abaixo (HRA) is a biological research station in an area of *montado*; and Castro Verde is a municipality (about 7,500 inhabitants) with a farming area of cereal steppes. (See the map in figure 4.1.)

Scales and study cases at each scale were chosen to balance relevance to users, availability of data, and coverage of different ecosystems (table 4.1). This constraint reduced the possibility of "nestedness" of the different case studies within each other (this was achieved, however, by integrating the case studies within the ecosystems at the national scale). It should also be noted that the case studies are not fully representative of the country but instead are designed to provide particular insights into ecosystem changes at each scale.

Figure 4.1.

Geographic situation of Portugal (mainland and islands) and of the case studies of the Portugal Assessment. Basin-level case studies: Mondego basin (1) and Mira basin (2). Local-level case studies: Sistelo (3), Quinta da França (4), Herdade da Ribeira Abaixo (5), and Castro Verde (6).

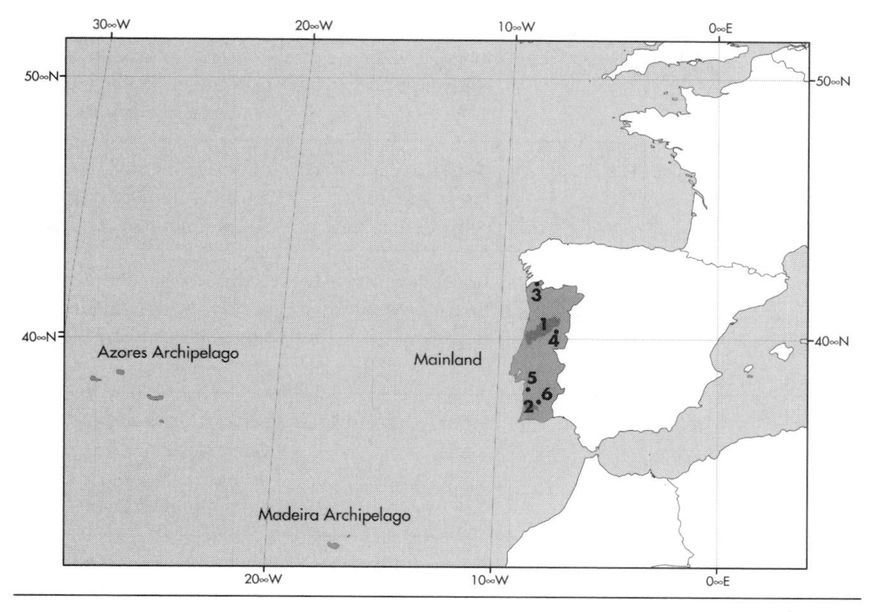

Adapting the MA Framework for the Portugal Assessment

The Portugal Assessment joins more than thirty-five scientists from the natural and social sciences, including subteams of two or three scientists for each study case. The scientists work together with the primary audience of the assessment: a group of ten users representing different societal sectors, including national and local government, nongovernmental organizations (NGOs), agriculture, and industry. However, the intended audience of the assessment is broader, including all professionals whose work depends on ecosystems or affects ecosystems as well as the general public. The assessment was also designed to provide feedback to the global MA and to validate the global MA results at a subregional scale. At the national level, most of the data were assembled from the literature and from expert opinion. This approach was also followed in the study cases, complemented with fieldwork by the assessment team at the local level.

Table 4.1
Case studies of the Portugal Assessment
(M = municipality, SM = submunicipality)

Case Study	Type	Area (km²)	Ecosystems	Justification
Mira	Basin	1,576	Coastal, inland water, forest, dryland, cultivated	Institute for Nature Conservation (national user) requested an estuary in Southern Portugal
Mondego	Basin	6,670	Coastal, inland water, forest, mountain, cultivated, urban	Intensively studied by a research group at University of Coimbra (Portugal)
Castro Verde	Local, M	567	Dryland, cultivated (pseudo-cereal steppe)	League for the Protection of Nature (national user) has a bird conservation program here
Sistelo	Local, SM	27	Mountain, cultivated, forest	National Park of Peneda-Gerês (local user) interested in protecting agricultural terraces
QF	Local, farm	5	Forest, cultivated	Pilot farm of ExtEnSity (national user)
HRA	Local, farm	2	*Montado*	Biological research station of CBA (leader of the assessment)

The MA approach consists of identifying the major direct and indirect drivers of ecosystem change, assessing the impacts of those drivers on biodiversity and ecosystem services, and establishing the linkages between ecosystems and human well-being. A *driver* is a natural or human-induced factor that directly or indirectly causes a change in ecosystem services (Petschel-Held et al. 2005). The condition of an ecosystem service is defined by ptMA as *the current capacity of the ecosystem to provide the service relative to the capacity at which the service could be maximized in a sustainable way* (Pereira, Domingos, and Vicente 2004). This definition emphasizes the status of the natural capacity of the ecosystem to continue to provide the service into the future, which can be equated with the economic "stock" in some instances.

The Portugal Assessment takes two different approaches to analyze the condition of ecosystem services at the national scale: (1) an analysis of the condition of a set of ecosystem services nationally and (2) an integrated analysis of each ecosystem. In a sense, the ecosystems approach is spatially nested within the services approach and could be seen as an intermediate scale of assessment between the national and the basin/local scale. However, in the services approach, the focus is on each ecosystem service, and on the most important areas for each service, while the ecosystems approach emphasizes how each ecosystem is functioning.

Socioecological scenarios are being developed by ptMA. Looking fifty years into the future at a spatial scale as small as Portugal means that the main drivers will likely have an external component. This can limit the usefulness of the scenarios for policy makers because national actors have little control of external drivers. Nevertheless, scenarios are potentially useful instruments for testing national policies. Early in the process, the most important and uncertain drivers of ecosystem change were identified. These drivers were to be used as axes for defining the scenarios, but it turned out that they were quite analogous to the axes of the global MA scenarios, which hinted to the assessment team that it was possible to use the work already done by the MA Scenarios Working Group. Adopting the global MA scenarios would also provide a regional calibration and validation of the global scenarios. Thus the assessment team decided to adopt the global scenarios as boundary conditions for the Portugal scenarios.

Because using global scenarios as boundary conditions can somewhat constrain the development of scenarios, a slightly different approach was followed at the local level in the Sistelo case study. The scenarios were first developed with the local stakeholders and are now in the process of being integrated by the research team into the national-scale scenarios.

We consider primarily responses as actions taken by people following an ecosystem change, or following a perception of threats and opportunities associated with an ecosystem change (Malayang, Kumar, and Hahn 2005). The purpose of a response, as an act, is to improve human well-being.

Bridging Epistemologies

The assessment's first epistemologies issue was how to bring scientists from very different areas of expertise to a common ground with user representatives. The latter group consisted of decision makers and officials with a technical background, so the technical gap between the users and the scientists was at most as wide as the gap between natural and social scientists. The actual gap that existed was more of a stakes gap (e.g., a representative from the paper and pulp industry has a different perspective from a conservation biologist), because a user and a scientist may place priorities on different ecosystem services. These divergences also occurred among the user representatives themselves, who represented different societal sectors.

Two exercises were key in bridging these gaps: the scenarios development and the qualitative assessment of conditions and trends. The emphasis placed on creating descriptive narratives for the scenarios allowed for easy communication. Further, the scenarios' "if . . . then . . . " approach allows for exploring different possibilities and trade-offs. Identifying these trade-offs is a good starting point for the open discussion of compensation mechanisms, including financial mechanisms. Surprisingly, it is the scientific basis of the narratives that often causes more disagreement. Socioecological systems are extremely complex, and we frequently lack the scientific knowledge to predict how they will evolve. Therefore, scientists may disagree on the future trajectory for a socioecological system, even when the evolution of such drivers as market regulation and society's behavior is well defined. These disagreements may be difficult to solve when the narratives are based on expert opinion (as in our case), which often mixes scientific knowledge with the experts' personal preferences and beliefs. We found that it is important to have team members with contrasting points of view, as long as the members are also flexible.

The qualitative assessment of conditions and trends also contributed to identifying and bridging gaps between participants. Scientists and users had to agree in plenary on the condition of each service in each ecosystem—on a scale of 1 (bad) to 5 (excellent)—and on the trend for that condition (decreasing, stable, or increasing). Our definition of *condition* emphasized the sustainability of the service, and we found that it is not simple even for two natural scientists to agree on how sustainable the management of a given service is. For instance, the sustainability of marine fisheries depends on the subset of the stocks being evaluated and on the criteria used to evaluate the condition of each stock. A further problem occurs when discussing trends. For example, water production has been increasing in Portugal, but the sustainability of the service has been decreasing.

In the Sistelo study case, the assessment focused on understanding the linkages between human well-being and ecosystem services. A participatory approach was used, in which people were asked in interviews and small discussion forums to assess their well-being, the importance of ecosystem services, and the linkages between the two (Pereira et al. 2005). Here, the challenge was twofold. On the one hand, a degree of trust needed to be established with the population so that people could freely express their opinions. On the other hand, communicating with the local people, who had little formal training, proved challenging. The local assessment team approached the first task by participating in community daily activities during the field visits. This also allowed the team to learn more about the community and add direct observation data to the assessment. The team addressed the second challenge by asking the local people, sometimes using pictures as an aid, to use their own terms and expressions to explain a given reality. Overall, this approach empowered the local people—that is, instead of having the researchers dominate the assessment and define the key issues, the local people played a strong role in defining what should be assessed and how.

Finally, let us give an example of how scales interact with epistemologies. A major challenge in an assessment is for experts at a given scale to contextualize their findings within a broader scale (for instance, to understand how biodiversity change in response to agricultural abandonment at the hectare scale will scale to the national level). Part of this difficulty is intrinsically related to not knowing enough about the situation at the larger scale: the detail obtained at the smaller scale is often lacking at the larger scale.

Findings of the Portugal Assessment: An Overview

The following drivers of ecosystem change in Portugal were identified as the most important: fire regime, land use changes (including abandonment of agricultural fields, afforestation, urban expansion, and development of transportation infrastructures), European Union (EU) common agricultural policy (CAP), global markets, and economic growth (table 4.2). Other important drivers include environmental legislation, social attitudes toward the environment, tourism, demographic change, and exotic species. Each driver's importance differs with the site at the local or basin scale. The different drivers interact and control conditions and trends in ecosystem services in Portugal.

Table 4.2
Drivers at multiple scales.

Driver	National	Basin		Local		
		Mondego	Mira	QF	HRA	Sistelo
Fire regime	X			X		
Land tenure and farm structure	X					X
Land use changes	X					
Tourism	X		X		X	
Exotic species	X	X				
Economic growth	X			X		X
Population distribution and migration	X			X		X
Environmental legislation and attitudes	X			X	X	
EU common agricultural policy and global markets	X			X	X	X

Indirect Drivers

Economic growth, first in the 1960s and then after the integration of Portugal into the EU in the 1980s and 1990s, made activities in the industrial and services sector increasingly attractive, which led to increased labor costs in agriculture—both hired labor costs, for agricultural companies, and opportunity costs, for farmers exploiting their farms directly. At the same time, entry into the EU's Common Market and changes in international trade agreements implied decreased agricultural prices (figure 4.2), which were only partially compensated for by subsidies.

Hence, maintaining economic viability of agriculture has required an increase in labor productivity, through either extensification (i.e., substituting labor by land) or intensification (i.e., substituting labor by machines and production inputs, such as water, fertilizers, and improved seeds).[1] If farm size is large and soil quality and water availability are low, as occurs in the South of Portugal, extensification occurs. This is observed in the Castro Verde and HRA study cases. If farm size is small, soil quality and water availability are high, and investment capacity and technical expertise are high, as occurs in the Coastal Center and North of Portugal, intensification will occur. This is seen in the Mondego basin study case, mainly with rice cultivation, leading to a high input of nutrients into estuarine waters. If neither of these conditions is fulfilled, as in many mountain areas in the North and Center (e.g., the Sistelo study case), abandonment will occur.

Figure 4.2

Illustration of interactions and feedbacks among drivers of Portugal ecosystems. Indirect drivers are on the left (the major one is economic growth), direct drivers are in the middle and to the right (mostly grouped as "land use changes"), and changes in ecosystem services are on the right.

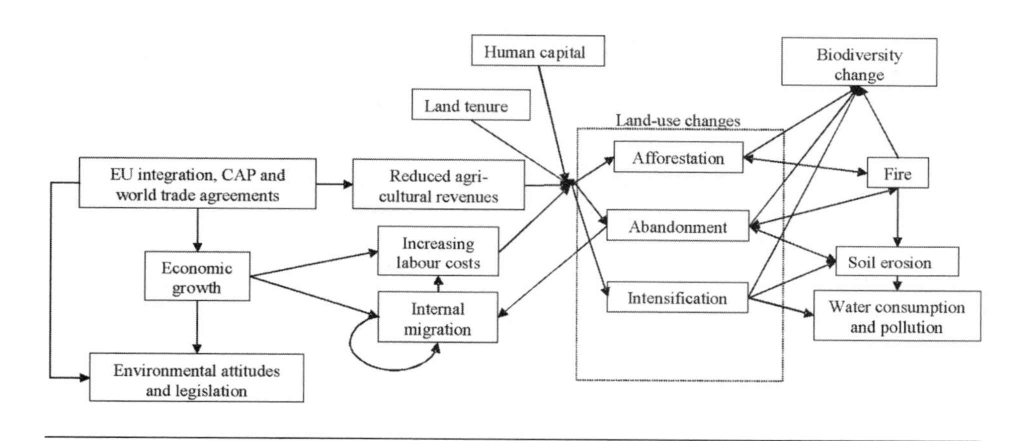

This set of effects is clearly seen for QF, in which the nominal price of sheep milk (the main production of the farm) has remained roughly constant since 1995 (implying a significant decrease in the real price) and maize, rye, and oats have shown significant decreases even in nominal prices. At the same time, labor costs have risen faster than inflation, in fact even faster than growth of the gross domestic product. The response has been intensification: increased mechanization (both for crops and for sheep milking), increased fencing, and increased stocking rates, through the replacement of natural pastures with sown ones. However, given the integration of QF in the project ExtEnSity (described below), intensification has been achieved in an environmentally friendly way.

Direct Drivers and Trends in Ecosystem Services

Intensification in general leads to increased water consumption and pollution risk. Extensification can mean a transition from arable crops to livestock production, but also to afforestation or simply abandonment. Abandonment leads to the establishment of shrubs corresponding to initial stages of the ecological succession, and to decreased landscape compartmentalization. One particular type of transition is related to fast-growing forest plantations: it is extensification in the sense of being a forest and decreasing labor input per unit area, while it is intensification in the sense of increased economic productivity per unit area. Most of this afforestation occurred with monocultures of maritime pine and, more recently, of eucalyptus, which, in contrast with the fire-resistant native oak forests (Nunez-Regueira, Anon, and Castineiras 1999), are very fire prone. The creation of vast plantations of these monocultures, together with the invasion of their surroundings by fire-prone scrubland in abandoned farm fields and the lack of forest management (partially because of the small average holding size), has led to a fast increase in wildfire frequency, with looming consequences for soil and water protection.

The importance of the industrial and services sectors also expanded the attractiveness of urban areas, which have evolved with little land planning. Additionally, the construction sector plays a disproportionate role in economic activity, a characteristic dating back to at least the 1960s. For example, within the EU, Portugal is second only to Luxembourg in per capita consumption of cement, this value being double the EU average (Vieira 2003), and the construction sector is the major driver of the increasing materialization (defined as the total material throughput per unit of economic output) of the Portuguese economy (Canas 2002). This gives the sector disproportionate political power, influencing land planning legislation and inducing heavy government investment in infrastructure. All this adds up to fast urban and infrastructure growth (e.g., highways). The former, taking place in coastal areas, places strong pressures on estuaries and coastal areas. The latter affects important terrestrial ecosystems.

The importance of construction carries over to a construction-based approach to tourism that intensively exploits coastal areas, decreasing their attractiveness, which is itself the basis for tourism (an effect identified by the global MA as the major negative effect of tourism; see Nelson et al. 2005). This leads to a downward spiral, with decreasing value added per tourist, leading to an ever-increasing need for additional tourists and to the "colonization" of new tourist areas in what might be called "slash-and-burn tourism." At smaller scales, however, as for QF and HRA, low-intensity, environmentally friendly tourism constitutes a major opportunity for the economic viability of landscape maintenance and biodiversity protection.

Scenarios

The two drivers chosen to delineate orthogonal axes for scenarios were society's attitude toward the environment and the evolution of agriculture. As explained above, these axes were integrated with the global MA scenarios. This process is ongoing, and here we discuss preliminary results for two scenarios: Order from Strength (a regionalized world with a focus on security) and Technogarden (a globalized world with a focus on environmentally sound lifestyles and policies). The other two scenarios are Global Orchestration (a globalized world with a focus on economic development) and Adapting Mosaic (a regionalized world with a focus on adaptive management of ecosystems).

In all scenarios, it is clear that external drivers have a predominant role. Because Portugal is a member of the EU, European policies and lifestyle contexts will largely determine the country's future. The global nature of the scenarios further reinforces the role of external drivers. In the Technogarden scenario, the EU's increased awareness of the problems associated with oil dependence leads to a push for renewable energies and training of environmental scientists and engineers. Marginal agricultural areas are abandoned, while farming practices are intensified in the best soils (two of the land use changes shown in figure 4.2). Biodiversity associated with agricultural habitats decreases as a result of intensification, but forest biodiversity increases in the forests growing on abandoned fields (see the right-hand side of figure 4.2). In contrast, in the Order from Strength scenario, Portugal's cultural traditions undergo a renaissance. The government promotes a policy of food self-sufficiency, which causes soil erosion, pollution of water resources, and biodiversity loss. Later in the scenario, the environmental impacts become so serious that the self-sufficiency food policy fails.

Contrasting Findings at Multiple Scales

In this section we perform a multiscale analysis of some of the assessment results, contextualizing national findings with global findings and complementing the national findings with data from the local scale.

Forest Policies and Biodiversity

Over 3.3 billion cubic meters of wood are delivered by forests, and numerous nonwood forest products figure significantly in the lives of hundreds of millions of people (Schvidenko et al. 2005). However, the world's forests as a whole are not managed sustainably, and the world's forest capital is being exhausted more rapidly than it is regenerated (Schvidenko et al. 2005).

In Portugal, as worldwide, forests have a high value: the annual economic value of ecosystem services of forest in 1998 was at least 939 million euros (Pereira, Domingos, and Vicente 2004). However, in contrast to the global situation, forests are not being overlogged in Portugal. Annual fellings for wood and pulp supply are smaller than the net annual increment of forests with the same main function (Direcção Geral de Florestas 1999). But historically this has not always been the case. After centuries of logging for fuel, timber, and agricultural expansion, by the nineteenth century forest cover in Portugal had decreased dramatically. As shown in figure 4.3, reforestation policies have more than quadrupled total forest cover over the past 120 years, bringing forest cover to 36 percent of the total land area (Direcção Geral de Florestas 2003).

Unfortunately, much of this expansion was achieved through eucalyptus and pine plantations. These two types of forest—in particular, the exotic eucalyptus—differ greatly from the native oak forest. When poorly managed, eucalyptus forests especially can experience soil erosion problems, excessive nutrient extraction, excessive soil tillage, and loss of hydrological regulation (Pereira, Domingos, and Vicente 2004) and have very low biodiversity (Blondel and Aronson 1999).

Interestingly, the importance of the native oak forest is often better recognized by local populations than by national authorities. A set of ecosystem services provided by native oak forest is particularly valued by the local population of Sistelo. In the Sistelo assessment, people ranked oak forest as the second most preferred landscape (immediately after agricultural fields), referring to such services as provisioning of high-quality timber, provisioning of fuel wood, aesthetic beauty, recreation and resting amenity, and contribution to a healthier environment (Pereira et al. 2005). Fuel wood production is particularly important at the local scale. Unfortunately, some uncertainty exists as to how much native oak forest will be a part of the future program of reforestation in Sistelo. As in the past, the focus of the national forest authority on timber production seems to favor pine plantations.

Drylands and Forestry: The Case of the Montado

While most drylands are experiencing desertification processes, such as vegetation destruction and soil degradation, which deteriorate the capacity of those ecosystems to provide goods and services (Safriel et al. 2005), the xerophytic Portuguese *montado* is performing relatively well (Pereira, Domingos, and Vicente 2004). The resilience of the xerophytic rangelands of the Mediterranean to

Figure 4.3.

Area of main forest types (by dominating tree species) through time.
(From Pereira, Domingos, and Vicente 2004.)

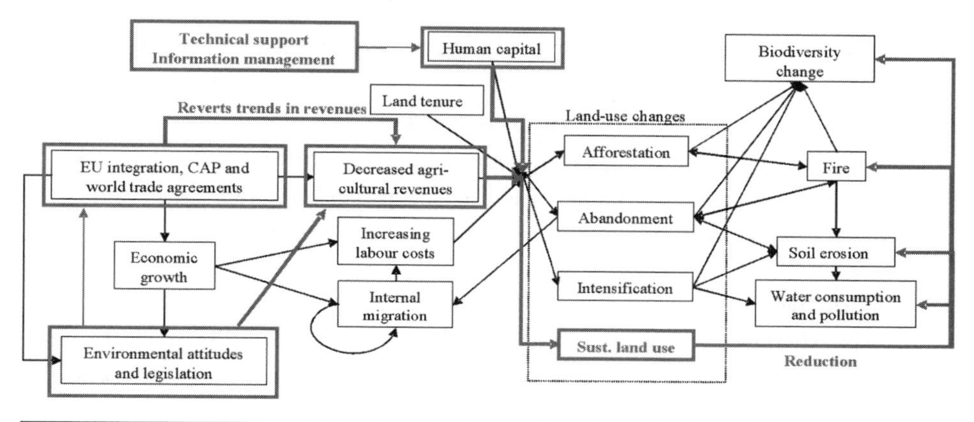

human impacts has been recognized in the global MA assessment (Safriel et al. 2005), but a full assessment of that ecosystem is being carried out only by ptMA.

The economic importance of *montado* is largely due to the production of cork, a renewable resource that is sustainably extracted. The cork is debarked from each tree in intervals of about ten years by skilled laborers. The national cork production is almost as valuable as all other Portuguese wood products together, at 222 million euros per year, with Portugal producing 50 percent of the world's cork (Pereira, Domingos, and Vicente 2004). Portugal had a cork oak *montado* area of 712,800 hectares in 1995–98 (Mendes 2005), which averages to a production value of about 300 euros per hectare per year. In the case study of HRA, cork production yields 768 euros per hectare per year (Pereira, Domingos, and Vicente 2004) and is the area's most important source of income.

Another important economic activity in the *montado* is bovine, swine, caprine, and ovine production, which is often improved by cultivating fodder plants under the tree layer or in the open spaces between the trees. The economic value of forage from the *montado* is estimated at 125 million euros per year (Pereira, Domingos, and Vicente 2004). These open spaces are also used to raise cereal crops, particularly wheat. Unfortunately, the "Wheat Campaign" of the midtwentieth century, aimed at ensuring the country self-sufficiency in staple cereals, led to a degradation of the *montado* due to a reduction of the tree density and the impacts of tillage on the root system of trees. The fragile soil, typical of dryland regions, was eroded and soil microbiota were affected, increasing susceptibility to insects and parasitic fungi. Fortunately, these damaging agricultural practices have since been abandoned in most areas.

The cultural services associated with the *montado* were studied at the local level in the HRA site. Cork oak *montado* provides important habitat for birds of prey and carnivore mammals, and its conservation value is recognized at the European level by the European Habitats Directive. Thus many students and researchers develop their fieldwork in the well-preserved HRA *montado*. Other activities, such as bird-watching and hiking, are also gaining importance.

Biodiversity Consequences of Abandonment versus Conversion to Agriculture: Contrasting Scale-Dependent Results and Perceptions

More land worldwide was converted to cropland in the thirty years after 1950 than in the 150 years between 1700 and 1850 (Cassman et al. 2005). This is not the case with Mediterranean forest, however; by 1950 only 30 percent of the original forest cover remained, but little change has occurred since then (Mace et al. 2005). Mediterranean regions show that agricultural habitats are not hostile to all biodiversity, particularly when species and agricultural habitats coevolved over millennia. In the Mediterranean countryside, extensive agriculture (e.g., agricultural terraces, cereal steppes, and agroforestry systems) and even, in some cases, intensive agriculture (e.g., rice fields) have been the preferred habitat of many species for centuries (Pereira, Domingos, and Vicente 2004).

The scenarios developed for Sistelo (Pereira et al. 2005) present an excellent example of the contrast between scales in the advantages and disadvantages of abandonment for biodiversity and other ecosystem services. One of the Sistelo scenarios can be broadly characterized by the abandonment of agricultural fields and the extinction of the farmer community. In this scenario, the agricultural terraces disappear and are replaced by oak forest and scrubland, with the loss of local provisioning services and of biodiversity associated with low-intensity farming (including hot spots of plant diversity, such as pastures and fallow fields). However, forest biodiversity, including such species as roe-deer, wild-boar, and wolf (Pereira et al. 2005), increases, and regulating services, such as climate regulation, improve. Some of these changes are also occurring at the national scale (see figure 4.2).

The other Sistelo scenario consists of rejuvenating the population while maintaining the agricultural terraces. In this scenario, the cultural landscape of the terraces is preserved and the local food products are sold as organic products with profitable margins (however, as established by figure 4.2, this requires some intervention from outside the system, e.g., agri-environmental measures).

For the local policy makers, the undesirable scenario is abandonment. However, in a national or global context, this scenario could be a part of the globalization trends in the Technogarden or Global Orchestration scenarios, which for many would be desirable scenarios. In contrast, the preferred scenario by local policy makers, the rejuvenation scenario, would occur if local communities were empowered, as in the Adapting Mosaic, or if self-sufficiency food policies were implemented, as in the national Order from Strength scenario. The latter would clearly have overall negative consequences for ecosystems and human well-being at the global scale.

Addressing Afforestation Threats on Biodiversity: The Cereal Pseudo-Steppe

The cereal pseudo-steppe is an agroecosystem consisting of arable crops in rotation with fallow land, occurring mostly in Castro Verde and created during the Wheat Campaign. It is environmentally problematic with respect to soil erosion and desertification, but very important for biodiversity, functioning as a preferred habitat of such vulnerable bird species as the great bustard (*Otis tarda*) and the lesser kestrel (*Falco naumanni*).

In the 1980s, this area was threatened by afforestation with eucalyptus. In response, the municipality of Castro Verde forbade the planting of forests with rapid-growth trees in about 85 percent of its area, to avoid agricultural abandonment and to maintain control of the land by the municipality's inhabitants. This measure led to a strategy for obtaining public funds to subsidize farmers adopting practices compatible with nature conservation. The strategy was first proposed by Nature Protection League (LPN), an environmental NGO, and then taken up by ERENA, a consulting company, and the local farmers' association, leading to the creation of the Castro Verde Zonal Plan (CVZP), financed by the agri-environmental measures of the CAP. This was the first, and is still the only, implemented plan of its kind in Portugal.

LPN continued its intervention by contributing to improvements in the zonal plan (e.g., by changing the dates of plowing, which were inadequate for bird nesting). This intervention was based on LPN's integration of information obtained by scientists working in this area. The CVZP has significantly improved the conservation of the target bird species (Pereira, Domingos, and Vicente 2004).

In parallel, from 1994 to 1997, LPN acquired farms at risk of eucalyptus afforestation using a 75 percent funding from the Life Program of the European Community, with the remaining support coming from individual donations and from the pulp and paper companies that originally owned the farms. The farms have been rented to local farmers on condition of their compliance with strict bird protection regulations.

However, since 2000, farmers have been progressively abandoning the CVZP because of a significant reduction in subsidies for areas larger than 100 hectares, precisely those areas that provide the most significant bird habitat (Pereira, Domingos, and Vicente 2004). The CVZP was also ineffective in preventing soil erosion and desertification. However, recent research suggest that subsoiling, injection of wastewater sludge, and use of direct seeding can reduce erosion by up to 90 percent (Pereira, Domingos, and Vicente 2004).

This response has clear multiscale and multiuser components: the threat to conservation at the regional scale (the Castro Verde municipality in Alentejo) of important bird species (which provide a global-scale biodiversity service) led to the acquisition of farms at a local scale by a national-scale environmental NGO (LPN), using funds at the EU (Life Program) and national (corporate and individual donors) scales.

Acting on Multiple Drivers: ExtEnSity

The shortcomings of the Castro Verde response, together with the current trends in drivers of ecosystem change in Portugal, have led to the development of the project ExtEnSity (Environmental and Sustainability Management Systems in Extensive Agriculture; see figure 4.4). ExtEnSity is a demonstration project financed by the Life Program that involves multiple types of organizations at different societal levels:

- research organizations
- NGOs, including farmer associations, LPN, and a consumer association
- private companies
- Ministry of Agriculture agencies.

The major thrust of ExtEnSity is the creation of a prototype for "sustainable land use" in the agricultural sectors of extensive livestock production and arable crops, with three main components: (1) optimized irrigation, (2) no tillage, and (3) biodiverse, legume-rich pastures with increased animal

Figure 4.4

Intervention by ExtEnSity on the causal structure in figure 4.2. (Thick boxes and lines indicate intervention.)

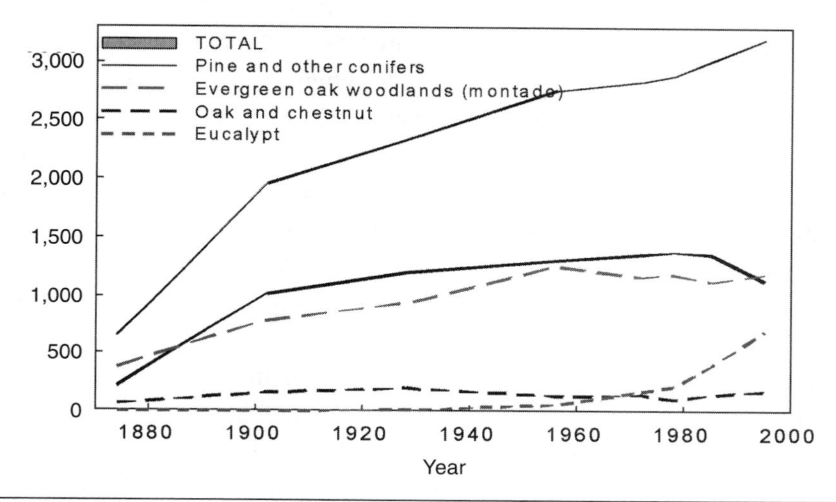

stocking rates. Components 2 and 3 lead to increased soil organic matter, thereby reducing soil erosion and increasing water retention. All three components reduce water consumption and nitrate leaching, also leading to reduced soil erosion and reduced water pollution.

Components 1 to 3 all lead directly to increased economic viability of agricultural activities, thereby promoting "sustainable land use" instead of afforestation, abandonment, and intensification. This transition is also promoted by compensating low human capital with technical support and information management.

Additionally, the transition to "sustainable land use" is addressed by counteracting "reduced agricultural revenues" through two interventions associated with an increase in rewards for the private (food quality and safety) and public (other ecosystem services) goods provided by "sustainable land use." The project directly increases the rewards for private goods by promoting the commercialization, at higher market prices, of the project farms' products. This addresses the environmental attitudes component of the "environmental attitudes and legislation" driver. Increasing the rewards for public goods requires a higher scale of intervention. The Ministry of Agriculture is in charge of developing and proposing to the EU agri-environmental plans. Thus ExtEnSity is collaborating with the Ministry in developing a specific agri-environmental measure for "sustainable land use." This addresses the CAP component of the "EU integration, CAP and world trade agreements" driver.

ExtEnSity is an example of an integrated response that addresses degradation of ecosystem services across multiple systems simultaneously, explicitly includes objectives to enhance human well-being, occurs at different scales and across scales, and uses a range of instruments for implementation (Brown et al. 2005). ExtEnSity has a strong multiscale, multiuser, multiknowledge system approach. Three of the project's partners are involved in managing some of the pilot farms, ensuring a smooth flow of knowledge from the local to the national level. Also, two different approaches are taken for interacting with farmers: direct interaction and through a local farmers' association. ExtEnSity acknowledges multiple forms of knowledge through its integration of scientific and "civil society" actors. Traditional knowledge is integrated mostly for a single region, through the local farmers' association, which has a large number of older farmers. The younger farmers do not have traditional knowledge, although they do have extensive local knowledge.

ExtEnSity addresses first the farm scale and builds on this to influence large scales through policy intervention. This is done through the three national level NGOs, which belong to "umbrella" organizations at the EU level. In the early stages of the project, these NGOs will bring information from their umbrella organizations. At the dissemination and policy-influencing stage of the project, they will transmit to the EU level the results of the project. The participation in the project by two government agencies also enhances this wider policy influence. In this way, ExtEnSity aims to solve a problem with integrated responses: they are often deemed successful at a small scale or in a particular locality, but their effectiveness is limited when constraints are encountered at higher levels, such as in legal frameworks and in government institutions (Brown, Mackensen, Viswanathan et al. 2005).

Conclusions

Performing a multiscale assessment requires a large team (roughly proportional to the number of case studies), an engagement of users at different scales, and more funds (Zermoglio et al. 2005).

The relationship between the national findings of ptMA and the global MA findings demonstrates the importance of having a larger-scale context. For instance, many of the drivers of ecosystem change are exogenous to the Portugal scale, and the analysis of the condition of Portugal ecosystems is best understood in the regional context of the Mediterranean. The global scenarios laid out the context for the development of the Portugal scenarios, a context that was particularly important given the long-term nature of the scenarios, which implies an analysis of the evolution of global drivers.

Local-scale assessments are powerful case studies of processes or systems relevant at the national scale. Data at the local scale can be collected by the assessment team directly. This is not possible in most instances for data at the national scale, where data are usually assembled from existing sources. For instance, in the Sistelo study case, field surveys showed the importance of the native oak forest for the well-being of the local population. Assessing the effectiveness of responses such as agroenvironmental measures and sustainability certification that are implemented at the local level requires local assessments.

Distributional and equity issues become apparent when comparing scenarios at different scales. What is a good scenario at a given scale when impacts on well-being are averaged across that scale may be an undesirable scenario at a smaller scale for a given portion of the population. The importance of the drivers differs across scales, and responses targeting a driver should be implemented at the scale where that driver is most important.

In conclusion, at any given scale, socioecological systems are open systems, with fluxes from scales above and below. A full understanding of those fluxes can be best obtained by using a multiscale approach.

Acknowledgments

This work was made possible by the commitment of the research team and the users representatives of the Portugal Assessment: H. Adão, C. Aguiar, F. Andrade, M. Araújo, C. Carmona Belo, F. Borges, M. J. Burnay, H. Cabral, H. Freitas, F. Salas Herrero, M. A. Loução, J. C. Marques, R. T. Marques, A. Mendes, E. Pereira, J. Santos Pereira, S. Pio, R. Rebelo, E. Sequeira, F. Brito Soares, H. Carvalho, A. Correia, M. Ferreira, J. Marques Ferreira, F. Fonseca, P. Marques, C. Marta, H. Martins, M. J. Martins, R. Pinto, V. Proença, C. Queiroz, M. Santos-Reis, T. Avelar, P. Canaveira, H. Carvalho, I. Guerra, S. Pio, C. Rosas, and N. Sarmento. Financial support was provided by the Millennium Ecosystem

Assessment, Caixa Geral de Depósitos, and Universidade de Coimbra. H. M. Pereira was supported by FCT fellowship BPD/9395/2002. T. Domingos was supported by Life Program grant LIFE03ENV/P/505

References

Blondel, J., and J. Aronson. 1999. *Biology and wildlife of the Mediterranean region.* Oxford: Oxford University Press.

Brown, K., J. Mackensen, K. Viswanathan, S. Rosendo, L. Cimarrusti, K. Fernando, C. Morsello, et al. 2005. Integrated responses. In Millennium Ecosystem Assessment, *Ecosystems and human well-being: Policy responses,* Vol. 3, 425–65. Washington, DC: Island Press.

Canas, A. 2002. Análise da Intensidade de Utilização de Materiais na Economia. Master's thesis, Instituto Superior Técnico.

Cassman, K., S. Wood, P. S. Choo, H. D. Cooper, C. Devendra, J. Dixon, J. Gaskell et al. 2005. Cultivated systems. In Millennium Ecosystem Assessment, *Ecosystems and human well-being: current state and trends,* Vol. 1, 745–94. Washington, DC: Island Press.

Direcção Geral de Florestas (DGF). 1999. *Anuário Florestal 1999.* Lisboa: DGF.

———. 2003. *Inventário Florestal Nacional.* 3rd rev. Lisboa: DGF. http://www.dgf.min-agricultura.pt/ifn/index.htm (accessed October 2, 2003).

Mace, G., H. Masundire, S. Baillne, T. Ricketts, T. Brooks, M. Hoffman, S. Stuart et al. 2005. Biodiversity. In *Ecosystems and human well-being: Current state and trends,* Vol. 1, 77–122. Washington, DC: Island Press.

Malayang, B. S., III, P. Kumar, and T. Hahn. 2005. Responses to ecosystem changes and to their impacts on human well-being: Lessons from sub-global assessments. In Millennium Ecosystem Assessment, *Ecosystems and human well-being: Multiscale assessment,* Vol. 4, 205–28. Washington, DC: Island Press.

Mendes, A.M.S.C. 2005. Portugal. In *Valuing Mediterranean forests,* ed. M. Merlo and L. Croitoru. Wallingford, U.K.: CAB International.

Millennium Ecosystem Assessment. 2003. *Ecosystems and human well-being: A framework for assessment.* Washington, DC: Island Press.

Nelson, G. C., E. Bennett, A. A. Berhe, K. G. Cassman, R. DeFries, T. Dietz, A. Dobson, et al. 2005. Drivers of change in ecosystem condition and services. In Millennium Ecosystem Assessment, *Ecosystems and human well-being: Scenarios,* Vol. 2, 173–222. Washington, DC: Island Press.

Nunez-Regueira, L., J.A.R. Anon, and J. P. Castineiras. 1999. Design of risk index maps as a tool to prevent forest fires in the northern coast of Galicia (NW Spain). *Bioresource Technology* 69 (1): 23–33.

Pereira, E., C. Queiroz, H. M. Pereira, and L. Vicente. 2005. Ecosystem services and human well-being: A participatory study in a mountain community in Northern Portugal. *Ecology and Society* 10 (2): 14.

Pereira, H. M., T. Domingos, and L. Vicente, eds. 2004. *Portugal Millennium Ecosystem Assessment: State of the assessment report.* Millennium Ecosystem Assessment. http://www.ecossistemas.org (June 1, 2005).

Petschel-Held, G., R. Lasco, E. Bohensky, T. Domingos, A. Guhl, J. Lundburg, M. Zurek, and G. Nelson. 2005. Drivers of ecosystem change. In Millennium Ecosystem Assessment, *Ecosystems and human well-being: Multiscale assessment,* Vol. 4, 141–70. Washington, DC: Island Press.

Safriel, U., Z. Andeel, D. Niemeijer, J. Puigdefabregas, R. White, R. Lal, M. Winslow, et al. 2005. Dryland systems. In Millennium Ecosystem Assessment, *Ecosystems and human well-being: Current state and trends,* Vol. 1, 623–62. Washington, DC: Island Press.

Schvidenko, A., C. Barber, R. Persson, N. Byron, P. Gonzalez, R. Hassan, P. Lakida, et al. 2005. Forest and woodland systems. In Millennium Ecosystem Assessment, *Ecosystems and human well-being: Current state and trends,* Vol. 1, 585–622. Washington, DC: Island Press.

Vieira, P. A. 2003. *O Estrago da Nação.* Lisbon: Publicações Dom Quixote.

Zermoglio, M. F., A. van Jaarsveld, W. Reid, J. Romm, O. Biggs, T. X. Yue, and L. Vicente. 2005. The multiscale approach. In Millennium Ecosystem Assessment, *Ecosystems and human well-being: Multiscale assessment,* Vol. 4, 61–84. Washington, DC: Island Press.

CHAPTER 5

A Synthesis of Data and Methods across Scales to Connect Local Policy Decisions to Regional Environmental Conditions

The Case of the Cascadia Scorecard

CHRIS DAVIS

Worldwide, the expansion of urban development poses a growing challenge to the goals of sustainable development. The transformation of land and ecological processes resulting from development is a driving force behind the lost ecological services that concern the Millennium Ecosystem Assessment. A quick glance at just a few global trends suggests why urbanization has quickly become an issue in need of deeper understanding.

Whereas less than 30 percent of the global population lived in urban settings in 1950, nearly 50 percent does so today. For the first time in history, more global residents live in urban areas than not. In the United States, 80 percent of citizens live in urban and suburban areas (Blair 2004). More than half live in coastal counties, where 27 million additional inhabitants are expected in only the next fifteen years (Beach 2002). In South America, 84 percent of all residents are expected to live in urban settings by 2010, completing a remarkable transition that will put the distribution of urban residents on a level equal to that of Northern Europe (Population Reference Bureau 2004).

Challenges posed by geographic and temporal scale underlie the problems researchers face in measuring this dramatic trend of urbanization. Measuring urban sprawl is a highly scale dependent undertaking. Whether or not a region sprawls very much depends on the extent, scale, and resolution of the analysis. An adequate policy response has been slow in coming because complicated scale questions muddy our understanding of how, and in response to what forces, urbanization occurs. Some scale questions are technical: can we isolate patterns of sprawl, to distinguish them from other forms of urban growth and to assess their rate of change? Others focus on determining an appropriate unit of analysis. For instance, development can transform the landscape by parcel and tract, or it may happen by subdivision and by river valley when roads are punched in through native vegetation; it may also occur at the level of an entire forested drainage being cleared for development.

Still other scale questions involve social relations: the benefits of development may be legally protected and transferable among individuals or corporations while the costs accrue across space and time, such as when a forest is clearcut to make room for new homes produces excessive erosion and flooding problems over years in downstream communities.

This chapter attempts to characterize the influence of geographic and temporal scale in measuring urban sprawl effectively. It summarizes the findings of a regional assessment at multiple scales of analysis, with a focus on the influence of geographic scale on an analysis of urbanization in the fast-growing metropolitan centers of the U.S. Pacific Northwest and southwestern Canada. The analysis used three separate methods to quantifiably measure sprawl. Each method was then evaluated and compared to the others to determine how it improves understanding of the patterns of urbanization in the region and how it overcomes challenges posed by geographic and temporal scale.

Background: The Challenge Posed by Scale in Urbanization Studies

The impact of urban growth on the landscape—in particular, the impact of its most corrosive form, urban sprawl—is the subject of a large and growing body of literature (Chin 2002; Gustafson 1998). Yet, little consensus regarding definitions and measurement methodologies is evident. Geographers and other researchers of urban form have not arrived at a widely accepted means of measuring the effects of these trends on the physical environment (Davis and Schaub 2005; Chin 2002; Theobold 2001; Fulton et al. 2001; Torrens and Alberti 2000; Daniels 1998).

Some studies have defined sprawl in terms of the relationship between population growth and built surface as mapped from remotely sensed imagery

(Sudhira, Ramachandra, and Jagadish 2003; Beach 2002; Imhoff et al. 2000). Others rely heavily on census data, comparing population growth to the extent of census-defined urban areas (UAs) (Kolankiewicz and Beck 2001).

Of the considerable number of sprawl analyses that appear in the literature, few explicitly consider geographic scale and its influence on results. But scale is clearly an influential factor, particularly in comparative studies that attempt to assess the performance of one metropolitan area to another. In studies using remotely sensed imagery, low-resolution data may be blind to scattered development, while high-resolution imagery may produce excessive "noise" problems created by the natural heterogeneity that characterizes the spectral signature of built surfaces. Data resolution is also critical to understanding scale-related influences. Bian (1997), for instance, showed that the r^2 value of a regression between biomass and elevation changed by an order of magnitude when the resolution of the input imagery went from one to seventy-five pixels.

Many sprawl studies have relied on changes in population density, focusing on the relationship between population growth and the associated expansion of the urban footprint delineated from one of any number of methods (Sudhira, Ramachandra, and Jagadish 2003; Beach 2002; Fulton et al. 2001; Imhoff et al. 2000). A scale problem arises here because a measurement of density clearly implies explicit reference to a standard spatial unit. And the scale of that unit—a city block, census tract, city, or UA—has a clear impact on results: can neighborhoods sprawl while the larger metro region contains growth? If density—the relationship between the number of residents and the quantity of land required to accommodate them—is a useful metric, what is the appropriate scale of aggregation for comparing metropolitan regions, especially in transboundary regions? Research has made clear that the unit of aggregation may alter results, but no single geographic scale has emerged as the most accepted unit for measuring urban sprawl using population density (Torrens and Alberti 2000).

To make the point more explicitly, Theobold (2001) demonstrates how studies that utilize census-defined UAs and consider population change may aggregate population at too coarse a scale to measure development at the rural fringe. This presents a crucial problem since it is often at this fringe that low-density development—the common denominator in most urban sprawl definitions—occurs at the fastest rate in many global, midsized cities (Montgomery et al. 2003).

Measuring sprawl strictly through changes in population density leads to an additional problem. Sprawling cities, defined as those growing less dense over the study period, may not be major consumers of new land. Instead, population density may be declining in certain areas or at the scale of an entire metropolitan center due to outmigration. Constraining the geographic scale of analysis to a static and potentially arbitrary spatial unit in a time change analysis may lead to the undesirable conclusion that many shrinking towns are actively sprawling. This is the case in several midwestern cities that ranked as major sprawlers in some studies, despite their comparatively low growth rates over the study period (Fulton et al. 2001).

Research Questions: Multimethod, MultiScale Approach

Since scale- and resolution-related issues have affected the results of much sprawl-focused research, an overarching goal of this study was to determine whether multiple analysis methods characterized by different strengths and weaknesses could offer a more nuanced understanding of urbanization patterns than could be obtained from a single method.

In the words of Zermoglio et al. (2005), would the research improve the definition of the problem and offer "improved understanding of scale-dependent processes"? Would the use of several methods shed light on the challenge of determining an appropriate unit of analysis—one that captures growth in the suburban-rural fringe and is indifferent to national and local jurisdictional boundaries?

Less technically, can multiple methods help address the question Wilbanks raises in chapter 2 of this volume: is the scale of decisions linked to the scale at which processes appear to transform the landscape? Ecologists and geographers have argued that a common ingredient in many global environmental problems is the disconnect between the scale of analyses that reveal the problem and the scale of decision making that affects it (Hobbs 1998; Lee 1993). Therefore, an additional goal of this research was to explore how more or less granularity and resolution may illuminate the connections between day-to-day policy setting: would it provide a better understanding of causality, as Zermoglio et al. (2005) argue is possible with multiscale analyses?

Figure 5.1
The Pacific Corridor, including major cities, Puget Sound, the Georgia Strait, and the Cascade Mountains.

Urban Area

Major Public Lands

Water

Figure 5.2
New impervious surface in the Puget Sound region, 1988–99.

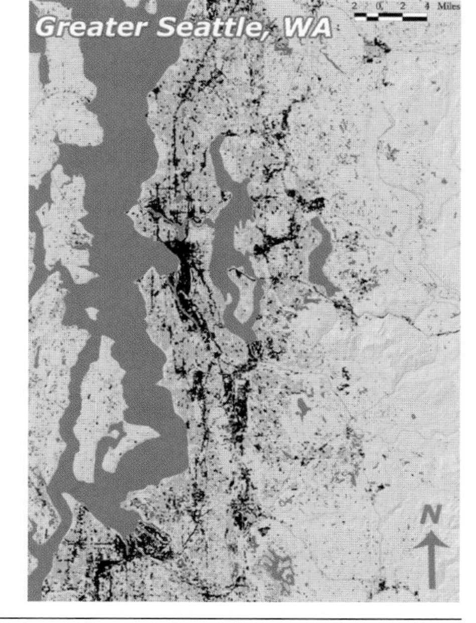

New Impervious Surface

Complete Impervious Surface

Partial Impervious Surface

Figure 5.3.
New impervious surface in the Portland metro region, 1989–99.

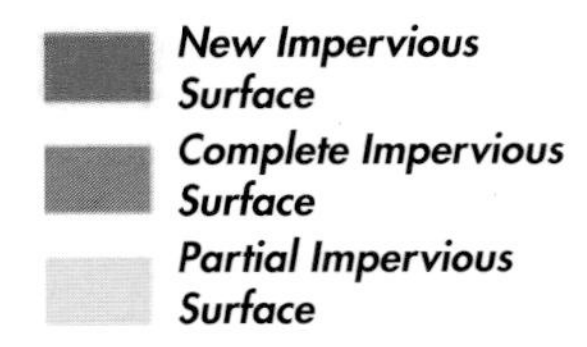

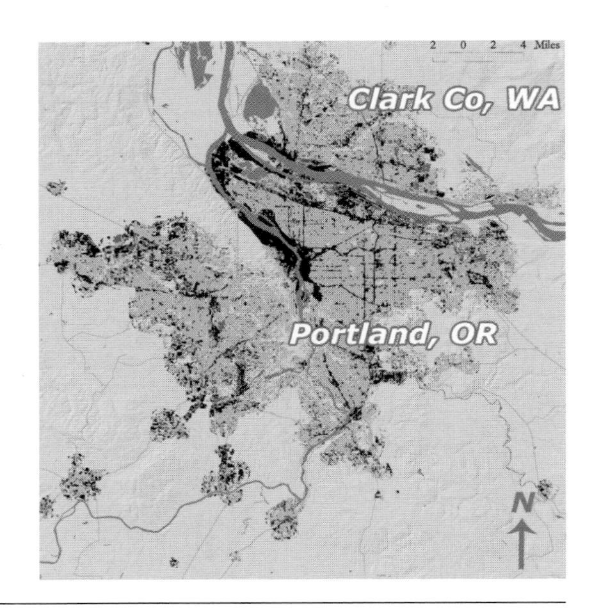

Figure 5.4
Neighborhood metric–population density in the Puget Sound region, 2000.

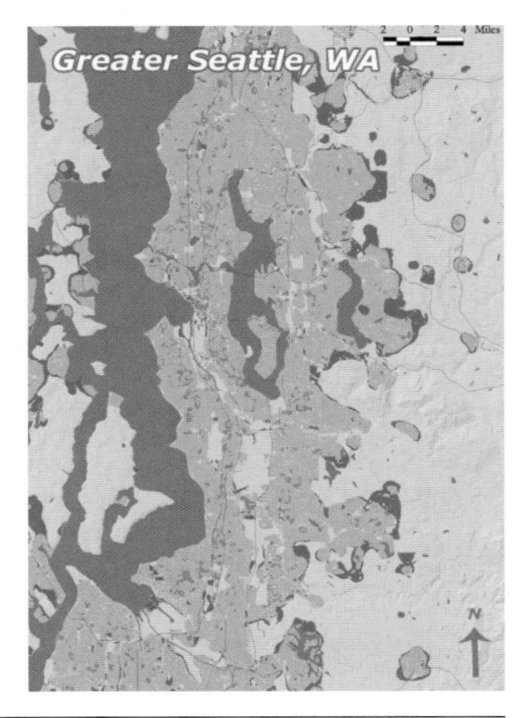

Figure 5.5

Neighborhood metric–population density for Vancouver, British Columbia, 1996.

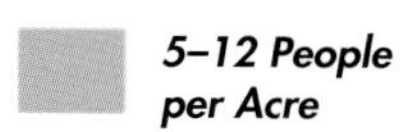

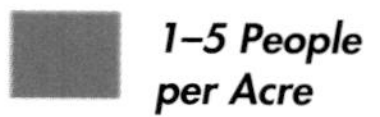

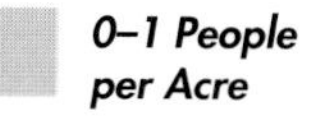

N

0 5 10 KM

Figure 5.6

Results of permit metric analysis, Puget Sound, 1991–2001.

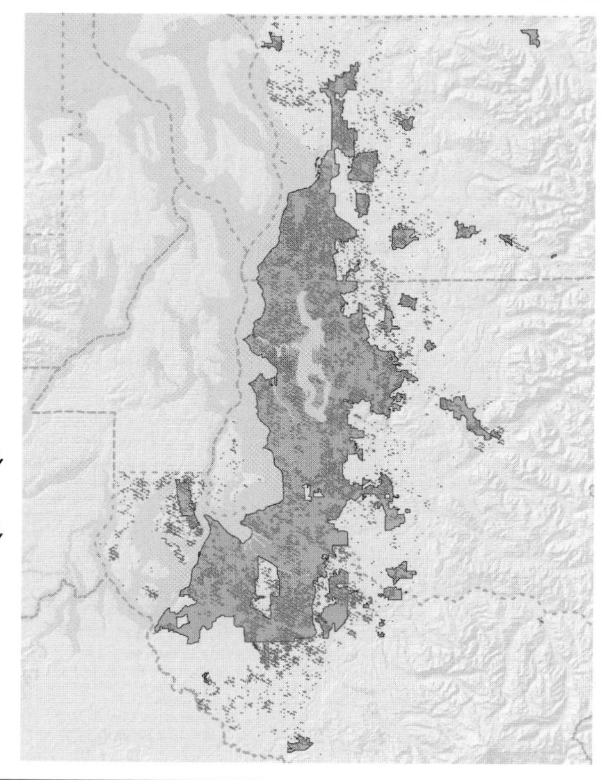

One Dot per Permit, inside UGB

One Dot per Permit, outside UGB

Water

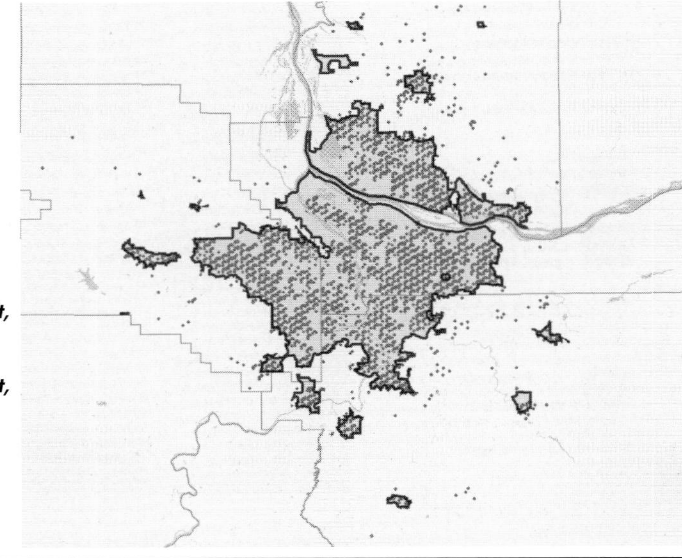

Figure 5.7
Results of permit metric analysis, Greater Portland, 1995–2001.

One Dot per Permit, inside UGB

One Dot per Permit, outside UGB

Water

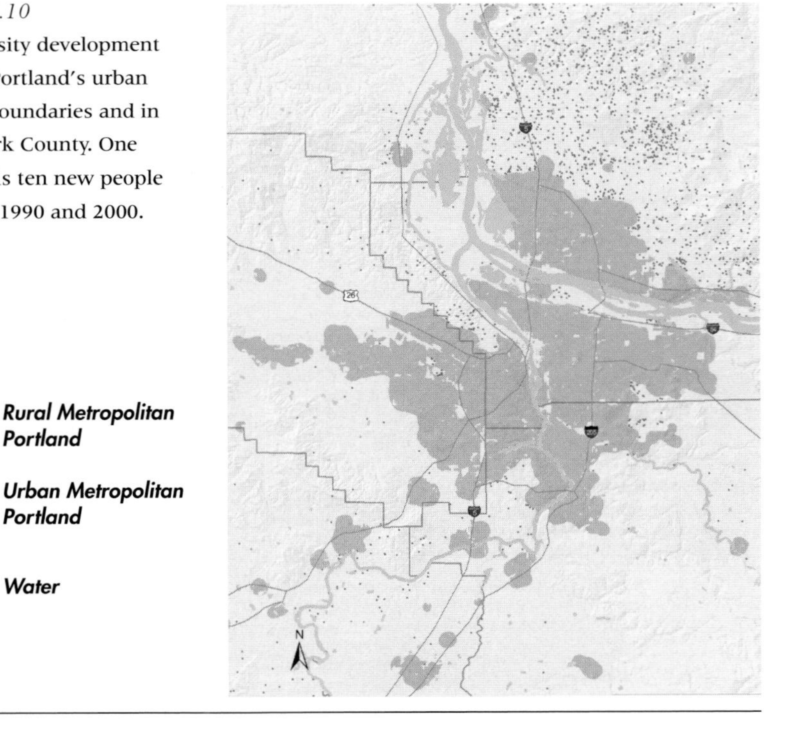

Figure 5.10
Low-density development outside Portland's urban growth boundaries and in rural Clark County. One dot equals ten new people between 1990 and 2000.

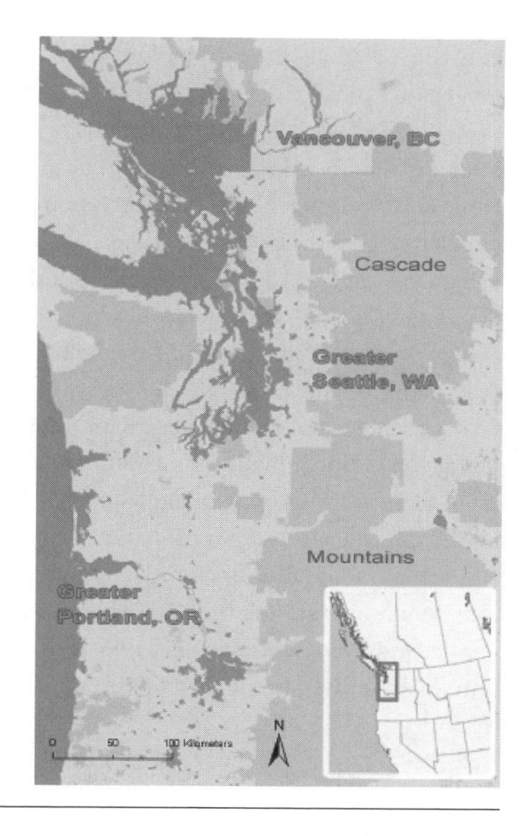

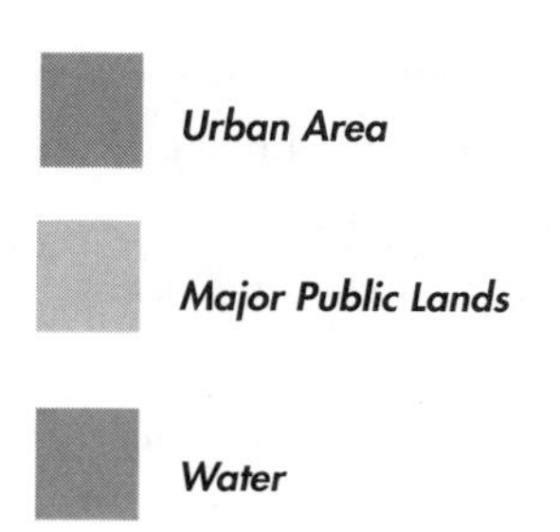

Figure 5.1
The Pacific Corridor, including major cities, Puget Sound, the Georgia Strait, and the Cascade Mountains.

Study Area: North American Transboundary Region

The challenges identified in the background section above influenced the methods chosen for this analysis. The research was aided by the fact that the three metropolitan areas are comparable in population size. Other attributes, however, made a comparative sprawl analysis difficult. The transboundary region of Puget Sound and the Georgia Basin in the Pacific Northwest of the United States and southwestern Canada spans two nations, three state or provincial governments, and dozens of cities (figure 5.1; see also color insert). Thousands of kilometers of unincorporated land connect the region's major cities—Portland, Oregon; Seattle, Washington; and Vancouver, British Columbia—along 450 kilometers of inner coastline. About fourteen and a half million people make their home in the region (Lewis 2001).

If viewed as a single region, this area was a global leader in population growth in the 1990s. Along the corridor formed by Interstate 5, which connects each of the region's major cities along the west coast, the total population of the major cities doubled since 1965. Puget Sound (referring to Greater Seattle and its surrounding suburbs and cities of Everett, Tacoma, and Olympia, Washington), Portland, and Vancouver saw relatively similar, though extraordinary, rates of population growth: 19, 27, and 26 percent, respectively (Vancouver is measured using data dating to 1985 because of differing census schedules in the United States and Canada). On a global measure, this puts Vancouver and Portland just behind Karachi, Pakistan, and New Delhi, India, and just above Cairo, Egypt, in growth rates of world cities for the same period (Durning et al. 2002).

Geographic constraints exacerbate the land use pressures created by dramatic population growth. From Vancouver, British Columbia, to the southern tip of the Seattle-Tacoma-Olympia metropolis, the region is bounded by Puget Sound and the Georgia Strait, natural barriers that limit growth to the west. To the east, some fifty to eighty kilometers from the urban centers, the Cascade Mountains stretch from British Columbia to Oregon. Comprising mostly public lands and alpine terrain, they form an eastern barrier that helps contain urban growth in the lowland trough.

Methods: Single-scale Assessment with Multiple Scales of Analysis

After considering a variety of analytical approaches, three methods were selected. Each offered different strengths and weaknesses, each posed separate challenges in the area of spatial and temporal scale, and each put particular focus on a separate type of what O'Neill and King (1998) call "grain," the smallest temporal or spatial intervals of an observation set. Combined, they addressed each of the research questions; individually, none was sufficient to overcome all the issues these questions raise. The methods are summarized below.

Impervious Metric

This approach started from the assumption that urban sprawl is fundamentally defined as a relationship between population and the built environment. Human development typically converts native vegetation to impervious surface, which has been implicated in a variety of ecological ills, including the degradation of stream and bird habitat, the pollution of surface waters, and the raising of air and water temperature (Blair 2004; Booth 2000; Booth and Jackson 1997). Ecologically efficient—or nonsprawling—growth would minimize the amount of impervious surface created with the influx of new residents to a region.

Remote sensing analysis is frequently used when calculating landscape metrics, and increasingly in measures of urban sprawl (Sudhira, Ramachandra and Jagadish 2003; Clapham 2003). Use of satellite imagery, in particular, is common because data are available in most areas of the urbanizing world, over common time frames and in highly consistent data formats (Vande Castle 1998). For this study, new impervious surface was compared to change in population as recorded by census data. The goals were (1) to understand the spatial distribution of new impervious surface and (2) to associate this transformation with population change to calculate the amount of built surface per capita. Sprawling regions would be those adding relatively more impervious surface per capita than their counterparts (Davis and Schaub 2005).

Neighborhood Metric

The neighborhood metric was designed to take advantage of the simplicity of population density measurement while avoiding the problems created by selecting an aggregation unit, as outlined above. This was primarily achieved using an analysis technique known as *dasymetric mapping,* an approach that may allow analysts to more accurately "see" the distribution of the mapped phenomena within enumeration units (Holloway, Schumaker, and Redmond 1997; Theobold 2001).

A second analytical step took these more highly resolved population data and used them to calculate changes in density in a way that both overcame the aggregation problem and added policy context. Using spatial analysis tools available in a geographic information system, neighborhoods of predefined density were dynamically delineated. For each grid cell in an urban area, local population density was calculated as the density of the smallest circle that contained at least five hundred residents—a rough proxy for a neighborhood. The number of people per acre was then calculated for that neighborhood, providing a measure of neighborhood density for every location on the map.

The resulting spatial data set was then classified in four categories. The categories were determined by population density thresholds shown to affect the viability of public transit (Newman and Kenworthy 1989). In North America, sprawling communities are car-dependent communities. Therefore, a sprawl measure that reveals the extent and distribution of car-dependent communities was deemed a useful approach to mapping sprawl. In addition to maps that display the change in "transit-friendly" development over time, statistical summaries allowed for an explicit comparison of the major metropolitan regions to determine which are characterized by growth in neighborhoods incapable of supporting public transit. The method also allowed us to report change at a dynamically defined unit of aggregation: the neighborhood as defined by its position in the Newman and Kenworthy (1989) classification scheme, rather than the static census block or tract, which may mask sprawl in low-density areas.

For the neighborhood metric, U.S. and Canadian census data were mapped at the block level and then converted into grid data for subsequent analysis. For the United States, input included data from the decennial censuses 1990 and 2000 for the Seattle-Tacoma region and for the Greater Portland region. In British Columbia, census data were gathered at the block level for 1986, 1991, 1996, and 2001.

Permit Metric

The third approach provides the highest resolution, or finest grain, and consequently the most direct measure of growth. Most of the metropolitan areas in the study area are subject to growth management regulations. Jurisdictions at both the state/provincial level and the local/county level are responsible for setting policy and implementing strategies that contain new growth within established urban growth boundaries (UGBs). UGBs are subject to revision over time but nonetheless provide a distinct geographic reference point for measuring how well growth is being channeled. The permit metric evaluates the percentage of annual residential building permits for new construction authorized outside established UGBs. More than any of the previously described metrics, the permit metric speaks to the impacts of day-to-day decision making and the local scale of neighborhoods and communities.

In both the United States and Canada, building permit data are collected by regulatory agencies at the local level responsible for overseeing construction standards. In the Portland and Seattle regions, permit data were gathered from the Regional Data Center at Metro and from the Puget Sound Regional Council (PSRC), respectively. Attribute data varied from year to year in both regions, but after cleaning data to ensure records accounted only for new home development (as opposed to other permit activities such as remodels) and completed projects (jurisdictions managed the distinction between applied and completed permits in different ways), time series were assembled for each region. In the

Table 5.1

Summary results of each sprawl metric by metropolitan area

	Annual Population Growth	Permit Metric % New Permits inside UGB		Neighborhood Metric % Residents in Compact Communities		Impervious Metric Open Space Converted to Development (square km)
		1995	2001	1990	2000	
Puget Sound	1.9%	78%	88%	21%	24%	138
Greater Portland	2.7%	94%	95%	20%	25%	120
Greater Vancouver	2.6%	NA	NA	51%	62%	67

Portland area, permit data from 1995 through 2001 were used; in Puget Sound, PSRC data covered the period from 1991 to 2001.

No unified regional data were available for the Greater Vancouver area. Although the Vancouver Regional District is the analogous regional entity, it does not make a policy of collecting and monitoring the construction activity of its constituent jurisdictions. Time and cost prohibitions precluded collecting data from each of the individual cities in the Greater Vancouver area. Consequently, no permit metric was calculated for Vancouver.

Results: Multimethod, Multiscale Approach

Table 5.1 summarizes the results across each of the metrics for each metropolitan area.

Impervious Metric

The results of the impervious analysis provide a view of new built development in each of the study areas at a high spatial resolution over comparable time scales. The analysis successfully isolated areas of increased built surface, allowing a comparison of development patterns resulting from population growth among metropolitan areas.

Puget Sound, the metro region with the largest developed "footprint," converted 156 square kilometers of undeveloped land to some level of imperviousness. The new development that occurred in the region was scattered and disconnected (figure 5.2; see also color insert). Some occurred along the fringes of existing developed areas, but much took place in previously undeveloped areas of the map.

Figure 5.2
New impervious surface in the Puget Sound region, 1988–99.

New Impervious Surface

Complete Impervious Surface

Partial Impervious Surface

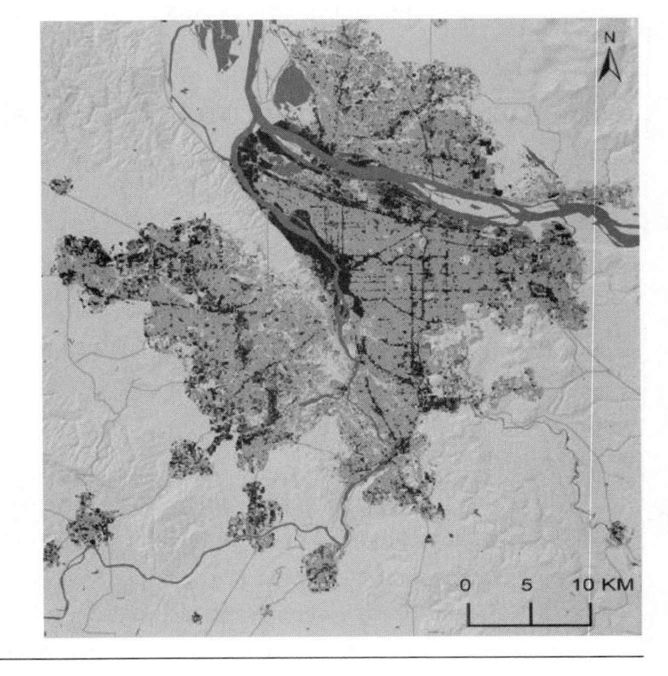

Figure 5.3.
New impervious surface in the Portland metro region, 1989–99.

New Impervious Surface

Complete Impervious Surface

Partial Impervious Surface

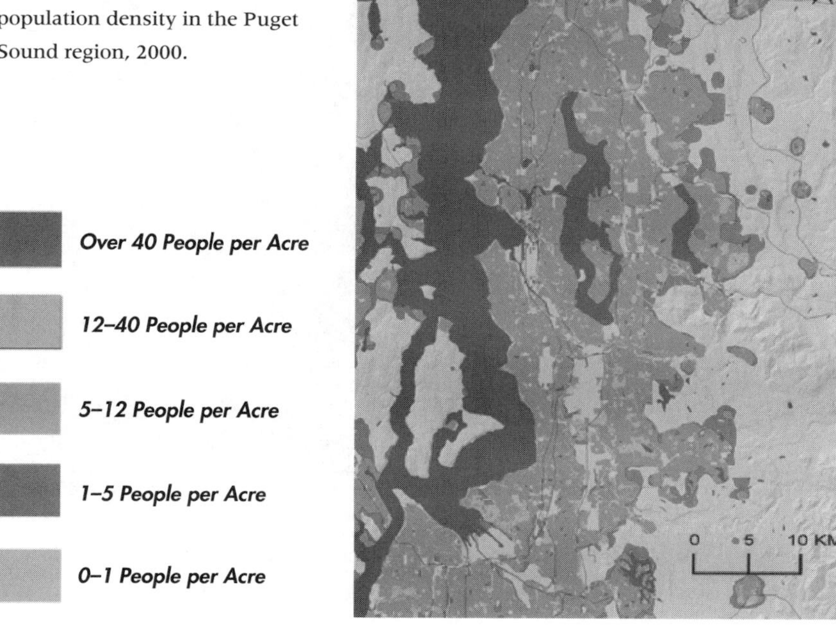

Figure 5.4
Neighborhood metric–population density in the Puget Sound region, 2000.

The Portland metro region added impervious surface closer to the already compact centers of its urban cores (figure 5.3; see also color insert). Mapping the resulting data suggests that through the time frame of the study, Portland's suburbs remained separated from one another by largely undeveloped land. Nevertheless, new impervious surface consumed 120 kilometers of open space, most of it within the bounds of the region's defined UGBs.

Based on the spectral mixing analyses (SMAs), Vancouver, British Columbia, set the standard for the region. Despite taking in the greatest percentage of new residents, Vancouver added the least amount of new impervious surface (67 square kilometers).

Neighborhood Metric

In Puget Sound, a comparison of density maps from 1990 and 2000 reveals that 55 percent of the new growth, or 253,000 new residents, settled in low-density areas with fewer than twelve people per acre. Figure 5.4 (also in color insert) reveals a picture of scattered, low-density development punctuated by

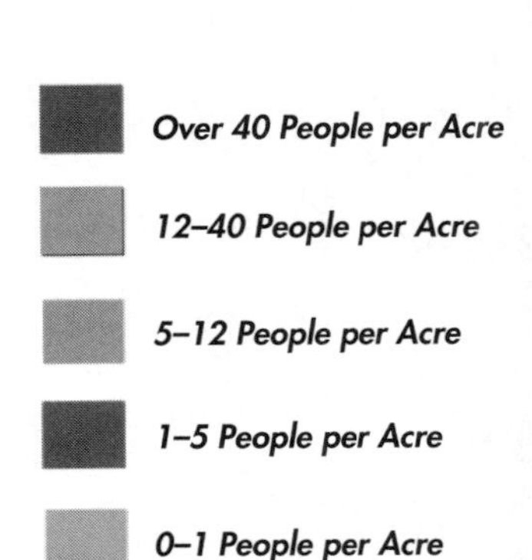

Figure 5.5
Neighborhood metric–population density for Vancouver, British Columbia, 1996.

concentrations of residents throughout the nearby suburban and rural lands. By the end of the decade, only one in four Puget Sound residents lived in compact communities.

By contrast, Vancouver managed its astounding 50 percent population growth over the fifteen years considered here with notably different results. Figure 5.5 (also in color insert) confirms that Vancouver's two million residents occupy far less land and reside in much more consistently compact neighborhoods than their counterparts in the Puget Sound region. By 2001, more than 60 percent of the city's inhabitants lived in transit-friendly areas.

In the Portland metro area, similarly rapid growth reshaped the landscape. Like Vancouver, it experienced population growth that put it near the top of the list of world cities in rate of expansion. But growth in compact neighborhoods in Portland doubled that in Puget Sound.

Permit Metric

In Puget Sound and Portland, building permit data were gathered for the years of 1991 through 2000. Forty-six thousand permits were issued outside the UGBs in the Puget Sound region over the study period (figure 5.6; see also color insert). Twenty-two thousand new permits were issued outside the boundaries after their establishment, which was ratified by law in 1995,

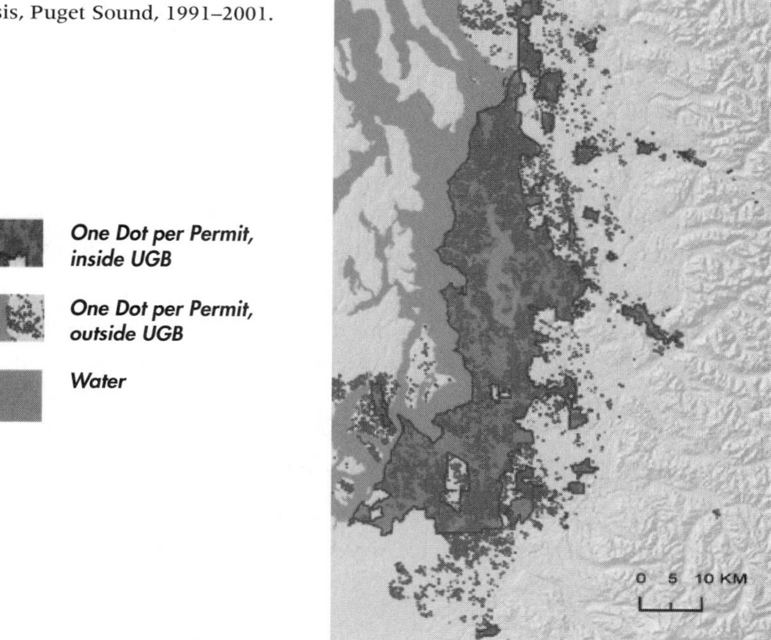

Figure 5.6
Results of permit metric analysis, Puget Sound, 1991–2001.

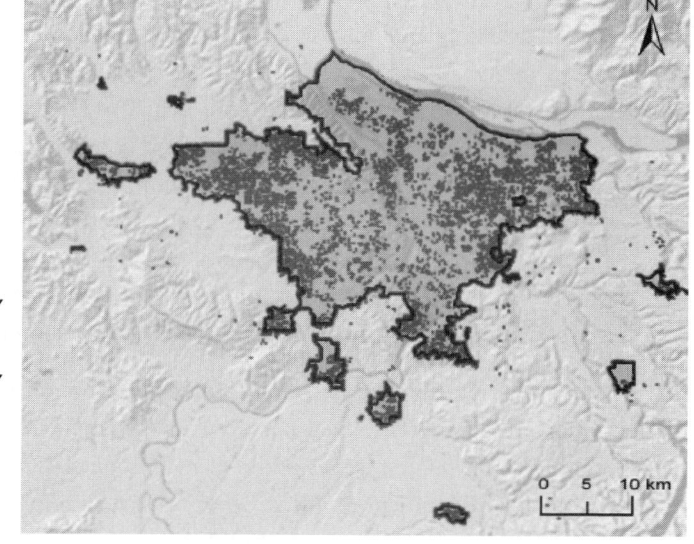

One Dot per Permit, inside UGB

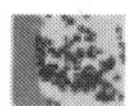
One Dot per Permit, outside UGB

Water

Figure 5.7
Results of permit metric analysis, Greater Portland, 1995–2001.

halfway through the period for which permit data were analyzed. By 2001, 88 percent of the permits issued in Puget Sound were inside the UGBs, more than doubling the number in 1991.

The Portland metro appears to have outperformed the Puget Sound region on this final metric. A lower percentage of permitted development went into low-density, car-dependent communities in the Portland metro area than in Puget Sound. Approximately 95 percent of new residential permits in Portland were issued within the UGBs, compared to the 88 percent in Puget Sound. Figure 5.7 (also in color insert) powerfully illustrates the success achieved by Oregon counties in managing Portland growth. However, data for Clark County, located in Washington State but within the range of the Portland metropolitan region, are lacking. This constitutes a significant problem, as addressed below.

Discussion: Multimethod, Multiscale Approach

The considerable literature on urban sprawl measures suggests that several issues have hindered the emergence of a consistent approach to measuring and monitoring urban sprawl.

- Research has not arrived at a consistent physical description of the sprawling landscape. Consequently, traditional landscape metrics used in landscape ecology to characterize land cover patterns have not been useful in standardizing an approach to measuring sprawl.
- Urban sprawl appears to be a highly scale dependent phenomenon—that is, whether a region is sprawling depends heavily on the scale of observation.
- Because of the limits of the data resolution typically used in many sprawl studies, measurements may be blind to land use changes in areas of low density or in the rural fringe, precisely the areas where development with the attributes of sprawl often occurs.
- Measurement methods focused on landscape form may not illuminate links between policies and their influence on development patterns.

The use of multiple metrics for the three metropolitan centers in the transboundary research completed here provided results that overcame most of these and other related issues while providing data that may help in benchmarking future studies. Findings were relatively consistent across each of the metrics, with each approach ranking the three areas in the same order

Table 5.2

Comparison of each metric's performance against the evaluation criteria

Scale-related Challenges	Characterizing Sprawl	Resolution/Scale Problems	Direct Policy Linkages
Impervious Metric	Consistent spectral signature across region; ignores international border	High resolution captures full heterogeneity of impervious surfaces, making classification and accuracy assessment difficult	Impervious surface is not regulated the same; abstract concept
Neighborhood Metric	Population density metric easily calculated from data available in both countries	Overcomes resolution issues with dasymetric methods and dynamically delineated, density-based neighborhoods instead of relying on census blocks or tracts	Limited connection between population density patterns and policy decisions; transit classes are not well known
Permit Metric	Locally collected data in variety of formats; various attributes, time scales, and levels of reliability	Provides a high spatial resolution if able to georeference and reconcile data from multiple jurisdictions	Directly related to decisions and local policies of planning departments and regional leadership

(excepting the permit metric, which omitted Vancouver, British Columbia). Table 5.2 summarizes some of the major pros and cons of each of the metrics.

Impervious Metric

The impervious metric offered reasonably high resolution (thirty-meter pixels) in a consistent data format available in each of the study areas. Because the entire study area lies within a reasonably uniform ecotone characterized by similar vegetation and precipitation patterns, the spectral characteristics of various relevant land classes is consistent. Focusing the analysis on the physical transformation of the landscape in response to population growth exploited the principal strength of satellite imagery: the data format ignores international borders, so it is a strategic choice in a study area that covers multiple political jurisdictions.

Somewhat complicating the applicability of this approach is that impervious surface is notoriously difficult to measure with remote sensing methods. The image components that comprise impervious urban landscapes—rooftops, parking lots, streets, and sidewalks—are marked by exceptional spectral variety, making it difficult to consistently define paved surfaces using automated methods. The SMA technique provided considerable help in solving this problem (a complete technical explanation of SMA and its contributions to measuring impervious surface can be found in Davis and Schaub 2005).

The problem of identifying the full variety of impervious surface types that appear in a satellite image is exacerbated by high-resolution imagery that captures greater heterogeneity in the landscape. Given the opportunity to take a measurement every thirty meters, the analyst must still select a scale at which to aggregate the measures of imperviousness to avoid being overwhelmed. The impervious metric successfully related changes in land form to population growth and to patterns formed in and out of designated urban growth areas to answer the scale of analysis problem.

We found that impervious area mapping from satellite imagery is also susceptible to criticism because of low certainty at large geographic scales and the likelihood of misclassification of pixels with similar spectral signatures. Problems with registration of images from multiple years may hamper attempts to capture fine-grained land use change at the fringe of the urban-rural interface.

Neighborhood Metric

The neighborhood metric characterized sprawl in the three regions in a clear, policy-relevant way: by tracking the growth in communities sufficiently dense to support public transit. Areas of low or middle density were not overlooked or masked out by large aggregation units used to summarize the area of development. Instead, the boundaries of the spatial unit of analysis were determined dynamically by the attribute being measured: transit-related population densities. The result was a higher resolution map of the spatial extent of density patterns that explicitly captures the edges between communities where growth is potentially transforming the landscape.

Using the dasymmetric mapping technique also helped ensure that spatial data provided as accurate a representation of the geographic distribution of residents as possible and contributed to the method's strength of capturing population density change at the rural fringe.

Carried out with census data widely available in both countries, the method benefited from its reliance on population density analyses, which are readily understood and easy to calculate. The structure of the data lends itself to time-series analyses, as does the availability of historical data. The processing technique used to dynamically define the neighborhood boundaries also supports future change analyses that might otherwise be complicated by revised census boundary definitions.

Some limitations emerge in conducting the analyses across international borders in two countries with different census schedules and sampling methods. However, these are largely surmountable and do not by necessity significantly affect the comparability of results across borders.

Permit Metric

Measuring sprawl by analyzing the spatial distribution of new building permits helps connect the abstract phenomena of land conversion and scattered development to the day-to-day policy decisions that drive them. Population-density patterns may seem out of the control of planners and land use agencies. Similarly, few jurisdictions have any mechanism for regulating impervious surfaces beyond rules for controlling stormwater runoff at construction sites. But building permit records provide data on new construction activities at very high spatial and temporal resolution.

The permit metric sought to identify or corroborate the patterns of sprawl revealed in the prior analyses by (1) tallying the number of permits for new residential units within and outside of UGBs and (2) summarizing the distribution of new residential permits in each of the population density bins used in the neighborhood metric analysis. In doing so, it captured growth in the rural fringe as well as in and outside areas designated for development at high spatial and temporal resolution. It was also useful in disaggregating the patterns of growth to understand how public policy differences across a metropolitan region may influence development patterns.

The challenge with the permit metric approach, as is often the case with high-resolution data, is the task of managing the volumes and varieties of data across the broad geographic extent of the study area. Given the cost of acquiring, managing, and reconciling high-resolution data across multiple jurisdictions, it may be that we often acquire the benefits of higher resolution by sacrificing spatial or temporal extent.

Figure 5.8

Neighborhood metric: proportions of population growth channeled into transit-friendly development, Portland and Puget Sound, 1990–2000, and Vancouver, British Columbia, 1991–2001.

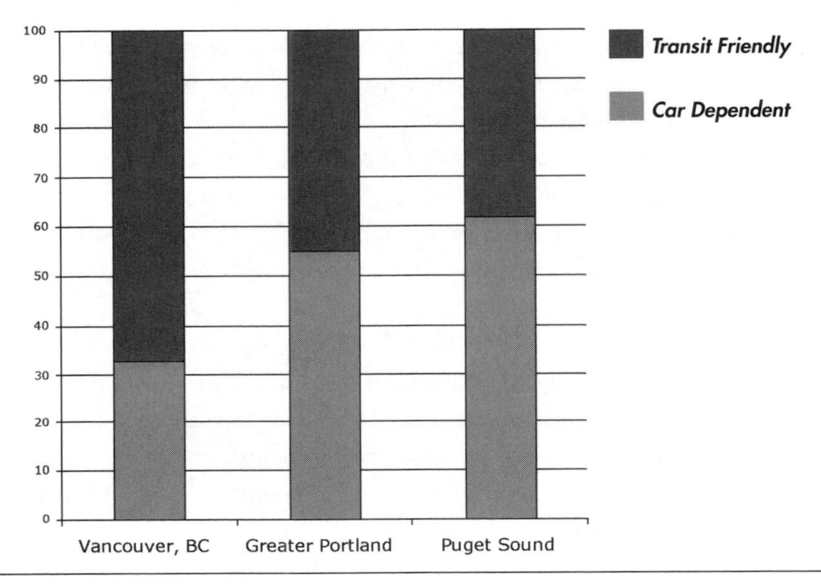

Building permit records are a common data set that most metropolitan planning agencies in North America maintain. However, multiple problems arise in implementing a multicity analysis relying on these data. Local jurisdictions gather varying types of data with permits. The agencies that gather data may not be the same ones responsible for documenting, archiving, and distributing them, leading to erratic gaps between the collection of data and the time that it becomes available for analysis. Additional issues also arise in reconciling the meaning of data from various organizations. Does the existence of a permit confirm the project was actually built? Can new residential and commercial projects be easily and systematically separated from add-ons or remodels that do not consume open space? Are the permits accurately georeferenced, or can they be from accompanying data?

Efficient analysis across many regions depends on the existence of a regional entity that gathers and formats data from local jurisdictions, helping researchers overcome these problems. Unfortunately, this was not the case in Vancouver, British Columbia, where the Greater Vancouver Regional District did not have such a policy in place.

The Power of Synthesis

Each of the methods conferred certain advantages with respect to the scale-related problems outlined above. By combining the results of the different methods, findings that were invisible to any one of them alone emerged.

For instance, figure 5.8 summarizes findings of the neighborhood metric in the three metro regions. It makes clear that by the yardstick of transit-friendly development, Vancouver excelled while the Puget Sound region failed to create communities dense enough to support the widespread public transportation necessary to curb sprawl.

With the benefit of additional measures that refine the temporal scale of analysis, however, a changing picture emerges. The communities of Puget Sound were forced to take regulatory measures to address sprawl only in the mid-1990s. Consequently, the permit rate within the UGBs increased significantly during the latter half of this study period. The higher temporal resolution of the permit data detected shifts in the development trends that arise from new policies that are invisible to the neighborhood metric, which relies on decennial census data.

This suggests that finer-resolution data would seem to offer an opportunity to isolate the relationship between policy and the landscape. There is, therefore, a compelling interest in disaggregating high-resolution data to the spatial scale of the decision making, wherever possible. Disaggregating high-resolution data to combine it with the results of other metrics is revealing, particularly in the Portland metro region.

The Portland metro area includes Clark County, Washington, on the north bank of the Columbia River. Unlike the three Oregon counties included in the Portland metro area, Clark County communities are subject to the more recently established and less stringent growth management regulations of Washington State. As the results of each of the metrics suggest, Portland grew more efficiently than Puget Sound, in spite of a faster-expanding population. However, significant portions of that new growth were accepted by Clark County (figure 5.9).

We suspected that Clark County, with its less restrictive regulatory environment, sprawled to accommodate Portland's growth. Permit data were available for the Oregon portion of Greater Portland but proved unreliable for Clark County, so this hypothesis could not be tested using that metric.

Instead, we modified the neighborhood metric data to map one dot per every ten new residents relocating to rural areas of the Portland metro region between 1990 and 2000. Figure 5.10 (also in color insert) shows the results.

Figure 5.9

Neighborhood metric: proportions of population growth channeled into transit-friendly development in Clark County, Washington, and the Oregon counties of the Greater Portland region, 1990–2000.

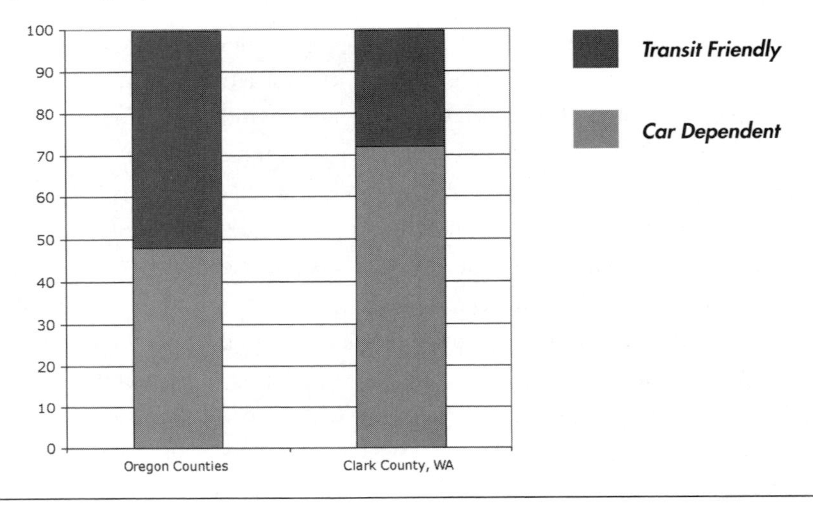

Not only did Clark County, Washington, accept a disproportionately large share of the Portland metro's new residents, it located them in highly inefficient, low-density communities at a rate that eclipsed rural land consumption on the Oregon side. Although Portland attained admirable achievements between 1990 and 2000, channeling most of a 2.1 percent annual growth rate into compact neighborhoods, the metro region might have performed far better but for Clark County's poor performance. Disaggregating the metro data to the constituent counties is essential to understanding the policy impact on land use efficiency in the Portland metro area. Indeed, it provides stark insights on the policy differences shaping land use across the state boundaries.

Conclusions

The conversion of natural landscapes to human-focused uses, particularly urban development, is a problem complicated by the need to explicitly consider geographic and temporal scale.

For researchers seeking policy solutions, a disconnect often arises between the scale of analyses that reveal the problem and the scale of decision making that

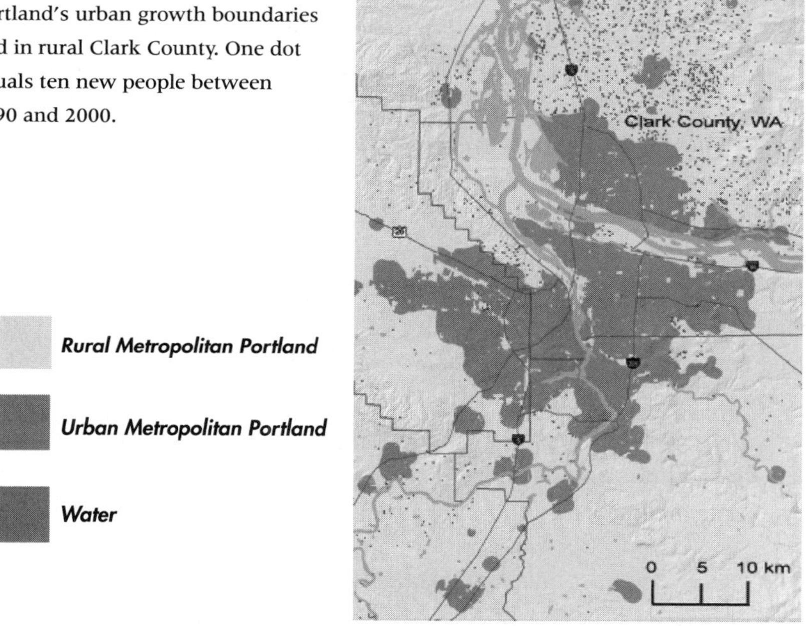

Figure 5.10
Low-density development outside Portland's urban growth boundaries and in rural Clark County. One dot equals ten new people between 1990 and 2000.

affects it. Ecologists and geographers have been observing this problem for years and interpreting it as a fundamental cause of many environmental problems (Hobbs 1998; Lee 1993). We see urban sprawl transforming lands in the geographies between metropolitan centers. By its nature, this is a problem involving multiple political jurisdictions. We recognize it as a regional phenomenon shaped by the interaction among multiple centers of growth. But more often than not, policy-setting institutions are not set up to facilitate or regulate these relationships.

The findings underscore at least two of the benefits of multiscale assessments identified by Zermoglio et al. (2005). First, they illustrate the improved analysis that can be attained with scale-dependent processes. Urbanization occurs at several spatial scales. Effectively measuring it to inform policy may require multiple scales of analysis aimed at different levels of organization, as the Clark County–Portland example makes clear.

Second, the results provide a better understanding of causality. The impervious metric captures the morphology of urbanization without necessarily isolating the drivers behind it. But governments regulate development.

Increasingly, they mandate density. The metrics used here provide indicators that are sensitive to specific decisions and policies at the local level.

As Kates (2001) and others have pointed out, theories aimed at bringing sustainable development into policy-setting realms must be cognizant of linking definitions of sustainability to explicit scales. Portland, Oregon, is cited ubiquitously for its far-reaching growth management and transportation planning successes. Many hold its record of forward-thinking policy setting as a model for other similarly sized cities to emulate. But as this study clearly shows, Portland's record at the scale of its decision making and its record at the scale of metropolitan growth are two separate things.

As a dynamic landscape form, urban sprawl is like other patterns scrutinized in landscape ecology: its character and shape are highly dependent on the spatial and temporal scale at which it is studied. Using three distinct analytical metrics, this study revealed some examples of how spatial and temporal scale may influence the interpretation of analytical results and the potential of multiple methods to enhance that interpretation.

Acknowledgments

Analysis received support from Northwest Environment Watch (NEW), in whose publication *Measuring What Matters* (2002) some of the initial findings of this study were published. Clark Williams-Derry of NEW carried out initial analysis of population data. Thanks to Tim Schaub of CommEn Space and Eugene Martin of the University of Washington Geography Department for input on impervious surface analysis methods.

References

Beach, D. 2002. *Coastal sprawl: The effects of urban design on aquatic ecosystems in the United States.* Arlington, VA: Pew Oceans Commission.

Bian, L. 1997. Multiscale nature of spatial data in scaling up environment models. In *Scale in remote sensing and GIS,* ed. D. A. Quattrochi and M. F. Goodchild, 13–27. Boca Raton, FL: Lewis.

Blair, R. 2004. The effects of urban sprawl on birds at multiple levels of biological organization. *Ecology and Society* 9 (5): 2. http://www.ecologyandsociety.org/vol9/iss5/art2/ (accessed July 17, 2005).

Booth, D. B. 2000. Forest cover, impervious surface area and the mitigation of urbanization impacts in King County, Washington. Seattle: King County Water and Land Resources Division.

Booth, D. B., and C. R. Jackson. 1997. Urbanization of aquatic systems: Degradation thresholds, storm water detection and the limits of mitigation. *Journal of American Water Resources Association* 33 (5): 1077–90.

Chin, N. 2002. *Unearthing the roots of urban sprawl: A critical analysis of form, function and methodology.* Working Paper 47. London: Centre for Advanced Spatial Analysis, University College, London. http://www.casa.ucl.ac.uk/working_papers/Paper%2047.pdf (accessed March 28, 2005).

Clapham, W. B., Jr. 2003. Continuum-based classification of remotely sensed imagery to describe urban sprawl on a watershed scale. *Remote Sensing of Environment* 86:322–40.

Daniels, T. L. 1998. The struggle to manage growth in the metropolitan fringe. In *Proceedings of the 1998 National Planning Conference Revolutionary Ideas in Planning,* ed. B. McClendon and B. Pable. N.p.: American Planning Association.

Davis, C., and T. Schaub. 2005. A transboundary study of urban sprawl in the Pacific Coast region of North America: The benefits of multiple measurement methods. *International Journal of Applied Earth Observation and Geoinformation* 7:268–83.

Durning, A., C. Williams-Derry, E. Chu, and E. dePlace. 2002. *This place on earth 2002: Measuring what matters.* Seattle: Northwest Environment Watch.

El Nasser, H., and P. Overberg. 2001. A comprehensive look at sprawl in America. *USA Today,* February 22, 2001.

Ewing, R. 1994. Causes, characteristics, and effects of sprawl: A literature review. *Environmental and Urban Issues* 21 (2): 1–15.

Fulton, W., R. Pendall, M. Nguyen, and A. Harrison. 2001. *Who sprawls most? How growth patterns differ across the U.S.* Washington, DC: Center on Urban and Metropolitan Policy, The Brookings Institution.

Gustafson, E. J. 1998. Quantifying landscape spatial pattern: What is the state of the art? *Ecosystems* 1:143–56.

Handy, S. L., and D. A. Niemeier. 1997. Measuring accessibility: An exploration of issues and alternatives. *Environment and Planning* A29:1175–94.

Hobbs, R. J. 1998. Managing ecological systems and processes. In *Ecological scale: Theory and applications,* ed. David L. Peterson and V. Thomas Parker. New York: Columbia University Press.

Holloway, S. R., J. Schumaker, and R. L. Redmond. 1997. *People and place: Dasymetric mapping using ARC/INFO.* Missoula: University of Montana.

Imhoff, M. L., C. J. Tucker, W. T. Lawrence, and D. Stutzer. 2000. The use of multisource satellite and geospatial data to study the effect of urbanization on primary productivity in the United States. *IEEE Transactions Geoscience and Remote Sensing* 38 (6): 2549–56.

Kates, R. 2001. Core questions of science and technology for sustainability: Long-term trends and transitions. http://sustsci.harvard.edu/index.html (accessed October 1, 2004).

Kolankiewicz, L., and R. Beck. 2001. Weighing sprawl factors in large cities: Analysis of U.S. Bureau of the Census data on the 100 largest urbanized areas of the United States. http://www.sprawlcity.org/studyUSA/index.html (accessed September 13, 2004).

Lee, K. N. 1993. Greed, scale mismatch and learning. *Ecological Applications* 3:560–64.

Lewis, M. 2001. Census 2000: I-5 corridor drives population increase. *Seattle Post-Intelligencer.* http://seattlepi.nwsource.com/local/grow272.shtml (accessed March 27, 2001).

May, C. W., R. Horner, J. R. Karr, B. W. Mar, and E. B. Welch. 2000. Effects of urbanization on small streams in the Puget Sound ecoregion. *Watershed Protection Techniques* 2 (4): 483–94. Also in Center for Watershed Protection, *The practice of watershed protection* (Ellicott City, MD: Center for Watershed Protection, 2000).

Montgomery, M., R. Stren, B. Cohen, and H. E. Reed, eds. 2003. *Cities transformed: Demographic change and its implications in the developing world.* Panel on Urban Population Dynamics, Committee on Population, Commission on Behavioral and Social Sciences and Education, National Research Council. Washington, DC: National Academies Press.

Newman, P.W.G, and J. R. Kenworthy. 1989. *Cities and automobile dependence.* Brookfield, VT: Gower Technical Press.

O'Neill, R. V., and A. W. King. 1998. Homage to St. Michael. In *Ecological scale: Theory and applications,* ed. D. L. Peterson and V. T. Parker, 3–16. New York: Columbia University Press.

Population Reference Bureau Staff. 2004. *Transitions in World Population* 59 (1).

Puget Sound Regional Council. 2000. Regional review: Monitoring change in the Puget Sound region. http://www.psrc.org/datapubs/index.htm (accessed October 18, 2004).

Rickets, T., and M. Imhoff. 2003. Biodiversity, urban areas, and agriculture: Locating priority ecoregions for conservation. *Conservation Ecology* 8 (2): 1.

Sabol, D. E. 1995. Monitoring land-use over time using spectral mixture analysis. In *Elements of change 1995,* ed. J. Katzenberger and S. J. Hassol, 79–81. Aspen, CO: Aspen Global Change Institute.

Sabol, D. E., D. A. Roberts, J. B. Adams, and M. O. Smith. 1993. Mapping and monitoring changes in vegetation communities of Jasper Ridge, CA using spectral fractions derived from AVIRIS Images. *Proceedings of the 4th Annual JPL Airborne Geoscience Workshop.* Washington, DC: Jet Propulsion Laboratory.

Sudhira, H. S., T. V. Ramachandra, and K. S. Jagadish. 2003. Urban sprawl: Metrics, dynamics and modeling using GIS. *International Journal of Applied Earth Observation and Geoinformation* 5:29–39.

Sutton, P. 2003. A scale-adjusted measure of "urban sprawl" using nighttime satellite imagery. *Remote Sensing of Environment* 86:353–69.

Theobold, D. M. 2001. Land-use dynamics beyond the American urban fringe. *Geographical Review* 91 (3): 544.

Torrens, P. M., and M. Alberti. 2000. Measuring sprawl. Working Paper 27. Centre for Advanced Spatial Analysis, University College, London. http://www.casa.ucl.ac.uk/measuring_sprawl.pdf (accessed August 30, 2004).

Torrey, B. B. 2004. Urbanization: An environmental force to be reckoned with. http://www.prb.org/Template.cfm?Section=PRB&template=/ContentManagement/ContentDisplay.cfm&ContentID=10186 (accessed August 30, 2004).

Vande Castle, J. 1998. Remote sensing applications in ecosystem analysis. In *Ecological scale: Theory and applications,* ed. D. L. Peterson and V. T. Parker, 271–88. New York: Columbia University Press.

Zermoglio, M. F., A. van Jaarsveld, W. Reid, J. Romm, O. Biggs, T. X. Yue, and L. Vicente. 2005. The multiscale approach. In Millennium Ecosystem Assessment, *Ecosystems and human well-being: Multiscale assessment.* Washington, DC: Island Press.

CHAPTER 6

Scales of Governance in Carbon Sinks

Global Priorities and Local Realities

EMILY BOYD

Global environmental problems are often complex and interconnected, with effects at different scales, local to global. An increasing number of environmental issues exhibit such linkages, both in effect and in driving force, including loss of biodiversity, land degradation, and climate change. It is typically recognized that management of all commons, including the global atmosphere and forests, requires robust institutions to coordinate and cooperate at different scales (Ostrom and Ahn 2003). This involves interactions among institutions both horizontally (spatially) and vertically (across levels of organization), from the global to the local.

In recent years, global environmental agreements have proliferated. There have been some two hundred global environmental agreements and protocols, including the United Nations Framework Convention on Climate Change (UNFCC) and its Kyoto Protocol. The Kyoto Protocol is the first legally binding commitment by nations to curb greenhouse gas emissions to 5 percent below 1990 levels. Under the Protocol, the so-called flexible mechanisms have been established to combat greenhouse gas emissions cost-effectively. One of these mechanisms, the Clean Development Mechanism (CDM), allows developed countries to offset emissions through energy or forest projects that mitigate carbon dioxide from the atmosphere and allows developing countries to voluntarily participate in efforts to reduce greenhouse gases in return for payments from developed countries. The CDM is considered by many developing countries an important and attractive opportunity to receive compensation for taking paths of lower emissions (for example, Costa Rica has been a critical advocate of land management projects).

Forests gained an important platform in the climate debate, brought into focus in Kyoto in 1997, with the realization that the world's forests, including tropical forests, were a net absorber of carbon dioxide (Adger and Brown 1995). In its Third Assessment Report (TAR), the Intergovernmental Panel on Climate Change (IPCC) estimated that deforestation (primarily in the tropics) accounts for about one-quarter of annual global emissions of carbon dioxide (Intergovernmental Panel on Climate Change 2001). While unmanaged forest stands absorb carbon dioxide, it is their destruction and resultant massive carbon dioxide emissions that are of most concern.

Climate policy discussion on land use change and forestry has revolved around the uncertainties of accurate monitoring of carbon emissions, given already limited information on deforestation rates and amounts of standing biomass. While proven scientific methods exist and have been used to quantify biomass and deforestation rates, such methods were ignored by many policy makers and other real challenges were not recognized—for example, leakage and methods of setting baselines were the primary technical challenges while local social issues were also paramount. The political process has rapidly adopted the challenge of incorporating land use change and forestry into its agenda without a sound understanding of their scientific, technical, and social challenges. Four main approaches exist to sequester and sink carbon or prevent the emissions of carbon through forest systems. These include (1) to maintain existing carbon pools (slowing deforestation and degradation), (2) to expand existing carbon sinks and pools through forest management, (3) to create new carbon sinks and pools by expanding tree and forest cover, and (4) to substitute renewable wood–based fuels for fossil fuels (Intergovernmental Panel on Climate Change 2000).

The development and push for land management in the CDM has come from both developing and developed countries. In particular, some countries in Central and South America (for example, Bolivia, Belize, Costa Rica, Chile, Colombia, and Guatemala) were strong proponents, with the perspective that the critical service their forests provide to the planet is a service that deserves compensation. The push against land management in the CDM was spearheaded by northern environmental nongovernmental organizations (NGOs) with support from the European Union, some developing countries, and some small island states. Ultimately, implementing such schemes under the Protocol calls for unprecedented levels of international cooperation, maybe even signaling a paradigm shift in the way that sovereign states interrelate, particularly regarding land (Fogel 2002, 2004).

The main aim of this chapter is to explore land management in the context of the CDM negotiations and how this policy plays out on the ground. The first section of the chapter draws on twenty elite interviews, extensive participant observation, and informal discussions with policy makers, NGOs, scientists, and other actors at the UNFCCC negotiations. The chapter then reviews a number of IPCC reports, specifically the IPCC (1990) Summary for Policy Makers of the IPCC Response Strategies Working Group, the IPCC (2000) Special Report on Land Use Change and Forestry (SRLUCF), the IPCC (1995) Second Assessment Report, and the IPCC (2001) TAR. The chapter then analyzes the local story lines, conflicts, and institutional dynamics in two pilot projects in Bolivia and Brazil. The final section of the chapter discusses the theoretical and policy implications of these findings relating to scale.

Definitions: Scale, Institutions, and Discourse

One way to advance thinking on global assessments is to deconstruct and reconstruct problems to reach a synthesis. This chapter focuses on the concern that solving problems through centralized controls and global blueprints tends to create its own vulnerabilities in the long term (Adger 2003). The theoretical framework adopted assumes three basic premises. First, the effectiveness of global treaties at the community level requires addressing multiple-scale assessment and multiple levels of decision making (Berkes 2002; Young 2002). Second, institutions represent the numerous ways in which society is held together that give it a sense of purpose and enable it to adapt. Third, institutions adopt and promote their own beliefs and values, which are manifest as discourse or narratives. Unraveling the global scientific and political discourses surrounding land management and the CDM is a useful way to understand the construction of policy choices that make up global institutions (Hajer 1995; Dryzek 1997; Adger et al. 2001).

Scale

Scale matters because actors and stakeholders in the global commons coexist at different spatial and temporal locations. Cash and Moser (2000), citing Holling (1978), suggest that meaningful understanding of systems can be fully reached only if the driving and constraining forces are addressed at different levels simultaneously. Scale is also important in terms of assessments. Complex systems require forging links between fine details and large outcomes in a manner that allows predictability. This requires addressing multiple levels of analysis simultaneously (Ahl and Allen 1996, 11). Multiple-level analyses may include the individual, household, community or village, district or municipality, state or province, and national and international levels.

Institutions

Institutions and institutional analysis incorporate a range of concepts and tools explored in the discipline of political science (Crawford and Ostrom 1995) and more recently applied to new institutional economics (Paavola 2005). Institutions are the entities from which collective action is taken for a variety of resource management activities—for example, water level control, tree harvesting, and health hazard mitigation—to achieve social or economic goals (Gunderson, Holling, and Light 1995). Ostrom, Schroeder, and Wynne (1993) explain that the structure of an institutional arrangement also includes analyzing which participants are involved, what their stakes and resources are, and how they are linked to one another and to outcomes in the world. This analysis, I argue, could be widened to include examining the *meanings* attached to and the constructs of environmental problems. These constructs are embedded in the narratives or story lines adopted by organizations and actors and subsequently manifest in institutions. Critics of global institutions argue that policy makers employ a *discourse* that focuses only on the global nature of problems.

Environmental Discourse

The understanding of discourse may be shared by a small or large group of people across different levels (Adger et al. 2001). Hajer (1995) argues that "environmental problems are ostensibly constructed through fragmented and contradictory discourses within and outside the environmental domain" (Hajer 1995, 15) and simplified into simple metaphors or symbols to ensure the successful transmission of a story line through the political realm. Adger et al. (2001), for example, explain global environmental discourses in terms of a dominant managerial story line and localist counternarrative. The former represents a blueprint, technocratic worldview, while the latter consists of a cultural or traditionalist view of peoples as victims of external intervention. Bringing together these strands of theory allows us to examine the way environmental debates are framed by global scientific assessments that concentrate on reducing greenhouse gas emissions over more localized intervention to reduce vulnerability or to build adaptive capacity on more locally defined terms. This clearly has implications for global-scale assessments.

Global Discourse

We examine two global discourses relevant to the study of carbon sequestration: global deforestation discourse and the global simplification of nature.

Global Deforestation Discourse

Land management surfaced late in the negotiating process in the run-up to Kyoto and was typified by poor scientific understanding and definitions and inconsistencies in national positions on land management (Fry 2002). Some suggest that the issue of tropical deforestation was brought into the debate because of the lack of money generated by the United Nations Convention on Biological Diversity, which is reportedly constrained by limited financial backing and an overloaded work program, while others point to the failure of the United Nations Forum on Forests process to bring about a convention on forests (Fogel 2002).

The politics of land management and the CDM are underpinned by two important story lines, namely the "alarmist" deforestation discourse of the 1980s and Hardin's "tragedy of the commons" (1968). Regarding the first story line, Fogel (2002) argues that the discourse inaccurately labels the South as culprits of deforestation and the North as victims of environmental externalities. She points out that despite alarmist suggestions by scientists such as Norman Myers (1989), scientific claims of deforestation rates are based on incomplete data and information. Forsyth (2003) also argues to allow some degree of local determination of what is considered environmental—for example, increasing forest cover may indeed be a facet of local environmental management, but it does not always follow that this is exclusively positive, as some forms of deforestation may also be considered acceptable if the resulting land cover is still sustainable for various uses.

Looking back to the early work of the IPCC and the Response Strategies Working Group in 1990, a number of forest-related recommendations are provided. One that stands out is that to address the pervasive forest crises, agriculture as well as people's need for employment and income needs to be addressed. In the words of the IPCC (1990, xlii): "Deforestation will be stopped only when the natural forest is economically more valuable for the people who live in and around the forests than alternative uses for the same land." Property rights strongly influence the adoption of the dominant development paradigm in Latin America and reportedly contribute to the existing pattern of deforestation into the frontier zones of the region (United Nations Development Programme 2001). Offsetting carbon dioxide emissions from one developed country in another developing country creates an implicit change in property rights between the investor and the land owner, or at least the rights to those forests are made acceptable (Brown and Adger 1994).

Fogel (2002) suggests that the IPCC has embraced an approach to halting deforestation that gives precedence to the establishment of property rights over other approaches, such as reducing population growth or reducing human overconsumption. In contrast, the "localist" discourses encountered in side events of the climate negotiations have focused on the developmental and rights-based aspects of forest management. For instance, a number of indigenous organizations highlight that the market approach to managing the commons could result in exacerbating existing inequalities between north and south and would do little to address the root of the climate change problem, namely industrial development.

This discourse is transposed into the story lines on land management in the CDM made evident in the rhetoric of a national delegate at UNFCCC COP-6 promoting a carbon sequestration project, and in a subsequent interview with the company investing in the project. First, the delegate wrote: "Deforestation resulting from indigenous people's settlement practices *creates a need* for alternative ways of improving the quality of life for such communities" (G77/delegate, personal communication, 2000; emphasis added). And later: "We are helping poor people in the tropics to change *centuries old practices* towards more sustainable lifestyles, such as alternatives to slash and burn" (personal communication, 2002).

These two statements overlook the role of international and national institutions in contributing to deforestation and the impact that it has on forests in the context of weak institutions and law enforcement. It also risks painting a misleading picture of the culprits of deforestation in a context in which the political economy has contributed to current rates of deforestation and where poor people have governed territories forested for over a hundred years, despite economic alternatives. Increasingly, evidence suggests that the poor are not necessarily perpetrators of environmental change but are actually important contributors to the management of the commons (Ostrom et al. 2002; Dolšak and Ostrom 2003). Skeptics of environmental orthodoxies also point to the benefits that local people have brought to some local environments—for instance, increasing forest cover and, in some cases, managing complex human-ecological systems (Forsyth 2003; Fairhead and Leach 1996).

Having said this, there is a shift in thinking among policy makers and scientists regarding the value of local and indigenous knowledge systems. Evidence of this shift prevails in the global Millennium Ecosystem Assessment (MA) (Millennium Ecosystem Assessment 2003). In contrast to the MA, the IPCC has been criticized for depicting the "real world" as sectoral, single-scale, with a single epistemology, and with no validation of knowledge outside of the peer-reviewed science, while in fact there are multiple actors as well as winners and losers across scales.

Global Simplification of Nature

Meanwhile, global managers, including policy makers, scientists, and conservation NGOs, have tended to frame carbon sequestration in terms of simple constructs of cause and effect. In reviewing the scientific discourse of the IPCC, the body that governs the science on land management and CDM, nature has been simplified into units and models. In 1990, the IPCC Response Strategies Working Group report recommended investing in plantations in developing countries, estimating two hundred million hectares of land to this option (Intergovernmental Panel on Climate Change 1990). Subsequently, in the 1995 Second Assessment Report, the IPCC took this thinking further in saying that protection, sequestration, and substitution of carbon dioxide globally (but predominantly in the tropics) could reduce atmospheric carbon by approximately 83 to 131 gigatons of carbon by the year 2050 (60 to 87 GtC in forests and 23 to 44 gigatons of carbon in agricultural soils) (IPCC 1995).

When the IPCC's SRLUCF came out in 2000, it was criticized by indigenous groups, NGOs, and scholars for its simplified portrayal of terrestrial systems and lack of information on the socioeconomic, political, and institutional consideration of carbon sequestration (World Rainforest Movement 2000). Land tenure was an issue of particular contention at the time, and it was only taken up much later in the negotiations on small-scale afforestation and reforestation in the CDM. The IPCC TAR in 2001 continued this line of simplifying complex systems in estimating that the global potential of biological mitigation options could reach an order of magnitude of 100 gigatons of carbon (cumulatively) equal to about 10 to 20 percent of potential fossil fuel emissions by 2050 (IPCC 2001) on proviso that the *appropriate organizations are available.* Achieving such a level is dependent on land and water availability as well as the rates of adoption of different land management practices. Furthermore, what "appropriate organizations" entails is not entirely clear.

Fogel (2002) suggests that the governments had to simplify complex forest ecosystems into objects in order to define, standardize, and universally agree on their carbon content. It has also been suggested that the simplification of classification systems, by government administrators and scientists, is done in order to know and to govern systems from a distance (Scott 1998; Latour 1997). Yet, at the same time, the IPCC acknowledges the overwhelming challenges that remain to this option. Methods of financial analysis and carbon accounting are still incomparable, in many instances the cost calculations do not cover costs for infrastructure, and appropriate discounting is absent. Further implementation challenges include monitoring, data collection and implementation costs, opportunity costs of land and maintenance, and other recurring costs, which are often excluded or overlooked. The IPCC also acknowledges that if projects are implemented inappropriately they may result in negative impacts: loss of biodiversity, community disruption, and groundwater pollution (IPCC 2001).

To its credit, the IPCC (1995) and its SRLUCF (2000) do point out that providing greater public participation in decision making may contribute to new approaches to sustainability and equity. The issue of participation and equity should, however, not be used lightly in the context of carbon offsets in the tropics. It seems unmerited of the IPCC to highlight equity to those countries where the major mitigation option is slowing or halting deforestation without paying mention to the issues of carbon price differentials (cost estimates reported to date of biological mitigation vary greatly from $.01 to $3 per ton of carbon in several tropical countries and from $20 to $100 per ton of carbon in nontropical countries), the role of the bigger emitters such as the United States, or the responsibility of governments in managing deforestation.

To conclude, this section has summarized the two key story lines surrounding carbon sequestration. At the global scale, land management and the CDM are underpinned by two important story lines: "alarmist" deforestation discourse and the tragedy of the commons, which tends to perpetuate the perception of poor people as culprits of environmental change. We have also brushed on another important element of carbon politics—the way that science has been used to simplify nature as an object. Policy on land management and the CDM appears to be underpinned by a simplified portrayal of complex natural ecosystems and the human dimensions of global environmental change.

We now examine two pilot projects in Bolivia and Brazil. We present an overview of the projects, a chronology, and a description of the stakeholders, followed by analysis of findings and comparisons of the institutional context.

Observations from the Field: Bolivia and Brazil

This section examines local story lines, conflicts, and institutional dynamics in two projects: the Noel Kempff Mercado Climate Action Project in Bolivia and the ONF/Peugeot Land Rehabilitation Project in Brazil. This section helps to contrast the global story lines discussed earlier with the local narratives to help identify conflicting institutional priorities. The analysis has also found that horizontal institutional dynamics (between organizations) is important to the way that carbon sequestration projects are played out at the local level. If this type of cross-scale approach were applied to global-scale assessments, it could help to inform policies better suited to the local context.

These pilot projects seek to provide local sustainable development benefits as well as to reduce greenhouse gas emissions of global impact (May et al. 2004). They also are entitled to claim under the non-Kyoto market, such as the Chicago Climate Exchange, and provide examples of what could develop under the Kyoto Protocol. Research included more than sixty semistructured stakeholder interviews, participation observation, and a workshop in 2001.

The Context of Latin America

In the context of climate change, Latin American forests are crucial, primarily as a contributor to upholding the global climate system. Land conversion to pastures and agriculture in the tropics contributes an estimated 20 percent of global greenhouse gas emissions (IPCC 2000). Brazil alone derives approximately one-fifth of its carbon dioxide emissions from land conversion of the Amazon region across Latin America. The potential for regulatory measures to succeed in abating deforestation and protecting the environmental services that forests provide, such as carbon sequestration, have been limited. In response, some argue for the use of the CDM in the forest sector (Fearnside 1999). Innovative approaches to conservation and carbon sequestration are, however, emerging among civil society and producer organizations in many parts of Latin America. For example, Brazil has begun to make use of fiscal instruments for encouraging conservation and providing environmental services, such as the ecological value-added tax (May et al. 2002) adopted initially by the states of Paraná and Minas Gerais and implemented more recently in the Amazon.

Case Study: Noel Kempff Mercado Climate Action Project

Project Description

The Noel Kempff Mercado Climate Action Project (NKMCAP) is one of the largest pilot projects of its kind undertaken globally. The NKMCAP is situated in the Noel Kempff Mercado National Park in northeastern Bolivia, bounded by the Paragua/Tarvo and Itenez rivers to the west and north and by Bolivia's international frontier with Brazil to the east. The park is biologically diverse lowland forest with a bird list of more than 630 species and with about 130 mammals, including abundant populations of giant otter and freshwater river dolphin (U.S. Initiative on Joint Implementation 1996).

Driven by a partnership among the Nature Conservancy, a consortium of companies (American Electric Power, BP Amoco, PacifiCorp), and the Bolivian government, the NKMCAP is an emission-avoidance project that is predominantly conservation focused in character complemented by diminished agriculture encroachment on purchased land. In 1996, logging concessions were indemnified for $2 million by the consortium, and the park doubled in size to about 1.5 million hectares. Carbon generation was originally estimated at 14 metric tons of carbon over thirty years, while recent monitoring and verification of the stands indicate the figure is 4.4 metric tons of carbon. The park is located in the municipality of San Ignacio de Velasco, within the department of Santa Cruz (800 kilometers from the provincial capital). Dispersed Chiquitano communities of the Bajo Paragua region have long used the forest that is now part of the expanded national park. The population is approximately 2,400.

Local Story Lines

Local villages resisted park expansion in the early days of the NKMCAP (Kaimowitz et al. 1998), and development assistance took off to a slow start:

"When the theme of protected areas and conservation appeared it was a big change for the communities because they were not ready to take on the norms and rules that the state was imposing and above all without consulting the communities or at least having workshops about how the new system was going to work" (headman Piso Firme, personal communication, 2003).

One key challenge was the issue of access and control of forest lands. The "alarmist" deforestation discourse adopted by the governing institutions initially excluded "people" from the mandate to protect forests and to prevent leakage from the destructive practices of local people. A former director of the national NGO noted that the project aimed primarily to "protect the Park to avoid leakage of carbon, deforestation, invasions, or timber extraction, and to restrict communities from entering to extract anything from the Park" (Adolfo Moreno, personal communication, 2001).

In contrast, the local story line emphasized the coexistence of humans with their natural environment for survival. For instance, in one community, people referred to the forest as their *supermarket,* from which they obtained animals, fruits, medicinal plants, and wood for construction and furniture. Locally, people took pride in their local knowledge, noted for example in their use of medicinal plants (Boyd, field notes 5, 2001). They also expressed concern about future opportunities that the forests would provide their children and envisioned forests as a means to generate income under controlled conditions (headman Piso Firme, personal communication, 2003).

In contrast to the state and the NGOs, community authorities were concerned about the impacts of new institutions on the existing way of life in the region and felt excluded from initial decision-making processes. The president of the Central Indígena de Bajo Paragua (CIBAPA), the community-based organization (CBO), wrote:

> [Because] everyone was not in agreement with the expansion of the park; they (the communities) didn't view the people responsible for the program with appreciation, they rejected them. [I] have learned from the process that above all the project should have consensus. There should be a participatory process in the communities, no? So, for a project to have success it should be done in a participatory way so that when it comes to project implementation everyone is in agreement, everybody knows and in this way work with responsibility and dedication. I see it

as a lesson learned from now on forward we have to take into consideration these things. (Ivar Vaca, personal communication, 2001)

The issue of land tenure was a central aspect of the NKMCAP. The park expansion zone was state-owned fiscal land, aside from a few small, private holdings. It was within this expansion zone that the state gave concessionary rights to forest harvesting companies. These rights, in turn, were indemnified by the project investors in order to clear the way for the state to officially declare the park's expansion area. Once the expansion zone was officially defined and the annual operating plan was in place, several "private property owners" within the expansion zone appeared out of the woodwork and demanded indemnification. These contested properties—which have a low chance of standing up in court—are currently holding up the process of fully clarifying the park's legal landholding rights.

The communal lands were also state-owned fiscal lands. After the park was expanded, the process of legally consolidating the indigenous lands adjacent to the park began in earnest with project funding. The convoluted Bolivian property rights system has allowed for two superimposed concessions within the community territory. Once this issue is resolved, the CBO will have legal ownership of its land holdings. This process continues to this day; once complete, it will mark the first time ever that the local communities of the Bajo Paragua hold legal titles to the land they have lived on for generations. Although the project has significantly contributed to strengthening local institutions, the process has emphasized the conflicts that may occur among international, national, and local institutions over entitlements to land or resources.

Institutional Power Dynamics

The institutional dynamics played out in the NKMCAP are important to understanding the barriers to implementation. In the design phase of the NKMCAP, institutional power dynamics figured predominantly between global and national institutions. The investor underscored bureaucratic government procedures as one of the weaker aspects of the collaboration, while the international NGO highlighted the different levels of knowledge required. Meanwhile, the state assumed credit for the existence of the scheme, stressing that without its capacity and knowledge the project would never have taken place. At the local level, the "Comite de Gestion," or management committee—an entity made up entirely of local actors, including the CIBAPA and the Municipality of San Ignacio—suggested changes that were taken seriously and generally incorporated into the planning and implementation cycle. Once the management committee approved the annual management plans, the project directors, government of Bolivia, and investors provided final approval for both the technical and financial aspects of the plan.

The management committee is an incipient body; while in theory it has say over the project, its members do not have the capacity or know-how to contribute much. Needless to say, with each passing year, the committee becomes increasingly savvy, attuned to the park's needs, and committed to offering their very best for the successful management of the project. Many of the issues that arose early on in the project had to do with the lack of capacity in the management committee. In the beginning, the CBO was not an equal partner in this sense—its members' lack of capacity meant that they were marginal to decision making.

Case Study: ONF/Peugeot Land Rehabilitation Project

Project Description

Brazil's official position in the run-up to the Marrakech Accords in 2002 was that projects that aimed to avoid deforestation should be excluded from the CDM (for more information on the politics of sinks in Brazil, see Fearnside 2001). Within this national context, the ONF/Peugeot project was established. It is a commercial project that sought an environmentally friendly image to counter the prevailing image of emission-intensive car manufacturers. Established in 1997, the project consists of a partnership among Peugeot, the Office Nacional de France (ONF, or French Forest Service), and Pronatura International, a Paris-based NGO with a Brazilian affiliate called Instituto Pronatura (IPN).

The project is located in the "arc of deforestation" of the Amazon basin between the municipalities of Juruena and Cotriguaçu, in northwestern Mato Grosso. In a 1991 census of Jurena (and Cotriguaçu), the total population was estimated to be just under six thousand. The area of the municipality is 33,688 square kilometers, with an extremely low population density of 1.38 persons per square kilometer. Both Juruena and Cotriguaçu are under increasing pressure from migration, cattle ranching, and gold mining. Peugeot reportedly invested approximately $10 million toward an initial aim to reforest degraded pasture with 10 million native trees on five thousand hectares, resulting in an estimated 2 million tons of carbon dioxide over forty years. Since the start of the project, the original carbon estimate has been reduced to 500,000 tons of carbon over one hundred years on two thousand hectares.

Mismatched Interests and Objectives

Peugeot's motivation, though largely publicity driven, was also to gain new technical knowledge about carbon monitoring, verification, and accounting (Marc Bocqué, personal communication, 2002). The company's self-interest is reflected in its market strategy for a green image and coincides with the timing of the installation of an industrial facility in Rio de Janeiro. Above all, Peugeot's aim was to have impressive results: "Our main aim was to promote the scientific process of controlling green house gases through such projects. We expect to encounter new knowledge about carbon measurements. [The] second motivation was our image—we feel that the environmental concern is something that is shared by all human beings today" (Marc Bocqué, personal communication, 2002).

Because of the project's ambitious aims and the company's sense of urgency, planting activities began prior to approval of environmental licensing and even prior to application for such a license at FEMA, the State Environmental Foundation. One year after the Peugeot project started, FEMA began to institutionalize single environment licensing—Licenciamento Ambiental Único (LAU)—for rural properties throughout Mato Grosso requesting deforestation approvals. The license was applied first to holdings above one thousand hectares and then lowered gradually to smaller rural enterprises. The São Nicolau ranch became one of the test cases for the LAU.

The indeterminate policy of the Brazilian government on carbon sequestration meant that there were no rules to guide the project, and the absence of guidelines produced uncertainties surrounding the project (Peter May, personal communication, 2001). The project became the object of regional criticism. Accusations linking use of herbicide by the project to the unexplained deaths of wild turtles and cranes found along the Juruena River, as well as accusations of smuggling native tree seeds to France, reached national media attention in November 1999. As a result, the land and environment committee of the state assembly, with the participation of the public prosecutor's office, mounted an official investigation commission to verify the facts. The project was able to keep the judicial issues at the local circuit court level. For a short time, it was also a diplomatic incidence but was resolved amicably. The government of Brazil is no longer concerned about the project being a sovereignty issue.

Meanwhile, the local NGO representatives in Brazil related a different intended vision. In their view, the aim of the project was to work with communities through promoting agroforestry systems—not just to focus on commercial reforestation. The original project feasibility study proposed a budget that included a considerable aspect of buffer zone work. Local municipalities anticipated that the project would create opportunities for local development through technical expertise and dissemination of potential know-how of carbon schemes and would generate an alternative vision of development in the area. In practice, a number of people were employed temporarily on the plantation, but the concept of carbon sequestration was not introduced to local farmers for fear of raising expectations. The development aspects were sidelined (IPN, personal communication, 2001), and locally the project came under scrutiny for lack of local integration.

Partnership and Power Dynamics

The designers of the Peugeot/ONF project chose a private property regime to ensure rapid implementation of project targets. IPN Brazil, a local NGO, was an important partner in acquiring the land but was sidelined after the land purchase was completed. Two specific issues are important to the institutional dynamic in this case. First, the type of land acquisition instigated a change in the dynamics of the institutional actors; second, the way authority (state or private company) exerted itself affected the relationship with local partners.

The dynamics among project stakeholders in the Peugeot/ONF pilot project as well as the interaction among partners involved in the design process are illustrated in figure 6.1.

Once Peugeot and ONF had secured their property, IPN became redundant to their objectives. Land was purchased by private investors, a regime that proved problematic as the private investors chose to bypass local institutions. Initially excluded from the Peugeot project committee, local scientists expressed a sense of exclusion, as did the local NGO that backstopped the project from Brazil (as described earlier). Following accusations by the state environmental agency, the institutional dynamics substantially changed. The ONF increasingly interacted with local government, local farmers, and IPN. Also, a scientific committee was established that consisted of predominantly Brazilian scientists.

Figure 6.1

Institutional dynamics in the Peugeot/ONF pilot project in Brazil. P = Peugeot, ONF = Office Nacional de France, ONF Brazil, IPN = Instituto Pronatura, SC = scientific committee, LG = local government, and F = farmers.

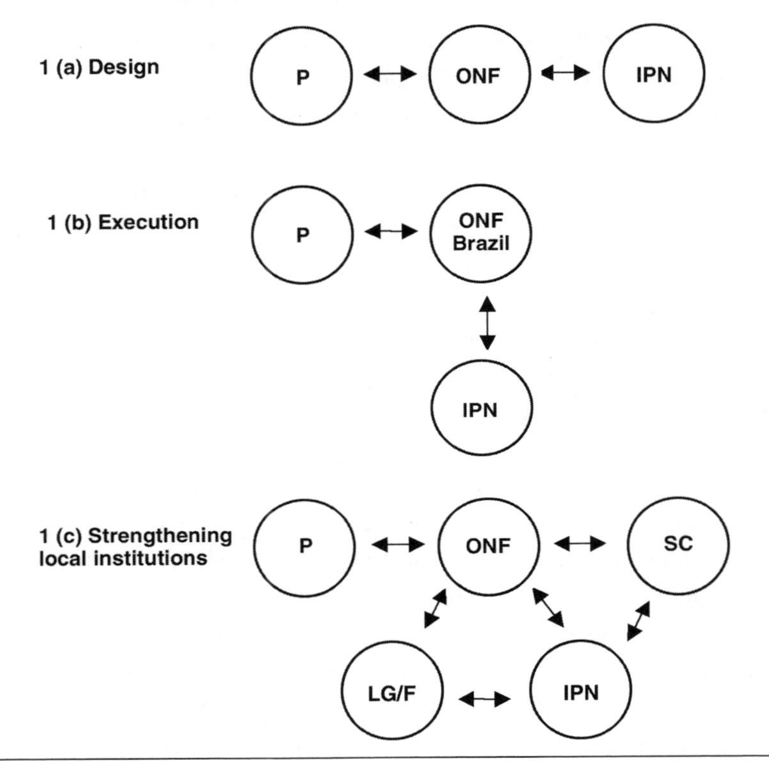

More recently, the project has gone into a process of "Brazilianization," as the ONF manager has departed and only Brazilians remain on site, with very remote "control" from France by the original site manger. Peugeot remains in the wings, using the project only for communication purposes and focusing instead on its primary aim in Brazil: promoting cars. The project mainly measures carbon and maintains a low-profile education and research program, although discussion revolves around longer-term forest management trials and conservation objectives to protect against encroachment from neighboring properties. The community tree planting activities have largely folded, as has IPN's "brittle" role in the region. Had ONF engaged more strongly with IPN or local producers, it may not have proved any more successful since the region is hampered by fluid land uses and local institutional weaknesses. Unlike in the NKMCAP, there is no local commons under sustained management; instead, there are private property and unclaimed public land with predatory occupation. ONF needed to show results fast, so its only option was the private property approach.

Discussion and Conclusion

At the global level, the politics of land management and the CDM display a tendency to seek blueprint solutions, while in practice these are implemented under conditions of scientific uncertainty and limited knowledge of the impacts on local well-being. Essentially, the CDM is based on the premise that mitigation should be done in an economically efficient manner—that is, adopt a property rights approach rather than a commons approach (Ostrom 1990). The Kyoto Protocol recognizes that there are externalities that go beyond national borders and that there are joint but differentiated responsibilities. This implies a shared responsibility for environmental goods that can be obtained by joining forces to introduce cleaner technologies. However, in the land use sector, this implies an attack on national sovereignty, and the commitment of land to permanent (or temporary) forest is seen as an unwelcome restriction on development. A distinct property rights challenge exists in the projects that involve dedicating large tracts of land to conservation rather than to management.

Potentially, community-based approaches to land management and CDM would be more appropriate than the Kyoto framework. Yet, it is unclear how to build credibility for such projects in the marketplace at a scale large enough to make a difference. There are initiatives such as the ICMS Ecologico in Brazil, which—although it has no carbon-based criteria—represents an innovative way to reward municipalities that have allotted land to conservation and taken it out of production. Efforts also exist to provide sources of funding for socially and environmentally friendly carbon projects under the Brazilian Environmental Fund and Biodiversity Fund (Peter May, personal communication, 2005).

Global Blueprint Meets Local Complexity

Adger, Brown, and Hulme (2005, 1) argue that "human responses to global environmental change have been driven on the one hand by underlying discourses of environmental management and control and of economic integration, and, on the other hand, by resistance to globalization and new perspectives on vulnerability and resilience." These observations fit with the way that land use and CDM have largely been conducted from the top down and driven by a set of global actors that largely subscribe to a managerial scientific and political discourse. From this chapter, we have seen that the impact of global policy includes cross-scale conflicts and entrenchment of global institutions against local worldviews perpetuated by myths (assumptions) and misinformation (media reporting and gossip), arguably resulting in adaptive learning processes but also in wasted opportunities for collective action and potentially compacting brittleness of local institutions in the long term (i.e., loss of resilience while increasing institutional efforts to control information and action) (Holling 1973).

It seems curious that the complexity of human-environmental interactions is poorly reflected in the global discourse on land management and CDM despite evidence of advancement in knowledge of these concepts (Millennium Ecosystem Assessment 2003). Concepts of land, spaces, dynamics, embedded identity, and complex social structures remain poorly understood in the best of circumstances and are deemed entirely irrelevant by some government administrators and scientists. Local values appear to be closely associated with development and land tenure, jobs, autonomy, and political leverage, while administrators and scientists lay claims to rights to carbon and conservation. However, the two scales are not irreconcilable: cooperation and increasing local involvement in the management of projects so as to incorporate local demands, rights, and privileges from the start are an important source of hope for improvement.

Participation versus Central Design

This chapter has helped to illustrate the risks of centrally designed projects. Clear, consistent messages will be required to reach people in the local context, as noted by the farmers' association in Brazil and by the local authorities in the NKMCAP. Increasingly, scholars suggest that concepts of equity require further notice in global environmental change research (Adger, Brown, and Hulme 2005). These concepts are highly relevant to implementation of global policies if they are to benefit those people they are aimed at helping (Adger et al. 2004). If future Kyoto compliant or noncompliant land management projects are to fulfill their sustainable development objectives, they will have to address issues of fairness—that is, who benefits—as well as processes and participation, which may initially require institutions to finance the development of local institutions. Boyd, Gutierrez, and Chang (2005) suggest that projects will have to adapt to local organizations, although they also recognize that this might be a barrier to project development in many locations.

At present, signals from "the top" have been construed as inconsistent and misunderstood. Such messages are likely to contribute to the brittleness of adaptive institutions and could enhance existing social, political, and institutional weaknesses. The NKMCAP case emphasized differences in priorities, and some actors stood to benefit more from the project than others. The Peugeot/ONF case highlighted the institutional barriers that exist within networks of like-minded groups of scientists, NGOs, and local officials. Analysis of these dynamics has helped to highlight the importance of communication among different stakeholder groups or actors.

Role of State Institutions

In touching on the interactions between global and local institutions, this chapter has highlighted the pivotal role of the state in directing and implementing global policy. The pilot projects discussed here show that a pivotal role for government agencies and devolved administrations exists in these partnerships but also that roles require clarification of responsibilities. In the NKMCAP, the state acted as a partner in designing and developing the project yet held a distinct position of authority in comanaging the project and the park. It also laid claim to the national park as a public good (49 percent of the potential carbon credits) and to control over financial resources. By creating a carbon land management project in a national park, the government will be involved either in controlling the project or in managing the resources. Meanwhile, the municipality played a more marginal role.

In contrast, the Peugeot/ONF project took place outside of a formal regulatory framework and relied on a local NGO as a linking institution to reach low-income small holders. In bypassing national authority, the project suffered from the uncertainty of rules and the lack of basic standards. These findings concur with Vogler (2000), who argues that there is an important role for the state in implementing global environmental policy but that it requires cooperation between levels of governance.

The government has a key auditing role in avoiding negative project impacts on local communities or the environment. Nevertheless, without local representation and participation of the communities and user groups that inhabit the local commons, such projects remain beneficial only on paper. The conflicts encountered may have been avoided with better guidance, greater transparency, and communication. This awareness, linked with accountability of discourse coalitions, networks, and organizations, might be a way to connect the global and the local.

Future Prospects

At the global level, government administrators and scientists have a responsibility to ensure that global standards are compatible with heterogeneous and diverse local institutions and to acknowledge that local institutions are not only diverse but also require capacity to develop or reestablish resilience. Policy on land use and carbon trading is at the early stages of development, but pilot projects have taught us that scale, institutions, and discourse play an important role in the outcomes of implementation.

References

Adger, W. N. 2003. Building resilience to promote sustainability. *International Human Dimensions Programme IHDP Update* (February): 1–3.

Adger, W. N., T. A. Benjaminsen, K. Brown, and H. Svarstad. 2001. Advancing a political ecology of global environmental discourses. *Development and Change* 32:681–715.

Adger, W. N., and K. Brown. 1995. Policy implications of the missing global carbon sink. *Area* 27 (4): 311–17.

Adger, W. N., K. Brown, and M. Hulme. 2005. Redefining global environmental change. *Global Environmental Change* 15 (1): 1–4.

Adger, W. N., M. J. Mace, J. Paavola, and J. Razzaque. 2004. Justice and equity in adaptation. *Tiempo: A Bulletin on Climate and Development* 52 (July): 19–22.

Ahl, V., and T.F.H. Allen. 1996. *Hierarchy theory: A vision, vocabulary, and epistemology.* New York: Columbia University Press.

Berkes, F. 2002. Cross-scale institutional linkages: Perspectives from the bottom-up. In *The drama of the commons,* ed. E. Ostrom, T. Dietz, N. Dolšak, P. Stern, S. Stonich, and E. Weber, 293–322. Washington, DC: National Academy Press.

Boyd, E., M. Gutierrez, and M. Y. Chang. 2005. Adapting small-scale CDM sinks projects to low-income communities. Working Paper 71. Tyndall Centre for Climate Change Research.

Brown, K., and N. W. Adger. 1994. Economic and political feasibility of international carbon offsets. *Forest Ecology and Management* 68:217–29.

Cash, D. W., and S. C. Moser. 2000. Linking global and local scales: Designing dynamic assessment and management processes. *Global Environmental Change* 10:109–20.

Crawford, S.E.S., and E. Ostrom. 1995. A grammar of institutions. *American Political Science Review* 89 (3): 582–600.

Dolšak, N., and E. Ostrom. 2003. The challenges of the commons. In *The commons in the new millennium: Challenges and adaptations,* ed. N. Dolšak and E. Ostrom, 3–34. Cambridge, MA: MIT Press.

Dryzek, J. S. 1997. *The politics of the earth: Environmental discourses.* Oxford: Oxford University Press.

Fairhead, J., and M. Leach. 1996. *Misreading the African landscape: Society and ecology in a forest-savanna mosaic.* Cambridge: Cambridge University Press.

Fearnside, P. 1999. Forests and global warming mitigation in Brazil: Opportunities in the Brazilian forest sector for responses to global warming under the "Clean Development Mechanism." *Biomass and Bioenergy* 16:171–89.

———. 2001. Saving tropical forests as a countermeasure: An issue that divides the environmental movement. *Ecological Economics* 39 (2): 167–84.

Fogel, C. 2002. *Greening the earth with trees: Science, storylines and the construction of international climate change institutions.* PhD diss., University of California, Santa Cruz.

———. 2004. The local, the global and the Kyoto Protocol. In *Earthly politics: Local and global in environmental governance,* ed. S. Jasanoff and M. Long Martello. Cambridge, MA: MIT Press.

Forsyth, T. 2003. *Critical political ecology: The politics of environmental science.* London: Routledge.

Fry, I. 2002. Twists and turns in the jungle: Exploring the evolution of land use, land-use change and forestry decisions with the Kyoto Protocol. *Review of European Community and International Environmental Law* 11 (2): 159–68.

Gunderson, L. H., C. S. Holling, and S. Light. 1995. Barriers broken and bridges built: A synthesis. In *Barriers and bridges to the renewal of ecosystems and institutions,* ed. L. H. Gunderson, C. S. Holling, and S. Light. New York: Columbia University Press.

Hajer, M. 1995. *The politics of environmental discourse: Ecological modernisation and the policy process.* Oxford: Clarendon.

Hardin, G. 1968. The tragedy of the commons. *Science* 162:1243–48.

Holling, C. S. 1973. Resilience and stability of ecological systems. *Annual Review of Ecology and Systematics* 4:1–23.

———. 1978. *Adaptive environmental assessment and management.* London: Wiley.

Intergovernmental Panel on Climate Change (IPCC). 1990. *IPCC response strategies—Report of Working Group III.* San Francisco: Island Press.

———. 1995. *Impacts, adaptations, and mitigation of climate change.* Second Assessment Report. Cambridge: Cambridge University Press.

———. 2000. *Land use, land-use change, and forestry.* Cambridge: Cambridge University Press.

———. 2001. *The scientific basis: Contributions of Working Group I to the Third Assessment Report of the IPCC.* Cambridge: Cambridge University Press.

Kaimowitz, D., P. Pacheco, J. Johnson, I. Pavez, C. Vallejos, and R. Vélez. 1998. Local governments and forests in the Bolivian lowlands. Rural Development Forestry Network, Paper 24b. Overseas Development Institute, London.

Latour, B. 1997. *Science in action.* Cambridge, MA: MIT Press.

May, P., E. Boyd, F. Veiga, and M. Chang. 2004. *Local sustainable development effects of forest carbon projects in Brazil and Bolivia: A view from the field.* London: International Institute for Environment and Development.

May, P., F. Veiga Neto, V. Denardin, and W. Loureiro. 2002. Using fiscal instruments to encourage conservation: Municipal responses to the "ecological" value-added tax in Paraná and Minas Gerais, Brazil. In *Selling forest environmental services for conservation and development,* ed. S. Pagiola, J. Bishop, and N. Landell-Mills, 173–99. London: Earthscan.

Millennium Ecosystem Assessment. 2003. *Ecosystems and human well-being: A framework for assessment.* Washington, DC: Island Press.

Myers, N. 1989. *Deforestation rates in the tropical forests and their climatic implications.* London: Friends of the Earth.

Ostrom, E. 1990. *Governing the commons.* Cambridge: Cambridge University Press.

Ostrom, E., and T. K. Ahn, eds. 2003. *Foundations of social capital.* Cheltenham, U.K.: Edward Elgar.

Ostrom, E., T. Dietz, N. Dolšak, P. Stern, S. Stonich, and E. Weber, eds. 2002. *The drama of the commons.* Washington, DC: National Academy Press.

Ostrom, E., L. Schroeder, and S. Wynne. 1993. *Institutional incentives and sustainable development: Infrastructure policies in perspective.* Boulder, CO: Westview.

Paavola, J. 2005. *Adding resolution to institutional analysis: Interdependence, institutional design and transaction costs.* Prepared for the Institutional Analysis for Environmental Decision-Making Workshop, United States Geological Survey, Fort Collins Science Center, Colorado, January 28–29, 2005.

Scott, J. 1998. *Seeing like a state: How certain schemes to improve the human condition have failed.* New York: Yale University Press.

United Nations Development Programme (UNDP). 2001. Promoting biodiversity conservation and sustainable use of the frontier forests of northwestern Mato Grosso. Report for UNDP/government of Brazil.

U.S. Initiative on Joint Implementation. 1996. The Noel Kempff Mercado Climate Action Project Bolivia: A United States Initiative on Joint Implementation pilot project proposal (October 25).

Vogler, J. 2000. *The global commons: Environmental and technological governance.* Chichester, U.K.: Wiley.

World Rainforest Movement (WRM). 2000. Climate Change Convention: Sinks that stink. WRM bulletin. http://www.wrm.org.uy/actors/CCC/sinks.html (accessed February 11, 2006).

Young, O. R. 2002. *The institutional dimensions of environmental change.* Cambridge, MA: MIT Press.

BRIDGING KNOWLEDGE SYSTEMS

CHAPTER 7

What Counts as Local Knowledge in Global Environmental Assessments and Conventions?

J. PETER BROSIUS

Several years ago, Arturo Escobar raised the specter of a new regime of "environmental managerialism" wherein the "Western scientist continues to speak for the Earth" (1995, 194). In its very conception, however, the Millennium Ecosystem Assessment (MA) represents a challenge to such a business-as-usual approach, and the current volume clearly shows this. One defining characteristic of the MA is a concern to link scales of analysis by integrating local/indigenous knowledge into global scientific assessments. At the same time, it represents an effort to create a scientific assessment process designed to meet the needs of decision makers (Millennium Ecosystem Assessment 2003; Reid 2000). Taken together, these two characteristics present several challenges to those involved in the MA process and to those responsible for translating MA outputs into policy.

In this chapter, I explore these challenges by examining how "local knowledge" is constituted in global environmental assessments and conventions, and I argue for a more expansive conception. My argument assumes that bridging scales requires more than bridging epistemologies. Across a range of disciplines, the theoretical landscape today is defined by a concern with questions of power, and the boundaries between the epistemological and the political are not as clear as we once took them to be.

In making this argument, I follow two trajectories. First, I consider the constitution of the "local" and the politics of translation. Specifically, I examine how local perspectives are elicited and presented in various mediated forms.

Second, I consider the constitution of "knowledge," showing how scientists interested in local/indigenous knowledge have focused overwhelmingly on environmental knowledge and ignored other domains of knowledge that are salient in the effort to link scales of analysis

The 1980s and 1990s showed remarkable growth and proliferation in indigenous movements worldwide. Much of the momentum for this movement was built around opposing the presence of extractive industries on indigenous lands, and the indigenous movement forged alliances with, among others, the global rainforest movement. Somewhat later, as the term *indigenous knowledge* began to appear on international agendas, the issues of bioprospecting and intellectual property rights became central concerns around which indigenous activists organized (Brush 1993).

Shifts in the conservation field have been equally significant. As global environmental change proceeds at an unprecedented pace, conservation has emerged as a central element in civic and political debates in the nations of both the North and the South. Responding to these debates, new forms of conservation practice are continually emerging. In the early 1990s, we witnessed the proliferation of bottom-up models under the rubric of community-based conservation. Since then, the "requiem for nature" argument has questioned the effectiveness of community-based approaches and called for stricter enforcement of protected area boundaries (Terborgh 1999). Simultaneously, we are seeing a host of new strategic priority-setting approaches that fall under the rubric of ecoregional conservation. Taking developments such as these into consideration, I conclude by offering an alternative approach to integrating local/indigenous knowledge into global scientific assessments that is premised on distinguishing several forms of mediation of local perspectives, and that incorporates a more expansive definition of knowledge.

The Constitution of the Local and the Politics of Translation

What counts as "local" when we speak of "local knowledge"? I want to suggest that when we invoke the "local," we might in fact be speaking about two distinct things. On the one hand are the voices of peasants, farmers, fishers, or indigenous peoples, often living in out-of-the-way places, frequently marginalized politically and economically. These are people we have come to valorize as possessing richly detailed knowledge representing generations of observation and experimentation about medicinal plants, crop varieties, trees, the habits of animals, and much more.

On the other hand are the voices of those who *are delegated* to speak for local or indigenous communities in national and international fora. They are no less local—it is more the context in which we encounter them. We do not go to them; they come to us. These are actors who have much to say to the scientific community and to decision makers. In an effort to counter long histories of oppression and dispossession, they are forthright in challenging national and international conservation or development agendas, conventions, and assessments and in asserting their rights to lands and livelihoods. These are relatively new actors on the global stage. Though local and indigenous peoples have mobilized in many times and places over the centuries, it has really only been since the 1980s that we have witnessed the emergence of a *global* indigenous rights movement—what Friedman has referred to as "the rise of the indigenous voice" (Friedman 1998, 567).[1]

These are the people who have increasingly made such a dramatic impression at such international events as the Fifth World Parks Congress (Durban, South Africa, September 2003), the CBD/COP7 (Seventh Meeting of the Conference of the Parties to the Convention on Biological Diversity, Kuala Lumpur, Malaysia, February 2004), the MA Bridging Scales and Epistemologies conference (Alexandria, Egypt, March 2004), and the World Conservation Congress (Bangkok, Thailand, November 2004). Over the last few years at these and similar events, representatives from indigenous and local communities worldwide have appeared in large numbers. At plenary sessions, on panels, and in workshops, indigenous and local community representatives speak of conservation initiatives undertaken without their consent, and of exclusion from ancestral lands.

That indigenous issues are increasingly on the agendas of such international events is in no small part the result of extensive preparatory work by indigenous organizations and their allies. They have lobbied to secure prime speaking slots and seats on drafting committees for indigenous representatives, sought funding for indigenous participation, and coordinated regional and preparatory meetings. As a result, indigenous and local representatives have been well prepared to make their voices heard and to ensure they are included in the final outputs of these events.

What this indigenous presence represents is a challenge to many basic assumptions about conservation. Indigenous representatives are suggesting that conservation can be done without externally imposed models, management plans, or monitoring and evaluation. They are also challenging assumptions about the roles of both Western science and major conservation organizations, asserting that conservation goals can be accomplished outside circuits of transnational expertise. Their message is that indigenous and local communities must represent something other than a "transaction cost," that threat assessments that classify their land use practices as disturbances are unacceptable, and that participatory methods that define them as just one more category of stakeholder have no place in their vision of conservation.

Though both kinds of actors—"local locals" and local/indigenous advocates—get coded as "local" in international fora, important differences exist between them. When we consider how their words, their insights, and their knowledge move between scales in the process of translation, we must recognize that both are mediated, albeit in very different ways. Making an effective link between local knowledge and policy requires that we recognize these different forms of mediation.

For researchers interested in local or indigenous knowledge, it is those "local locals" that we usually work with the most. After all, these are the people who exist "on the ground" as repositories of the knowledge that interests us. The point, however, is that their knowledge enters circuits of global knowledge production in mediated form *through us*. Most of us who conduct research on local knowledge are able to do so because powerful institutions are interested in supporting our research, and because these institutions are increasingly interested in what we have to say about particular peoples and places. When they want to learn about local realities and local perspectives, they turn to the social sciences. This is what Gledhill was getting at when he reminded us that "intellectuals are contributing to new regulatory strategies being pursued by states and transnational agencies. There is a particular danger that anthropologists will reinforce a politics of containment where this offers a new market opportunity for peddling our services as experts on 'culture,' either to the national state as an employer of specialists in the administration of ethnic difference or to the wider world of transnational agencies and NGOs" (1998, 516–17).

The key to understanding this process of mediation lies in an understanding of the tools we use. As an anthropologist, I believe in the value of ethnographic research methods. Other social scientists rely on more rapid, formalistic, survey-based methods. Whichever methods we prefer, the social sciences get positioned as speaking for the local. In so doing, the danger is that the representations of those who possess expertise in making the local legible and intelligible to those working at other scales are conflated with local voices themselves. These are not the same thing; we must never conflate data provided by those who work at a local level with local voices themselves. We can offer our translations, our mediated accounts, and these can be very valuable, but we must never presume that we actually ever speak *for* the local.

The voices of indigenous advocates or representatives are mediated as well, albeit very differently. While they may be unmediated by social science conventions and formalistic methods, they *are* mediated by transnational discourses of indigeneity. While asserting locality and connection to place, they simultaneously speak with reference to global categories. This does not make their claims any less authentic: there is nothing inauthentic about the solidarity that is emerging from a recognition of shared histories of marginalization. Still, the fact that indigenous representatives are compelled to speak in global categories is a form of mediation.

There is yet another aspect of how the local is constituted that deserves our attention: the pervasive distinction made between local actors and "decision makers." This is achieved in part through what I have elsewhere termed the "topology of simple locality" (Brosius 1999c): a topology that defines the task of the ethnographer as one of inscribing and representing for an audience some actually existing place or set of places—our research sites, the communities in which we work. It is a kind of focalizing strategy, drawing our attention to particular places as the most significant loci for the production of knowledge, and diverting our attention from the ways in which those places articulate with other places or with actors working at other scales. The topology of simple locality suffers from the same shortcoming that has produced critiques of that other convention of anthropological writing, the "ethnographic present."

Anthropologists today are much more alert to the politics and histories that have shaped the communities they study.[2] However, the "ethnographic present" is a still-extant convention of ethnographic writing wherein an anthropologist describing a particular set of cultural practices writes about them in the present tense, even though their research may have occurred many years in the past and though much of what is described may no longer exist in the same form as it did when it was observed (Fabian 1983). Just as the ethnographic present acts as a distancing mechanism that relegates our research subjects to a timeless irrelevancy, immune from history and from the effects of our ethnographic presence, so too does the topology of simple locality create a coherent "there" that can be known and represented and kept in its place.

As Tawfic Ahmed and Reid (2002, 219) remind us, the Millennium Ecosystem Assessment is "designed to meet the needs of decision-makers." Unspoken here is an assumption about the inherent distance between local actors and decision makers and, therefore, about the relation between scale and hierarchy. Viewed in this way, indigenous knowledge is provided to those in the policy domain, but it speaks only in the passive voice of science rather than in the active voice of advocacy and it speaks from the subordinate position of knowledge solicited and translated up for the purpose of governance.

Whether our goals are purely instrumental (rendering local voices and local knowledge into forms useful in managerial terms) or emancipatory (rendering local voices into compelling narratives designed to secure rights), those local voices are situated in a subject position (Laclau and Mouffe 2001, 84).

The Constitution of Knowledge

I now turn to the question of what counts as "knowledge" when we speak of "local knowledge." As it is used by ethnoecologists and others, reference to indigenous or local "knowledge"—often referred to by the acronyms IK (indigenous knowledge) or TEK (traditional ecological knowledge)—is generally applied to knowledge of the natural world: what such groups know about the resources they exploit, how these societies cognize or interpret natural processes, and so forth. In short, when we speak of indigenous or local knowledge, what we generally mean is *environmental* knowledge.

That we are at last recognizing the value of local/indigenous knowledge, rather than dismissing it as anecdotal, irrelevant, or merely a lesser form of knowledge, is clearly a positive development. But that we limit our valorization of knowledge largely to that which pertains to the natural world yet again consigns that knowledge to the irrelevancy of the ethnographic present, destined forever to fill what Trouillot has termed the "savage slot" (Trouillot 1991), an epistemological backwater distinct from, and subordinate to, the forms of knowledge possessed by decision makers.

Let us, for a moment, consider the domains of knowledge that concern the

Millennium Ecosystem Assessment. One strength of the MA is not only that it is firmly science driven, dedicated to producing and synthesizing reliable scientific data, but that it goes beyond this to identify trends, scenarios, trade-offs, and response options (MA 2003; Reid 2000). Central to the MA vision is that it provide information that is not only scientifically credible but salient and legitimate as well. According to Reid: "Scientific information is *salient* if it is perceived to be relevant or of value to particular groups who might use it to change management approaches, behavior, or policy decisions. . . . It is *legitimate* if the process of assembling the information is perceived to be fair and open to input from key political constituencies, such as the private sector, governments, and civil society" (Reid 2000; emphasis added).

But what might happen, we may ask, if these three criteria were applied not only to objective scientific information but to local and indigenous knowledge as well? What if, when we went out to seek information from local people, we not only asked about their knowledge of the natural world but also sought their analyses of the political world? How might their analyses of drivers and their assessments of threats differ from our own? What if we asked them about trends, scenarios, trade-offs, and response options? In other words, instead of treating our informants as reservoirs of local/indigenous knowledge, what if we treated them as political agents with their own ideas about the salience and legitimacy of various forms of knowledge? And what if we made a more systematic effort to incorporate that into MA outputs?

A brief example illustrates what is at stake here. For several years in the 1980s and 1990s, I worked with various groups of Penan hunter-gatherers in the Malaysian state of Sarawak. As traditionally nomadic hunter-gatherers, Penan depend on the forest for virtually every aspect of their existence. They exemplify the depth and richness of environmental knowledge that indigenous peoples hold, with a remarkable knowledge of trees, plants, and animals and of the relations among them. Penan also possess a rich vocabulary for describing landscape and an extensive knowledge of places in the landscape they inhabit. This landscape is more than a reservoir of detailed ecological knowledge or a setting in which they satisfy their nutritional needs. It is also a repository for the memory of past events, a vast mnemonic representation of social relationships and of society. For Penan, landscape, history, and kinship—the bonds linking individuals to households to communities to generations past and future—are part of a larger whole.

In the late 1980s, the Penan became the focus of a high-profile transnational indigenous rights campaign concerned with logging. Since the 1980s, timber companies have expanded their reach throughout virtually every river valley occupied by Penan, and Penan have responded with intermittent blockades. During the first wave of blockades in 1987, images of Penan resisting the approach of logging companies traveled global environmental and indigenous rights circuits, producing an outpouring of support (Brosius 1997a, 1997b, 1999b, 2001a, 2001b, 2003a, 2003b). Penan continue to assert their rights to land using every tool of persuasion available to them, though their efforts have largely been futile. The official government view of Penan is that their way of life is little more than a form of vagrancy in which would-be subjects are able to evade the gaze of the state. The only way Penan can be heard, the only discourse audible to the state, is that of development. The overall effect of the campaign was that the government shifted the debate over logging in Sarawak from a focus on forest destruction and the rights of indigenous communities to an issue of sustainable forest management. The discursive contours of the debate were shifted away from the moral and political domain toward the domain of environmental management (Brosius 1999a). By the mid-1990s, the campaign's momentum had largely dissipated.

The question I want to pose is, in a "policy environment" characterized by dispossession, where the threats to local communities result from the actions of "decision makers," of what relevance is indigenous knowledge of nature by itself, divorced from its significance with respect to the making of claims? What is needed, I would argue, is a more expansive, less fixed notion of knowledge. What matters is not how much Penan know about the landscape they inhabit but how they position that knowledge, and themselves, within the broader contours of power.

Whether or not they are actively engaged in explicit acts of resistance, the topic of logging is one that consumes Penan and that they discuss endlessly. Their narratives recount confrontations between themselves and state authorities or company representatives: police, judges, government ministers, camp managers, and others. They recount the arguments put forth by themselves or others: why they decided to blockade, why they should not be blamed for those blockades, and who they believe to be ultimately responsible.

Any effort to understand Penan narratives of dispossession must begin with recognizing the variety of forms they take. Such narratives, and the forms of action they prescribe, exist on a continuum from the concrete to the aesthetic and oblique. Many of the concerns they express are practical in nature—for instance, the simple difficulty of making a living in a logged-over landscape. Penan describe in matter-of-fact terms the destruction of the forest and the hardship this has caused them. They speak of river siltation, the destruction of sago, rattan, and fruit trees, and the depletion of game. At times, Penan make direct claims: they speak of boundaries and of the need to prohibit the entry of outsiders onto their lands. At other times they speak movingly about the qualities of the forest and their life within it. They speak of the heat, dust, and desolation of logging against the coolness and cleanliness of the forest, the harsh sound of chainsaws versus the squeaking of trees rubbing in the wind. The words and images they employ are contrastive and tinged with nostalgia: what the forest was like before logging and after it. And they speak of loss and pain—at seeing valuable fruit trees destroyed and the graves of loved ones bulldozed.

What is further striking about Penan commentaries on landscape and forest destruction is the degree to which the arguments they put forth are about *locality* and *biography.* Penan do not talk about the need to preserve rainforests as a generic abstraction; they talk about the need to preserve particular watersheds "from which we eat." It is the transgression of that densely biographical and genealogical locality that Penan find to be such a great injustice.

Often too, Penan speak in metaphors—for example, linking the forest to a supermarket or a bank. Such arguments are meant to appeal to what Penan presume is a shared sense of justice and respect. The arguments that Penan are putting forth should be viewed not exclusively as acts of resistance but simultaneously as efforts at engagement. In making their arguments to loggers, civil servants, environmentalists, and others, Penan are attempting to speak across difference, to *familiarize* themselves, to frame their arguments in ways that they hope outsiders will recognize. Their purpose is to persuade.

In considering how Penan frame their struggle against logging, it is important to consider not merely the rhetorical elements of these narratives but the *forms* they take as well: letters addressed to government officials, verbal arguments with timber company managers, maps produced with the aid of local activists, videotaped interviews produced by Euro-American documentary filmmakers, and others. What happens when Penan claims are textualized in different ways? How do Penan conceptions of their audience condition the arguments they put forth and the forms of knowledge they deploy?

What this points to is the need to foreground notions of agency in narratives of landscape and dispossession. The questions of whom Penan believe to be responsible for their plight and whom they believe is in the best position to help them are as central to this whole domain of discourse as are statements about what is occurring and how it effects their everyday lives. These are as much narratives of culpability as narratives of place.

For instance, in asserting claims to land, arguing for the establishment of reserved areas, attempting to demarcate borders, or contesting the claims of timber companies, Penan—often with the help of nongovernmental organization allies—produce maps or written declarations. Penan see that loggers bring maps, show them official letters, and try to compel them to sign documents, and that all of these serve to validate company claims to Penan lands. Penan recognize that these methods are the single most effective way to assert their claims in a way that is meaningful to outsiders

At the same time they are asserting their own claims to land, Penan deny the validity of maps produced by others. One nomadic headman, referring to map-making practices, described timber companies as "stealing [land] from open places"—that is, from the outside. He declared government maps a lie because they are made from high above, showing only the shape of the land. The Penan see the fact that these maps are made from a distance as an indication of duplicity. Penan contrast the way companies make maps from a distance with the way they themselves do: by walking through and over every valley and ridge, by filling the place with names, and by sustaining themselves on resources that have been passed down for generations. As one nomadic Penan man sarcastically told me, he would ask loggers, "If this is your land, why do you always ask us the names of rivers? Do you know the names of places? You and your people are always asking—what is the name of this river?, what is the name of that river? If you don't know these, you don't belong here."

The Penan response to logging is a product not only of the tangible effects of environmental degradation but also of the way Penan perceive themselves to have been treated by those with an interest in its continuation. They are responding not only to logging as an activity that directly affects their lives but also to the *agents* of logging. When Penan discuss why they erect blockades, one theme arises more than any other: they say they blockade because "the government does not hear what we say," repeatedly describing the government and companies as being "deaf." Company and government officials do not listen to them, Penan assert, because the officials do not respect them, and they interpret this as a form of insult. Further, because they have made innumerable good-faith attempts at dialogue, any action they might then take—most often blockades—can no longer be considered their fault.

In recent years, we have observed a florescence of scholarship focused on indigenous conceptions of landscape (Basso 1984; Feld 1982; Hirsch and O'Hanlon 1995; Myers 1991; Povinelli 1993; Rosaldo 1980; Roseman 1991; Weiner 1991; Zerner 2003). This literature has alerted us to the rich variety of narrative forms through which societies inscribe their presence in places. Yet in listening to Penan statements about the forest and its destruction, we should be cautious about assuming that documenting Penan conceptions of landscape as some fixed entity—"indigenous knowledge"—is ever enough. Rather, we also need to try to discern how Penan conceptions of their audience condition the arguments they put forth.

Discussion: Local Knowledge, Indigenous Peoples, and Environmental Governance

In the past decade or so, it has become axiomatic to state that indigenous peoples "possess, in their ecological knowledge, an asset of incalculable value: a map to the biological diversity of the earth on which all life depends. Encoded in indigenous languages, customs, and practices may be as much understanding of nature as is stored in the libraries of modern science" (Durning 1992, 7). As self-evident as this may now seem, it does not provide much guidance with respect to how one moves between local knowledge and global science. That is the question that animates this volume and the Millennium Ecosystem Assessment as a whole.

In our efforts to bridge scales and epistemologies, we stand at a critical crossroad. For today we are confronted with two apparently contradictory trends in the domain of environmental governance. On the one hand, we have witnessed a trend toward valorizing indigenous/local forms of knowledge and mobilizing indigenous peoples. The Bridging Scales and Epistemologies conference, the outputs of the World Parks Congress, the World Conservation Congress, and the CBD/COP7—all of which express some form of support for indigenous priorities—are four manifestations of this trend.

On the other hand, in the last few years we have witnessed a decisive move by major conservation organizations toward cartographically enabled regional land use planning approaches under the rubric of ecoregional conservation (Olson et al. 2000, 2001; TNC 2000, 2001; World Wildlife Fund–US 2000). Along with this, we have witnessed the emergence of the field of "conservation finance" (Bayon, Lovink, and Veening 2000; Conservation Finance Alliance 2002; WWF-US 2001) and the proliferation of social science–based metrics and models designed to monitor and manage social and political processes in conservation (Brosius and Russell 2003). These three are linked discursively, strategically, and institutionally in a broader process of consolidation, and together they are reshaping the way conservation is conceptualized, planned, and administered. The comprehensive visions being promoted and the proprietary databases being produced in the emerging complementarities of spatial planning, investment, and social metrics have the potential to reshape the contours of the relationship between humanity and nature for generations to come.

Events such as the World Parks Congress and the World Conservation Congress can be seen, in essence, as exercises devoted to normalizing and reinforcing these complementary manifestations of consolidation. Attending these events, one is left with the impression that an enormous weight of managerialism has descended over conservation, much as it once did on development, and that this state of affairs is in large part due to the efforts of major conservation organizations to consolidate their authority over global conservation practices. They are achieving this consolidation by establishing administrative technologies in which they are taken for granted as methodological gatekeepers. Increasingly, conservation has become a gated community that one can enter only by accepting the methodological terms promulgated by major conservation organizations. This has occurred as tools or approaches that originated as emancipatory moves—stakeholder analysis, participatory mapping, community-based natural resource management—have been incorporated into the managerial apparatus of conservation. Once incorporated, they become tied to the imperatives of funding cycles, scaling up, accountability to donors, and more.

This situation is ironic, and it has major implications for integrating local and indigenous perspectives into conservation. Just when local voices and local forms of knowledge are being invoked as relevant to the setting of global conservation strategies and local conservation management, the institutional structures of global conservation that are now emerging are preventing them from being meaningfully included. What then becomes of alternative forms of conservation that are informed by local ways of knowing?

This consolidation of conservation practices by major conservation organizations is being achieved by the shift in scale that I have described. As Smith (1992) and Harvey (1996) remind us, scale is always political. Left unspoken in contemporary conservation is the relation between scale and hierarchy. Higher-level scales of visualization require higher-level structures of governance. Ecoregional conservation is fundamentally *about* scale—both enlarging the scale of environmental interventions and linking information created at different scales into a single strategic blueprint for the future at an extended temporal scale. All the talk about scaling up in conservation is accompanied by a concern for improving "efficiencies" and reducing "transaction costs." Indigenous knowledge, when it moves up the scale, becomes both simplified and embedded in a range of other agendas.

Earlier I drew a distinction between two forms of locality: that mediated by the research activities of social scientists and that articulated by local/indigenous activists and advocates. One speaks in the passive voice of science, translating indigenous ways of knowing into forms intelligible to practitioners and decision makers; the other speaks in the active voice of advocacy. Making this distinction draws our attention to the question of how local/indigenous perspectives and ways of knowing are elicited and translated between scales and how the link is made between this knowledge and the policy domain. The former reifies the distinction between local/indigenous peoples living their lives in particular places and policy makers who are making sometimes momentous decisions about those peoples' lives.

Local/indigenous advocates, on the other hand, are refusing that distinction. Making meaningful progress in the future will entail a willingness on the part of conservation scientists and practitioners to work with indigenous/local communities in new ways, ways in which the tools of Western science are offered in support of local conservation priorities. What that means for how conservation initiatives are planned, implemented, and governed is not yet clear, but it is an effort that we must take seriously. The challenge is to seek productive terms of engagement. We cannot afford to perpetuate the polemic that the goals of conservation and indigenous rights are at odds with each other, or that indigenous knowledge is something to be packaged and passed up to "decision makers." The fate of biodiversity rests in part on how the conservation community responds to the challenge posed by indigenous and local communities and whether it is able to embrace this as an opportunity to create new alliances for conservation.

References

Basso, K. 1984. Stalking with stories: Names, places and moral narratives among the western Apache. In *Text, play, and story: The construction and reconstruction of self and society,* ed. E. Bruner, 19–55. Washington, DC: American Ethnological Society.

Bayon, R., J. S. Lovink, and W. J. Veening. 2000. *Financing biodiversity conservation.* Washington, DC: Sustainable Development Department, Inter-American Development Bank. http://www.iadb.org/sds/doc/ENV-134FinancingBiodConservaE.pdf (accessed September 25, 2005).

Benhabib, S., ed. 1996. *Democracy and difference.* Princeton, NJ: Princeton University Press.

Brosius, J. P. 1997a. Endangered forest, endangered people: Environmentalist representations of indigenous knowledge. *Human Ecology* 25 (1): 47–69.

———. 1997b. Prior transcripts, divergent paths: Resistance and acquiescence to logging in Sarawak, East Malaysia. *Comparative Studies in Society and History* 39 (3): 468–510.

———. 1999a. Analyses and interventions: Anthropological engagements with environmentalism. *Current Anthropology* 40 (3): 277–309.

———. 1999b. Green dots, pink hearts: Displacing politics from the Malaysian rainforest. *American Anthropologist* 101 (1): 36–57.

———. 1999c. Locations and representations: Writing in the political present in Sarawak, East Malaysia. *Identities: Global Studies in Culture and Power* 6 (2/3): 345–86.

———. 2001a. Local knowledges, global claims: On the significance of indigenous ecologies in Sarawak, East Malaysia. In *Indigenous traditions and ecology,* ed. J. Grim and L. Sullivan, 125–57. Cambridge, MA: Harvard University Press and Center for the Study of World Religions.

———. 2001b. The politics of ethnographic presence: Sites and topologies in the study of transnational environmental movements. In *New directions in anthropology and environment: Intersections,* ed. C. Crumley, 150–76. Walnut Creek, CA: Altamira.

———. 2003a. The forest and the nation: Negotiating citizenship in Sarawak, East Malaysia. In *Cultural citizenship in Island Southeast Asia: Nation and belonging in the hinterlands,* ed. R. Rosaldo, 76–133. Berkeley: University of California Press.

———. 2003b. Voices for the Borneo rainforest: Writing the history of an environmental campaign. In *Imagination and distress in southern environmental projects,* ed. A. Tsing and P. Greenough, 319–46. Durham, NC: Duke University Press.

———. 2004. Indigenous peoples and protected areas at the World Parks Congress. *Conservation Biology* 18 (5): 609–12.

Brosius, J. P., and D. Russell. 2003. Conservation from above: An anthropological perspective on transboundary protected areas and ecoregional planning. *Journal of Sustainable Forestry* 17 (1/2): 39–65.

Brosius, J. P., A. Tsing, and C. Zerner. 1998. Representing communities: Histories and politics of community-based natural resource management. *Society and Natural Resources* 11 (2): 157–68.

———. 2005. *Communities and conservation: Histories and politics of community-based natural resource management.* Walnut Creek, CA: Altamira.

Brush, S. 1993. Indigenous knowledge of biological resources and intellectual property rights: The role of anthropology. *American Anthropologist* 95 (3): 653–86.

Conservation Finance Alliance. 2002. Conservation Finance Alliance. http://www.conservationfinance.org (accessed September 25, 2005).

Dirks, N., G. Eley, and S. Ortner, eds. 1993. *Culture/power/history: A reader in contemporary social theory.* Princeton, NJ: Princeton University Press.

Dryzek, J. 1990. *Discursive democracy: Politics, policy and political science.* Cambridge: Cambridge University Press.

Durning, A. 1992. *Guardians of the land: Indigenous peoples and the health of the earth.* Worldwatch Paper 112. Washington, DC: Worldwatch Institute.

Escobar, A. 1995. *Encountering development: The making and unmaking of the third world.* Princeton, NJ: Princeton University Press.

Fabian, J. 1983. *Time and the other: How anthropology makes its object.* New York: Columbia University Press.

Feld, S. 1982. *Sound and sentiment: Birds, weeping, poetics, and song in Kaluli expression.* Philadelphia: University of Pennsylvania Press.

Friedman, J. 1998. The heart of the matter. *Identities* 4 (3/4): 561–69.

Gledhill, J. 1998. Contesting regimes of truth: Continuity and change in the relationships between intellectuals and power. *Identities* 4 (3/4): 515–28.

Harvey, D. 1996. *Justice, nature and the geography of difference.* Oxford: Blackwell.

Hirsch, E., and M. O'Hanlon, eds. 1995. *The anthropology of landscape: Perspectives on place and space.* Oxford: Clarendon.

IUCN (World Conservation Union). 2003. *Message to the Convention on Biological Diversity, Vth IUCN World Parks Congress, Durban, South Africa, 8–17 September 2003.* Gland, Switzerland: IUCN. http://www.iucn.org/themes/wcpa/wpc2003/pdfs/outputs/wpc/cbdmessage.pdf (accessed September 25, 2005).

Laclau, E., and C. Mouffe. 2001. Hegemony: The genealogy of a concept. In *The new social theory reader: Contemporary debates,* ed. S. Seidman and J. C. Alexander, 76–86. London: Routledge.

Mascia, M., J. P. Brosius, T. A. Dobson, B. C. Forbes, L. Horowitz, M. A. McKean, and N. J. Turner. 2003. Conservation and the social sciences. *Conservation Biology* 17 (3): 649–50.

Millennium Ecosystem Assessment. 2003. *Ecosystems and human well-being: A framework for assessment.* Washington, DC: Island Press.

Mintz, S. 1985. *Sweetness and power: The place of sugar in modern history.* New York: Viking Penguin.

Myers, F. 1991. *Pintupi country, Pintupi self: Sentiment, place, and politics among Western Desert Aborigines.* Berkeley: University of California Press.

The Nature Conservancy. 2000. *Designing a geography of hope: A practitioner's handbook to ecoregional conservation planning.* Vols. 1 and 2. Washington, DC: The Nature Conservancy. http://conserveonline.org/docs/2000/11/GOH2-v1.pdf (accessed September 25, 2005).

———. 2001. *Conservation by design: A framework for mission success.* Washington, DC: The Nature Conservancy. http://nature.org/aboutus/howwework/files/cbd_en.pdf (accessed September 25, 2005).

Olson, D. M., E. Dinerstein, R. Abell, T. Allnutt, C. Carpenter, L. McClenachan, J. D'Amico, et al. 2000. *The global 200: A representation approach to preserving the earth's distinctive ecosystems.* Washington, DC: Conservation Science Program, World Wildlife Fund–US, Washington, D.C. http://www.gm-unccd.org/FIELD/NGO/WWF/Global200.pdf (accessed September 25, 2005).

Olson, D. M., E. Dinerstein, E. D. Wikramanayake, N. D. Burgess, G.V.N. Powell, E. C. Underwood, J. A. D'Amico, et al. 2001. Terrestrial ecoregions of the world: A new map of life on earth. *BioScience* 51 (11): 933–38.

O'Neill, J. 2001. Representing people, representing nature, representing the world. *Environment and Planning C: Government and Policy* 19 (4): 483–500.

Painter, J, and C. Philo. 1995. Spaces of citizenship: An introduction. *Political Geography* 14 (2): 107–20.

Povinelli, E. 1993. *Labor's lot: The power, history and culture of Aboriginal action.* Chicago: University of Chicago Press.

Reid, W. 2000. Ecosystem data to guide hard choices. *Issues in Science and Technology Online* 16 (3, Spring). Online journal. http://www.nap.edu/issues/16.3/reid.htm.

Rosaldo, R. 1980. *Ilongot headhunting, 1883–1974: A study in society and history.* Stanford, CA: Stanford University Press.

Roseberry, W. 1989. *Anthropologies and histories: Essays in culture, history, and political economy.* New Brunswick, NJ: Rutgers University Press.

Roseman, M. 1991. *Healing sounds of the Malaysian rainforest: Temiar music and medicine.* Berkeley: University of California Press.

Schneider, J., and R. Rapp, eds. 1995. *Articulating hidden histories: Exploring the influence of Eric R. Wolf.* Berkeley: University of California Press.

Smith, N. 1992. Geography, difference and the politics of scale. In *Postmodernism and the social sciences,* ed. J. Dougherty, E. Graham, and M. Malek, 57–79. London: Routledge.

Tawfic Ahmed, M., and W. Reid. 2002. Millennium Ecosystem Assessment: A healthy drive for an ailing planet. *Environmental Science and Pollution Research* 9 (4): 219–20.

Terborgh, J. 1999. *Requiem for nature.* Washington, DC: Island Press.

Trouillot, M. 1991. Anthropology and the savage slot: The poetics and politics of otherness. In *Recapturing anthropology: Working in the present,* ed. R. Fox, 17–44. Santa Fe, NM: School of American Research Press.

Wallerstein, I. 1974. *The modern world system: Capitalist agriculture and the origins of the world economy in the sixteenth century.* Orlando, FL: Academic Press.

Weiner, J. 1991. *The empty place: Poetry, space, and being among the Foi of Papua New Guinea.* Bloomington: Indiana University Press.

Wolf, E. 1982. *Europe and the people without history.* Berkeley: University of California Press.

World Wildlife Fund–US (WWF-US). 2000. *The global 200: A blueprint for a living planet.* Washington, DC: World Wildlife Fund–US. http://www.panda.org/about_wwf/where_we_work/ecoregions/index.cfm (accessed September 25, 2005).

———. 2001. *Center for Conservation Finance business plan: Building conservation capital for the future.* Washington, DC: Center for Conservation Finance, World Wildlife Fund–US. http://www.worldwildlife.org/conservationfinance/pubs/business_plan.pdf (accessed September 25, 2005).

Zerner, C., ed. 2003. *Culture and the question of rights.* Durham, NC: Duke University Press.

CHAPTER 8

Bridging the Gap or Crossing a Bridge?

Indigenous Knowledge and the Language of Law and Policy

MICHAEL DAVIS

In December 2002, Australia's High Court dismissed an appeal made by the Yorta Yorta Aboriginal people of Northern Victoria and New South Wales against an earlier federal court determination that had decided against their claim for native title under the Native Title Act 1993. These Aboriginal peoples' struggle for recognition of their enduring connections with their ancestral lands under Australia's native title laws had, in this hearing, depended solely on the outcome of complex legal deliberations regarding notions of tradition and custom.

The Yorta Yorta peoples' claim had been dismissed by a 1998 federal court decision on the basis, the judges held, that the "tide of history had washed away" the peoples' connection to lands and waters. The court's argument was that the traditional laws acknowledged and customs observed by Yorta Yorta today were not the same as they had been in the period before Europeans arrived. Laws designed to provide for indigenous peoples' rights and interests in land or native title, or for their participation in managing or protecting environment and biodiversity, incorporate terms and concepts intended to denote aspects of Aboriginal culture relevant to the particular law in question. Examples include *tradition, traditional knowledge,* and *law and custom.* Yet such terms are employed in legal texts in ways that present idealized, or fictive, notions of Aboriginal culture and society. They are derived not from indigenous ways of understanding and articulating the world but, rather, from Western intellectual worldviews and presuppositions.

This chapter explores some issues that flow from these problems in cultural translation by first examining and then challenging the often-held notion of a divide between indigenous knowledge and "Western" science. Although the term *Western science* refers in this context to all modes of knowledge and practice that form dominant epistemologies, have claims to truth or authority, and are said to be "derived from facts," this notion of scientific modes of knowing is as problematic as the construct of indigenous knowledge that is the subject of this chapter (Chalmers 1999).[1]

The idea of a divide between indigenous knowledge and Western science has been founded on a view that Western science and allied systems of knowledge have formed a dominant discourse that has obliterated, marginalized, or assimilated local, traditional, and indigenous traditions and discourses. In reviewing this divide, this chapter argues for a greater emphasis on the complexity, diversity, and plurality of indigenous knowledge and draws on some examples from the Australian literature to illustrate. The recognition of the "plurality of cultural systems and the diversity of environmental knowledge within and between cultures" (Grim 2001, liii) might also help with incorporating understandings of the dynamism and innovative and adaptive qualities of indigenous cultures into the dominant discourses of law, policy, and administration.

When advocating plurality in discourses and epistemologies, some caution is needed to avoid representing indigenous knowledge in law and policy either (1) as a set of essentialized or homogeneous entities that satisfy some stereotypical Western image or (2) as being utterly incommensurable, or radically other in an extreme relativistic position that renders cultural comparison untenable or negates any possibility of finding common ground or integrating different knowledge systems.

Indigenous knowledge and Western science are best regarded as complementary, or parallel, systems of knowledge, rather than as fundamentally incommensurable. As Turnbull points out, all knowledge systems can be regarded as localized, situated ways of making coherent systems of meaning from an array of heterogeneous, disorganized, and fragmented elements. The differences that can be observed cross-culturally among and between knowledge systems arise from their different power structures, modes of social and political organization, and the particular ways in which they seek to produce coherent systems (Turnbull 2000; also see Agrawal 1999).

Creating the Divide

Indigenous knowledge has historically been regarded in the dominant, Western society as inferior and marginalized, and as a devalued form of knowledge. This lowly status of indigenous knowledge is a result of the growth of dominant forms of knowledge concomitant with indigenous peoples' historical experiences of colonization and oppression. This marginalizing of indigenous knowledge also has resulted from the particular bureaucratic-administrative machinery of government, founded on the creation of hierarchies that privilege those forms of knowledge, such as science and law, that claim to purvey some truth and authority.

As Dei et al. (2000, 4) note: "The negation, devaluation, and denial of indigenous knowledges, particularly those of women, is the result of deliberate practices of establishing hierarchies of knowledge. . . . Institutions are not unmarked spaces of thought and action. Knowledge forms are usually privileged to construct dominance, and can be 'fetishized' so as to produce and sustain power inequities." Vandana Shiva (in Dei et al. 2000, vii) similarly asserts that "Western systems of knowledge in agriculture and medicine were defined as the *only* scientific systems. Indigenous systems of knowledge were defined as inferior, and in fact as unscientific."

Not only were indigenous knowledge systems seen as inferior, they were also "systematically usurped and then destroyed in their own cultures by the colonizing West" (Shiva, in Dei et al. 2000, vii). Within this framework of "knowledge hierarchies" (Dei et al. 2000), local and indigenous knowledge systems are rendered invisible or devalued by the dominant culture. This view is also seen in some conventional development approaches, wherein indigenous and local peoples are "developed" by those doing the developing. As a result, dependent relations are established and maintained through which indigenous systems of knowledge are usurped by the dominant developed discourses (Agrawal 1995; Antweiler 1993; Hobart 1993).

Knowledge systems and epistemologies may often be seen as jostling in apparent adversity and competition rather than striving for integration and mutual interdependence. There are many examples of competing systems, which are typically played out in contexts of claims for recognition. One example in recent years was the Hindmarsh Island case, in which Aboriginal women's knowledge relating to a certain place in South Australia was subordinated and denigrated by those advocating and supporting the proposed development of a bridge from the mainland across to Hindmarsh Island (Simons 2003).

The Tyranny of Dualism and Categories

Indigenous knowledges are subordinated not only through the formation of hierarchies but also by the perpetuation of binary oppositions of such categories as us/them, self/other, or we/they. The perceived dichotomies between "traditional" and "modern," and between "indigenous" and "nonindigenous," are further consequences of this pervasive dualism.

These dualities extend most significantly into discussions on modes of thought. In the history of anthropology and philosophy, a strand of debate has centered on the notion that differences exist between modes of thought of non-Western, "primitive" others and Western, "rational" modes of thought (Goody 1977). Allied to this is the Enlightenment idea of progress and the historically rooted shift from superstition to magic to religion to science. Indigenous peoples in this schema possess what have been regarded as exemplars of so-called primitive or irrational modes of thought. One problem in this debate over rationality and modes of thought is the specific categories that have been used to define and describe the binary oppositions flowing from us/them (Goody 1977; Lévi-Strauss 1966). Goody has noted that "the trouble with the categories is that they are rooted in a we/they division which is both binary and ethnocentric, each of these features being limiting in their own way." He goes on to suggest that "we speak in terms of primitive and advanced, almost as if human minds themselves differed in their structure like machines of an earlier and later design" (1977, 1).

Understanding different societies and cultures in terms of contrasts and binary oppositions is deeply embedded in European thought, both historically and institutionally. There persists in many discourses about indigenous involvement in and approaches to land, resource, and environmental management a perceived divide between "folk" systems of ecological knowledge, considered intuitive and informal, and scientific approaches, defined as rational, rigorous, and technically accurate. An example of how this kind of opposition has influenced interpretation and analysis is the use of fire for land management in Australia's Northern Territory. Aboriginal people had traditionally used fire as a management tool for maintaining or increasing natural resources. Fire is also used by cattle tenders for pastoral purposes, and by non-Aboriginal national park rangers in park management. Although Aborigines have in recent years become more involved in park management and ranger activities, perceived differences still exist in the worldviews of Aborigines, cattle tenders, and park rangers regarding burning practices (Lewis 1989).

Beyond Categories

Some of the literature on humans' knowing and interacting with landscapes and environments has emphasized or reinforced a divide between indigenous knowledge and Western science founded on the oppositional categories of indigenous/Western or indigenous/nonindigenous. However, the category of indigenous knowledge is formed from a complex intertwining of knowledge traditions and practices through the engagement of indigenous and nonindigenous peoples. Far from being considered a unitary, homogeneous entity founded in some perceived idea of indigeneity, indigenous knowledge must instead be understood as contingent, historically situated, and particular to the specifics of locality, group dynamics, place, and time. The term *indigenous knowledge* needs to be interrogated in order to shift from positing it as a reified, essentialized construct suspended in space and devoid of context, toward a more nuanced view. Simultaneously, the presumed sharp distinction between indigenous knowledge and other knowledge systems also needs to be reconsidered.

What is usually termed *indigenous knowledge* comprises complex interactions and relationships among peoples (indigenous and nonindigenous), situations, experiences, observations, and practices. In what way might we define a point at which "traditional" knowledge differs from, say, "new," "adapted," or "modernized" knowledge? There may be a continuum or spectrum of systems of knowledge across time, space, and locality, thus rendering difficult or irrelevant any attempts to create artificial distinctions or dichotomies between "indigenous" knowledge, "traditional" knowledge, and "science" (Agrawal 1995).

Ellen and Harris (2000, 2) are among those who have critiqued the sharp distinction between "indigenous" and "nonindigenous" knowledge systems claiming that such a distinction "has many highly specific regional and historical connotations which are not always appropriate to other ethnographic contexts." In this view, creating these distinctions makes comparative work difficult.

The Same and Yet Different

Indigenous and scientific systems of knowledge and practice share some common characteristics yet also reveal some important differences. One study illustrates some contrasts between the knowledge systems, or epistemologies, of Aborigines and pastoralists in the context of land management in the Kowanyama River catchment in Far North Queensland. Here, Strang (1997) has noted fundamentally different discourses on land and environment that appear to reflect contrasting worldviews. Discussing Aborigines' perceptions of and approaches to land management, she comments that "the most important point about Aboriginal land use is that economic interactions with country are never wholly divorced from social and spiritual interactions." She goes on to argue that "land provides a central medium through which all aspects of life are mediated, and economic considerations are merely part of an intimate, immediate, fundamentally holistic relationship" (p. 84).

Strang describes some stark differences between pastoralists' worldviews and those of Aboriginal peoples in this region:

> Aboriginal cosmology is typically presented as the foundation for a primarily mystical, spiritual interaction with the physical world, while in the European or white Australian cosmos, scientific rationalism and crass materialism are largely believed to have marginalized spiritual life.
>
> The Aboriginal groups and the pastoralists experience quite different kinds of physical and emotional interaction with the environment. The traditional Aboriginal economy demands intimate and highly detailed knowledge of the local ecology and geography, with an intense focus of attention on the indigenous flora and fauna. Being integrated with the spiritual and emotional aspects of Aboriginal life, it is part of a deep engagement with a particular landscape, encouraging a continual investment of value in the land. The interaction based on traditional activities—walking, fishing, collecting resources and so on—is a very immediate, tactile engagement, lending itself to qualitative and affective responses to the land. (Strang 1997, 237)

Highlighting the different ways in which pastoralists engage with the land and environment, Strang (1977, 280) observes that "the pastoralists are focused on the foreign elements they have imposed on the landscape: the Western technology, the infrastructure and the stock. Their attention is firmly engaged by, and therefore invested in, their economic activities. On a daily basis, their adversarial efforts to control the cattle and the land are largely mediated by technology, separating them from a more gentle, intimate interaction with the landscape."

Whereas Strang's study emphasizes difference and incommensurability, others stress integration and complementarity between knowledge systems. An example of this latter group is a comparative study of landscape classification and ecological knowledge of Anangu Aboriginal people in Central Australia, and scientific ecological approaches to land management (Baker and Mutitjulu Community 1992; Reid et al. 1992). This study shows that two quite distinct systems of taxonomy and classification of the natural world can be worked together toward the common goal of sustainable land and environmental management. It illustrates the ways in which indigenous and scientific systems of knowledge can find common ground and can be regarded as complementary or parallel systems. This complementarity can be explored further by examining what characteristics are shared by indigenous and scientific—indeed, by all—systems of knowledge. Slikkerveer (1999, 169) points out that both indigenous and what he terms "global" knowledge systems "are alternative pathways in the human/scientific quest to come to terms with the universe, and are the result of the same process of creating order out of disorder."

At the heart of both indigenous/local and scientific/global knowledge systems is the practice of making observations about local phenomena and interpreting patterns and trends. All knowledge systems, in their applications and techniques, consist of classifying the world and creating typologies, rules, and methods for understanding. They are based on experimentation and innovation. The practices, the techniques, and the applications are to be seen as somewhat distinct to the knowledge itself. All knowledge, in this sense, is concerned with the task of making sense of the world around us and of adapting to changes in the world or adjusting the world to achieve a balance between societies and their environments. The common elements underpinning all knowledge systems have been explored in some detail by Turnbull, who argues that "there is not just one universal form of knowledge (Western science), but a variety of knowledges" (2000, 1). Turnbull demolishes the notion of a hegemonic, authoritative Western science, proposing instead that the production of all kinds of knowledge is a process of assembling a vast array of heterogeneous components (2000, 4). He suggests that "all knowledge traditions, including Western technoscience, can be compared as forms of local knowledge so that their differential power effects can be explained without privileging any of them epistemologically" (2000, 6). Thus it is—in Turnbull's scheme—the particularized, localized social and spatial settings that we must look to if we are to engage in a cross-cultural exploration of differences between and among different knowledge traditions.

Agrawal (1999, 177) supports the view that different knowledge traditions are best understood by examining their contexts. He argues that relations of power are the critical factors to consider in different knowledge systems: "Most scholars have now come to accept that there are no simple or universal criteria that can be deployed to separate indigenous knowledge from western or scientific knowledge. Attempts to draw a line between scientific and indigenous knowledge on the basis of method, epistemology, context-dependence, or content, are intellectually barren and have produced little that is persuasive."

In considering the contextualized nature of knowledge systems, Agrawal argues that it is important to consider the social and political contexts of knowledge, and the relationships between power and practice, if the study of indigenous knowledge systems is to serve the interests of indigenous peoples themselves. Since these peoples are usually poor and marginalized, we must consider the problem in terms of how the "institutions and practices sustained by different forms of knowledge" contribute to their plight (Agrawal 1999, 178). The differences, therefore, between indigenous and scientific knowledge systems are to be found not as intrinsic properties of the systems themselves but, rather, in terms of how the systems are formed, practiced, and applied. It is in the social and political relations between and among knowledge holders and transmitters, in the distribution of power and authority, and—crucially—in the contexts in which these knowledge systems are formed, maintained, and presented that we might discern some comparative cross-cultural and cross-disciplinary distinctions as well as seek commonality.

Considering both the problem of comparative engagement between and among different knowledge systems and the need to find common ground, should the distinctive aspects of indigenous knowledge systems also be emphasized? If we are to highlight the distinctiveness of indigenous knowledge, one suggestion could be to highlight its "traditional" nature. Although, as this chapter discusses, using the term *tradition* in reference to indigenous knowledge is highly problematic, the Canadian-based indigenous organization Four Directions Council (cited in Posey 1999a, 4) makes a useful point about this notion of tradition: "What is 'traditional' about traditional knowledge is not its antiquity, but the way it is acquired and used. In other words, the social process of learning and sharing knowledge, which is unique to each indigenous culture, lies at the very heart of its 'traditionality.'"

Indigenous writer Laurie Anne Whitt (1999, 69) emphasizes the distinctly indigenous nature of indigenous knowledge by referring to its "intimate" relationship to land and to the natural world. Barsh (1999, 73), too, has proposed the features he believes distinguish indigenous systems of knowledge. While regarding the "traditional ecological knowledge" of indigenous and tribal peoples as "scientific in that it is empirical, experimental and systematic," he suggests some important differences. He states that indigenous knowledge differs in two respects from Western science:

> First, knowledge is highly localized. Its focus is the complex web of relationships between humans, animals, plants, natural forces, spirits and landforms within a particular locality or territory. . . . Second, local knowledge has important social and legal dimensions. Every ecosystem is conceptualized as a web of social relationships between a specific group of people (family, clan or tribe) and the other species with which they share a particular place.

In sum, the most distinctive feature of indigenous knowledge that sets it apart from scientific and other systems of knowledge is its holism, the way it functions as a complex set of interrelationships among the physical world, the world of humans, the natural world, and the unseen world of ancestors and cosmology.

Beware the Noble Savage

A growing recognition of the value of indigenous knowledge (Brush and Stabinsky 1996) provides a useful and much needed counterpoint to earlier discourses that denigrated such knowledge systems. However, it also brings with it a risk of constructing indigenous peoples as environmentalists par excellence. These noble-savage ecological warriors become, in some discourses, the saviors of the planet, standing as powerful symbols for those who oppose globalization and unfettered development (Sackett 1991). Ellen and Harris (2000, 1) note that "most of us will also accept that the claims made for the environmental wisdom of native peoples have sometimes been misjudged and naïve, replacing denial with effusive blanket endorsement and presenting an 'ecological Eden' to counter some European or other exemplary 'world we have lost.'" To avoid proliferating this kind of unexamined, essentialized view of indigenous knowledge, we must strive to develop plurality wherein a space is created for juxtaposing different systems of knowledge and actions in structures of complementarity rather than of competition and adversity—one that might also lead to a greater understanding of the complexity of indigenous knowledge systems and practices.

More informed, systematic understandings of indigenous knowledge, taxonomies, categories, and concepts may be gained through such rigorous, applied disciplines as anthropology, geography, and history. An example of such an endeavor is geographer Richard Baker's (1999) study of the Yanyuwa Aboriginal people around Borroloola in Australia's Northern Territory. Baker writes: "It is important to try and see Yanyuwa country through Yanyuwa eyes." He explains that "what can seem to European imagination to be an unproductive, strange and at times frightening landscape, is the known and bountiful home of the Yanyuwa" (1999, 45). Baker's study shows these Aboriginal peoples' environmental knowledge to be dynamic and responsive, changing and adapting over thousands of years through constant observation, experimentation, and transmission across the generations. Characterizing this type of innovative knowledge also helps refute the notion that what is often called "traditional" knowledge is fixed and immutable (Baker 1999, 45–50).

A better cross-cultural understanding of systems of thought and practice can also powerfully challenge the authority and hegemony of the dominant modes of thought, as Overing argues. She states (1985, 17): "An excellent antidote to the power of our Western hierarchical oppositions and the theory of knowledge upon which they ride is an acquaintance with other theories of knowledge and ontologies." Clearly, a need exists for greater understanding of other systems of knowledge and translation across categories and boundaries. However, this understanding should be approached with some degree of caution. Not all indigenous knowledge can or should be revealed to those outside the culture, or even to certain persons within the culture. It may be, in this sense, conceivable to appreciate the complexity and richness of a particular system of knowledge across cultural boundaries without having access to the details of that knowledge tradition. There is much that must remain confidential, and respect for the internal rules governing the management of knowledge in indigenous communities is an essential part of cross-cultural understanding.

Defining Indigenous Knowledge

Indigenous writer Winona LaDuke (1994, 127) has written that "traditional ecological knowledge is the culturally and spiritually based way in which indigenous peoples relate to their ecosystems." She states that "this knowledge is founded on spiritual-cultural instructions from 'time immemorial' and on generations of careful observation within an ecosystem of continuous residence." Many writers have grappled with the terminology and definitions of "indigenous traditional knowledge."

Acknowledging the difficulties of defining indigenous knowledge, Howden (2001, 60) suggests the following working definition: "[Indigenous knowledge] is a living system of information management which has its roots in ancient traditions. It relates to culture and artistic expression and to physical survival and environmental management. It controls individual behavior, as it does community conduct. In short, it is a concept that essentially defies description in Western terms, but which lies at the heart of Indigenous society."

In this view, the problem in understanding indigenous knowledge within Western discourses lies in the kind of categorization that these discourses use to separate such categories as "law," "culture," "heritage," and "religion" (as discussed above in terms of the Western preoccupation with hierarchies of knowledge). Howden (2001, 62) writes: "Indigenous knowledge systems are better understood as practical, personal and contextual units which cannot be detached from an individual, their community, or the environment (both physical and spiritual)."

Working definitions of indigenous or "traditional" knowledge have also been proposed by others, including Davis (1999, 1), who bases such a definition on certain identifiable characteristics said to be common to all types of indigenous knowledge. These include the following:

- The holding of communal rights and interests in knowledge
- A close interdependence among knowledge, land, and spirituality
- The passing down of knowledge through generations
- Oral exchange of knowledge, innovation, and practices according to customary rules and principles
- The existence of rules regarding secrecy and sacredness that govern the management of knowledge.

Although some analytical use lies in formulating a working definition of indigenous knowledge, the risk also exists that such defining and classifying returns to the very problem argued against in this chapter: the reifying and essentializing of indigenous categories and concepts. Formulaic definitions, once established in the literature, become vulnerable to appropriation by dominant discourses, thus perpetuating the very problem we address here. Another concern with definitions revolves around who is doing the defining and for what purposes. Finally, the formation of definitions places at risk the possibility of recognizing the diversity and plurality of indigenous knowledge. As Dei et al. (2000, 4) have explained this plurality: "All knowledges exist in relation to specific times and places. Consequently, indigenous knowledges speak to questions about location, politics, identity, and culture, and about the history of peoples and their lands." Is it possible then to represent such fluidity within a single definition? And even more important, what purpose would such definitions have and for whom?

Valuing Indigenous Knowledge

Indigenous knowledge has often been undervalued, or perceived to be of less worth than other forms of knowledge. This undervaluing has been discussed in the context of development. As Chambers and Richards (1995, xiii) point out: "In the past, indigenous knowledge was widely regarded among development professionals as an academic, if not dilettantish, concern limited largely to social anthropologists. Much of it was seen as superstition. In the dominant model of development, useful knowledge was only generated in central places—in universities, on research stations, in laboratories, then to be transferred to ignorant peasants and other poor people."

However, an increasing body of literature is recognizing the intrinsic value of indigenous knowledge systems and of the benefits of harnessing these systems toward sustainable development goals (Agrawal 1995).

Plurality, Complexity, and Understanding

Recognizing the value of indigenous systems of knowledge is a critical step toward greater appreciation of the plurality between and among different traditions. An appreciation of plurality rests on developing a sound comparative understanding across and within different cultural systems. Shiva (2000, viii) advocates a plural approach to knowledge systems, arguing:

> It is now generally recognized that the chemical route to strengthening agriculture and health care has failed, and must be abandoned. This provides us with an opportunity to re-evaluate indigenous knowledge systems and to move away from the false hierarchy of knowledge

> systems back toward a plurality. The pluralistic approach to knowledge systems requires us to respect different such systems—to embrace their own logic and their own epistemological foundations.

She elaborates (2000, viii–ix):

> It also requires us to accept that *one* system (i.e., the Western system) need not and must not serve as the scientific benchmark for all systems, and that diverse systems need not be reduced to the language and logic of Western knowledge systems.

If this plurality and complexity are better understood and respected, bridging the gap between different knowledge systems is more likely to occur.

Crossing the Divide

The divide—imagined, perceived, or invented—between indigenous and non-indigenous knowledge traditions can be crossed by considering different ways of thinking, talking, and writing about environmentally based practices. One such approach is "caring for country," a phrase that has been used to describe specific nurturing strategies and practices that "promote the well-being of particular types of ecosystems" (Rose 1996, 63). For Aboriginal people, caring for country might be considered a way of attaining a balance among environmental consciousness, pragmatic approaches to sustaining livelihoods, and spiritual or cosmological perspectives on food, living things, and being in the world.

However, the expression can also suggest a more thoughtful or considered way by which humans generally and collectively might approach the maintenance of the land and environment. In this way, a notion of "care" can be deployed as a metaphor for a regime of intercultural environmental ethics, practices, and epistemologies that are not derived from or dependent on specific historically or culturally based techniques and technologies. By promoting an "Aboriginal land ethic" (Rose 1988) and, more broadly, an "ecological ethic," it is possible to transcend divisive, conflict-based approaches to the environment and develop "attitudes of care, concern, respect, responsibility and perhaps awe for the value of all living things which compose the larger web of life" (Tully 2001, 150). The working together of multiple epistemologies—indigenous, "Western," scientific, and others—is central to such an approach.

The divide between so-called Western rational, instrumental, scientific discourses and actions and indigenous epistemologies has been based on a perceived dichotomy between the scientific approach—with its emphasis on pragmatic, rational, and logical actions founded in measurement, accuracy, and technology—and indigenous approaches, thought to be more integrative and to juxtapose the physical and the pragmatic with the spiritual and the religious. However, if we focus not on imposed presuppositions about an indigenous knowledge–Western science divide but, rather, on collective approaches to caring for, nurturing, and maintaining land and ecosystems, then we may be able to integrate or harmonize different traditions and epistemologies. In Rose's view, good ecological management is achieved by working together different ways of "caring for" or nurturing country, such as meshing the "conventional" fire management regimes employed by rangers with the systems used by Aboriginal people. In this way, she proposes, "the congruence of two knowledge systems . . . offers models for how ecological knowledge more generally can be managed on the continent, and for how Indigenous and settler Australians can share in the work of life" (1996, 63; see also Rose 2004). An appreciation and incorporation of culturally different concepts and categories when forming laws and policies can provide the grounds for implementing the policies more ethically.

Translating Concepts: Tradition and Custom

Translating concepts and categories between different cultural systems requires reexamining and rethinking some key concepts of law, policy, and administration. One such concept is tradition, which recurs often in discourses on native title and heritage in Australia. As anthropologist Peter Sutton (2003, xviii) observes: "The focus of native title in Australia is on the translation of customary and traditional rights in country into legal "rights and interests.'" The concept of tradition as articulated in the legal arguments is rooted in Enlightenment ideas of progress and finds expression in a traditional/modern dichotomy. This historically situated concept of tradition within discourses of modernity further complicates the position of indigenous peoples as exemplars of tradition. In this sense, "tradition" is often regarded as some imagined construct that posits an "authentic" or "truthful" set of beliefs, values, customs, and practices, rooted in antiquity and reinforced by ancient and enduring mythic charters. This "tradition" predates modernity or rests in opposition to it.

In the history of anthropological and ethnographic work in Australia and elsewhere, as well as in the forming and implementing of law and policy for indigenous peoples, there has been a tendency to search for or construct some perceived intangible, residual, and elusive "traditional culture" that is thought to underlie contemporary indigenous lives (Povinelli 2001). However, in the present argument concerning the dissolving of boundaries between "indigenous" and "Western" knowledges, it is more productive to posit a greater complexity in the relationships between "tradition" and "modernity." The tradition/modernity boundary can be blurred by adopting a view of "traditional culture" not as some immutable, fixed set of customs and practices but, rather, as a more malleable entity. Swain (1993, 178) provides a useful guide to this kind of approach: "The 'traditional Aborigine' is an academic fiction. We are dealing with an inherently dynamic ontological fabric, constantly being made relevant to an ever-changing world."

Traditions, argues Swain, are "entities which are forever becoming" (1993, 279). By taking this more pluralistic, dynamic understanding of tradition and extending it to suggest a multiplicity of traditions sharing a mutually compatible space, the traditional/modern dichotomy begins to fade. Instead, following Muecke (2004), there is a constant movement between ancient and modern wherein, if we equate the ancient with that which is "traditional," the ancient can be said to be always already present within the modern. If the concepts of "tradition" and "traditional culture" are deconstructed in this way, what then of "modernity"? Rather than positing a unitary or homogeneous modernity, which can be "understood as an attitude of questioning the present," Gaonkar (2001, 13–14) suggests it is useful to "think in terms of alternative modernities." Establishing a field containing a multiplicity of traditions and modernities creates a space wherein it becomes possible to reformulate relationships between and among different knowledge traditions.

Dissolving the binary opposition of tradition/modernity exposes the many levels of meanings, values, and contexts within which concepts such as "tradition" may be reexamined. The current use of terms and categories in legal, policy, and administrative discourses and practices has little to do with the historically, socially, and culturally situated actualities of indigenous communities. Such uses are generally divorced from the adaptive, dynamic processes of cultural systems in indigenous societies and reflect more the ideologies and presuppositions of the dominant legal and political machinery. The role of disciplines such as anthropology, grounded in field observation and close engagement with indigenous communities, is important to consider here, as such disciplines might provide a more nuanced and complex understanding of indigenous cultural systems (see Brush 1993; Davis 2001; Smith 2003).

Conclusion

This chapter has argued that although there may be innate, fundamental, a priori principles underlying all systems of knowledge and epistemology, the application and practices stemming from these systems differ across, between, and within cultures. In other words, common principles or core elements are perceived and sensed differently by different cultures, which then construct their own classifications and taxonomies to describe the environment in ways that accord with their cultural systems. Dominant legal and sociopolitical systems delimit and bound indigenous cultural and epistemological systems in artificially constructed categories and concepts that have more reference to bureaucratization and program management than to specific, localized, and particularized cultural knowledge and epistemological systems.

While national policies and legislation serve the interests of the nation-state by legitimizing its dominance over marginalized and minority peoples through the use of essentializing language, the potential for engagement with indigenous forms of knowledge and practice also occasionally arises. Despite the totalizing tendencies of national discourse regarding indigenous epistemologies, there nonetheless remains the scope for a deeper, more engaged understanding of the complexities, malleability, and adaptability of indigenous knowledge systems within national policy and legislative discourse, as well as for a plural approach to help different traditions and epistemologies work together. This may be achieved by creating a space within national laws and policies for inscribing indigenous forms of cultural practice as well as by using interdisciplinary and multifaceted approaches to legislative and policy development. Such approaches can benefit from applied disciplines, such as anthropology and cultural criticism, that attend to the complexities of indigenous cultural systems. They will also be greatly enhanced by a commitment to engagement with indigenous peoples wherein these peoples can participate in, and contribute meaningfully to, policy and legislative development.

References

Agrawal, A. 1995. Dismantling the divide between indigenous and scientific knowledge. *Development and Change* 26:413–39.

———. 1999. On power and indigenous knowledge. In *Cultural and spiritual values of biodiversity,* ed. and comp. D. A. Posey, 177–80. London: United Nations Environment Programme/Intermediate Technology Publications.

Antweiler, C. 1993. Local knowledge and local knowing: An anthropological analysis of contested "cultural products" in the context of development. *Anthropos* 93:469–94.

Baker, L. M., and Mutitjulu Community. 1992. Comparing two views of the landscape: Aboriginal traditional ecological knowledge and modern scientific knowledge. *Rangeland Journal* 14 (2): 174–89.

Baker, R. 1999. *Land is life: From bush to town, the story of the Yanyuwa people.* St. Leonards, New South Wales: Allen and Unwin.

Barsh, R. L. 1999. Indigenous knowledge and biodiversity. In *Cultural and spiritual values of biodiversity,* ed. and comp. D. A. Posey, 73–76. London: United Nations Environment Programme/Intermediate Technology Publications.

Brush, S. B. 1993. Indigenous knowledge of biological resources and intellectual property rights: The role of anthropology. *American Anthropologist* 95 (3): 653–71.

Brush, S. B., and D. Stabinsky. 1996. *Valuing local knowledge: Indigenous people and intellectual property rights.* Washington, DC: Island Press.

Chalmers, A. F. 1999. *What is this thing called science?* 3rd ed. St. Lucia: Queensland University Press.

Chambers, R., and P. Richards. 1995. Preface. In *The cultural dimension of development: Indigenous knowledge systems,* ed. D. M. Warren, L. Jan Slikkerveer, and D. Brokensha, xiii–xiv. London: Intermediate Technology Publications.

Davis, M. 1999. Indigenous rights in traditional knowledge and biodiversity: Approaches to protection. *Australian Indigenous Law Reporter* 4 (4): 1–32.

———. 2001. Law, anthropology, and the recognition of indigenous cultural systems. In *Law and anthropology: International yearbook for legal anthropology,* vol. 2, ed. R. Kuppe and R. Potz, 298–320. The Hague: Martinus Nijhoff Publishers.

Dei, G. J. Sefa, B. L. Hall, and D. Goldin Rosenberg, eds. 2000. *Indigenous knowledges in global contexts: Multiple readings of our world.* Toronto: University of Toronto Press.

Ellen, R., and H. Harris. 2000. Introduction. In *Indigenous environmental knowledge and its transformations: Critical anthropological perspectives,* ed. R. Ellen, P. Parkes, and A. Bicker, 1–29. Amsterdam: Harwood Academic.

Gaonkar, D. P. 2001. On alternative modernities. In *Alternative modernities,* ed. D. P. Gaonkar, 1–23. Durham, NC: Duke University Press.

Goody, J. 1977. *The domestication of the savage mind.* Cambridge: Cambridge University Press.

Grim, J. A., ed. 2001. *Indigenous traditions and ecology: The interbeing of cosmology and community.* Cambridge, MA: Harvard Press for the Center for the Study of World Religions.

Hobart, M., ed. 1993. *An anthropological critique of development: The growth of ignorance.* London: Routledge.

Howden, K. 2001. Indigenous traditional knowledge and native title. *University of New South Wales Law Journal* 24 (1): 60–84.

LaDuke, W. 1994. Traditional ecological knowledge and environmental futures. *Colorado Journal of International Environmental Law and Policy* 5 (1, Winter): 127–48.

Lévi-Strauss, C. 1966. *The savage mind.* Chicago: University of Chicago Press.

Lewis, H. T. 1989. Ecological and technological knowledge of fire: Aborigines versus park rangers in Northern Australia. *American Anthropologist* 91 (4, December): 940–61.

Muecke, S. 2004. *Ancient and modern: Time, culture and indigenous philosophy.* Sydney: University of New South Wales Press.

Overing, J. 1985. Introduction. In *Reason and morality,* ed. J. Overing, 1–28. London: Tavistock Publications.

Posey, D. A. 1999a. Introduction: Culture and nature—the inextricable link. In *Cultural and spiritual values of biodiversity,* ed. and comp. D. A. Posey, 3–16. London: United Nations Environment Programme/Intermediate Technology Publications.

———, ed. and comp. 1999b. *Cultural and spiritual values of biodiversity.* London: United Nations Environment Programme/Intermediate Technology Publications.

Povinelli, E. A. 2001. Settler modernity and the quest for an indigenous tradition. In *Alternative modernities,* ed. D. P. Gaonkar, 24–57. Durham, NC: Duke University Press.

Reid, J., L. Baker, S. R. Morton, and Mutitjulu Community. 1992. Traditional knowledge + ecological survey = better land management. *Search* 23 (8): 249–51.

Rose, D. B. 1988. Exploring an Aboriginal land ethic. *Meanjin* 3:378–87.

———. 1996. *Nourishing terrains: Australian indigenous views of landscape and wilderness.* Canberra: Australian Heritage Commission.

———. 2004. *Reports from a wild country: Ethics for decolonisation.* Sydney: University of New South Wales Press.

Sackett, L. 1991. Promoting primitivism: Conservationist depictions of Aboriginal Australians. *Australian Journal of Anthropology* 2 (2): 233–47.

Shiva, V. 2000. Foreword: Cultural diversity and the politics of knowledge. In *Indigenous knowledges in global contexts: Multiple readings of our world,* ed. G. Sefa Dei, B. L. Hall, and D. Goldin Rosenberg, vii–x. Toronto: University of Toronto Press.

Simons, M. 2003. *The meeting of the waters: The Hindmarsh Island affair.* Sydney: Hodder.

Slikkerveer, L. J. 1999. Ethnoscience, "TEK" and its application to conservation. In *Cultural and spiritual values of biodiversity,* ed. and comp. D. Posey, 169–259. London: United Nations Environment Programme/Intermediate Technology Publications.

Smith, B. R. 2003. "All been washed away now": Tradition, change and indigenous knowledge in a Queensland Aboriginal land claim. In *Negotiating local knowledge: Power and identity in development,* ed. J. Pottier, A. Bicker, and P. Sillitoe, 121–54. London: Pluto.

Strang, V. 1997. *Uncommon ground: Cultural landscapes and environmental values.* Oxford: Berg.

Sutton, P. 2003. *Native title in Australia: An ethnographic perspective.* Port Melbourne, Victoria: Cambridge University Press.

Swain, T. 1993. *A place for strangers: Towards a history of Australian Aboriginal being.* Cambridge: Cambridge University Press.

Tully, J. 2001. An ecological ethics for the present: Three approaches to the central question. In *Governing for the environment: Global problems, ethics and democracy,* ed. B. Gleeson and N. Low, 147–64. Basingstoke, England: Palgrave.

Turnbull, D. 2000. *Masons, tricksters and cartographers: Comparative studies in the sociology of scientific and indigenous knowledge.* Amsterdam: Harwood Academic.

Whitt, L. A. 1999. Metaphor and power in indigenous and Western knowledge systems. In *Cultural and spiritual values of biodiversity,* ed. and comp. D. A. Posey, 69–72. London: United Nations Environment Programme/Intermediate Technology Publications.

CHAPTER 9

Mobilizing Knowledge for Integrated Ecosystem Assessments

CHRISTO FABRICIUS, ROBERT SCHOLES, AND GEORGINA CUNDILL

The "truth" is elusive when dealing with complex, dynamic systems (Kay et al. 1999). Researchers, natural resource managers, and environmental practitioners face a number of challenges, including how to deal with information "fuzziness," how to reconcile seemingly contradictory data, how to smooth over geographic and spatial variability or "lumpiness," and how to consolidate information gathered at different spatial scales. One proposed solution has been to amalgamate different types of knowledge, such as by working across disciplines, combining qualitative and quantitative information, and linking formal and local knowledge in a complementary manner. But this approach is no panacea for ecosystem assessments involving complex systems, and new challenges arise when attempts are made to combine knowledge in this way. The techniques to combine different forms of knowledge and data from disparate sources, different spatial scales, and indeed different worldviews are neither well developed nor validated.

The Southern African Millennium Ecosystem Assessment (SAfMA, http://www.maweb.org) was undertaken at a variety of spatial scales, from the regional (with sub-Saharan Africa as the assessment area) to the local (at the scale of a village, single protected area, or microwatershed). Each of these scales had its own stakeholders and thus its own key topics of concern. These in turn defined the information needs for the assessment at that scale. We found that as the scale of assessment moved from regional to local, so the balance of information availability shifted from formal, documented data, typically

Table 9.1

Characteristics of knowledge along a formal–informal and a tacit–explicit gradient

	Formal	**Informal**
Explicit	Most but not all "scientific" knowledge is in this quadrant. The typical outputs of a conventional assessment are also here.	This knowledge is codified but neither collected nor tested in accordance with conventional scientific rules.
Tacit	Scientifically trained people have formal knowledge that is uncodified.	This knowledge is embedded in local customs, traditions, and memory and is transferred through oral history.

regarded as being in the "scientific domain," toward informal, tacit information contained in the life experience of local residents and in folklore transmitted by oral tradition, or perhaps documented but not in accordance with conventional scientific standards. We contend that the distinction between "formal" and "informal" knowledge is not as absolute as is often thought and that, at the level of broad principles, similar rules of use and validation apply, although the procedures may differ. Elements of both sorts of knowledge exist at all scales, although informal knowledge is generally more site specific and restricted by design and circumstances than scientific knowledge is.

Knowledge can be classified and defined in a variety of ways. Here we use "explicit" to mean knowledge that exists in a written (i.e., codified, including numeric or graphical) and categorical form. "Tacit" knowledge, on the other hand, is held in people's memories and is not documented. "Formal" knowledge has passed through a strict and universally accepted set of rules qualifying it for a particular use, whereas "informal" knowledge has been subject to local rules of validity (table 9.1). "Local" knowledge has a fine-grained perspective and is highly context specific as opposed to "universal" knowledge, which is more coarse grained and incorporates a variety of contexts.

The application of different types of knowledge can be depicted in two dimensions, with the "informal–formal" and "local–universal" gradients on the respective axes (figure 9.1). Local, informal knowledge is mostly reserved for customs, traditions, and local systems of resource utilization, whereas universal, formal knowledge often characterizes large-scale initiatives, such as international conventions, global change models, and space aviation programs. A particular set of rules pertains to the scientific method, and knowledge that satisfies these rules is "scientific" and usually also explicit.

Figure 9.1
The most common uses of different types of knowledge (local to universal), depending on perspective and formality.

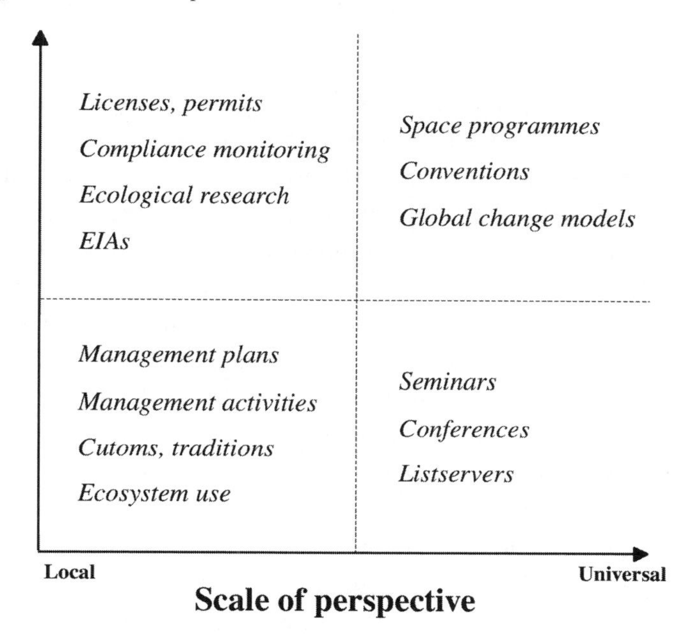

The SAfMA team faced a number of challenges when attempting to amalgamate these different types of knowledge across spatial scales. We confronted these potential challenges from the outset by proactively and, sometimes, reactively devising strategies for dealing with them. In the process we learned several lessons about knowledge amalgamation and sense making in complex assessments. This chapter shares the experience in SAfMA of soliciting (making explicit) and assessing (formalizing) traditional knowledge at the local scale and of making explicit the tacit knowledge from "scientific expert" sources at the regional scale. It then discusses the processes by which the assessment adds value to this input data, from whatever source it is derived.

Incorporating Informal, Local Knowledge Systems

Local ecological knowledge, also sometimes called "local knowledge," "informal knowledge," or "traditional ecological knowledge," is embedded in local customs, belief systems, and learning. Local knowledge is particularly relevant in ecosystem management, and its integrity is acknowledged in the Convention on Biological Diversity (Article 8j). The characteristics of local knowledge include the following:

- As with all types of knowledge, it constantly evolves through generations of hands-on experimentation and is carried over from one generation to the next in folklore, societal norms, management systems, and social memory (Berkes and Folke 1998). This adaptive process more often than not acts as a filter on the quality and validity of knowledge that is transferred.
- Local knowledge is very seldom documented (except through intermediaries, such as researchers, writers, and journalists) and is mostly tacit.
- Local knowledge is used in everyday situations. Its main value lies in helping local people cope with day-to day-challenges, detecting early warning signals of change, and knowing how to respond to challenges. It is extensively used by local practitioners to develop natural resource management strategies, to set rules that govern the use of ecosystem services, and to make day-to-day decisions, such as knowing which medicines to use, where to find food and water in times of crisis, and which plants and animals are best avoided or best to use.
- Knowledge is the backbone of local social institutions, which act as knowledge banks and mechanisms for knowledge transfer between individuals and over time. Social institutions convert knowledge into sets of rules, norms, and social behaviors, which then become local management systems (Folke, Berkes, and Colding 1998). Institutions are therefore the conduit that converts knowledge into management systems, strategies, and policies.

Local knowledge, and especially traditional knowledge, is seldom documented or "refereed." Traditional knowledge is often jealously guarded. Many scientists are skeptical of the validity of informal knowledge because of the lack of rigor, while traditional people may be skeptical about science, either because they do not understand it or because science has on some occasions been used to mask realities or manipulate the truth. Concerns about data integrity can mar the confidence in results based on knowledge amalgamation.

Drawbacks of Purely Formal, Scientific Knowledge

The principles and processes of scientific assessments are rooted in the "formal, explicit" quadrant of our knowledge classification. It helps to recognize the shortcomings of science as a knowledge system in order to work around them. Three are particularly salient here.

1. The scientific method tends to be highly compartmentalized and reductionist. It is evolving methods, such as systems modeling, to balance this tendency, but it remains generally discrete rather than integrated.
2. Scientific knowledge remains the domain of a small elite, even in developed countries. It is often either inaccessible or incomprehensible to the general public and even to highly educated policy makers. A consequence is that scientific knowledge is often patchy, with large spatial or subject gaps.
3. It struggles to engage usefully in problems that do not lend themselves to quantification and mathematical representation.

Why Include Local Knowledge in an Ecosystem Assessment?

Local and tacit knowledge can help address some of the shortcomings in formal, explicit knowledge in ecosystem assessments—if the knowledge can be moved into the explicit domain where such assessments reside. There is, however, a fear, especially among indigenous groups, that this could lead to the manipulation and co-option of local and traditional knowledge. Scientists must be perceptive to such sensitivities. Calls have intensified from various disciplines and institutions for broader approaches and solutions to environmental and societal problems as a whole (Berkes, Colding, and Folke 2003), emphasizing, among other things, decentralization and integrated conservation and planning that is sensitive to local cultural values and institutions (Mauro and Hardison 2000). In southern Africa, this has lead to policies that emphasize community participation and cross-sectoral integration—for example, the South African National Water Act (1998), which requires the devolution of authority to local catchment management forums; the community-based natural resource management program in Botswana (Madzwamuse and Fabricius 2004), which enables local communities to contribute to decisions about wildlife harvesting; and the National Forests Act (1998) in South Africa, which stipulates that local communities should participate in forest management.

Traditional knowledge, in particular, is increasingly being recognized as holding lessons for adaptive managers. Berkes, Colding, and Folke (2000), for example, suggest that traditional knowledge can be described as adaptive because it acknowledges that environmental conditions will always change, assumes in many instances that nature cannot be controlled, and assumes that yields cannot be predicted. Adaptive management is designed to improve on a trial-and-error basis, an attribute inherent in the social learning process, where learning occurs at the level of the group rather than that of the individual.

Local knowledge is an invaluable source of fine-grained, detailed information about local ecosystem services, especially (but not exclusively) in areas where little formal knowledge exists. At Mt. Coke in South Africa, local geographic knowledge was, for example, converted to formal maps with the aid of a geographic information system (GIS) (Bohensky et al. 2004). This provided new insights into such fine-grained information as the positive correlation between tree density and distance from the village, due to fuel wood depletion near villages. Knowledge about patterns of ecosystem change can be used to inductively develop and test models of ecosystem dynamics, as was done in the Gorongosa area in Mozambique (Lynam et al. 2004). Resource users possess detailed knowledge of fine-grained resource patches, such as fountains, sacred pools, caves, patches rich in soil nutrients, and fuel wood (Hendricks 2003; Fabricius and Cundill, forthcoming).

Local knowledge is often the only source of information about past patterns of ecosystem use, past land use, traditional customs, and the history of local politics, especially in communal areas where this information is mostly undocumented. In the Mt. Coke area, for example, local information about land boundaries and political events could be triangulated with historical records to produce a rich body of information about the drivers of the social-ecological system that would not otherwise have been available (Shackleton et al. 2003). Local people routinely adopt an integrated approach when assessing and managing ecosystems. Culture, natural resources, livelihoods, and management practices are viewed as part of the same system. Economic, political, and climatic drivers of change are assimilated in local knowledge systems, and the links between these causal factors are more obvious to local resource users than to scientific investigators. In the Macubeni catchment near Queenstown in South Africa, local groups were able to construct complex "problem trees" of the underlying causes of land degradation in a matter of hours. The causes included chronic poverty, past politics, national economic change, and human population density (Fabricius, Matsiliza, and Buckle 2003).

Local knowledge has, in many instances, coevolved with ecosystems. The feedbacks between ecosystem change and knowledge is evident in local customs, belief systems, and day-to-day adaptive management practices. In South Africa's Richtersveld National Park, for example, Nama pastoralists move their livestock in response to short-term and seasonal fluctuations in rangeland productivity and condition, and fuel wood collectors in the Great Fish River basin adapt their wood collection patterns in response to resource availability (Bohensky et al. 2004). Many of the flexible livelihood strategies observed in local societies are intended to reduce people's vulnerability to sudden change. The flexible social systems—such as the mobility, flexible leadership structures, and variable group sizes of the Basarwa people in the Okavango Delta—have evolved with highly dynamic ecosystems (Madzwamuse and Fabricius 2004).

Shortcomings of Local Knowledge

Local knowledge falls short where the rate of change in social-ecological systems is faster than the rate of knowledge evolution. Consistently high livestock densities in the Great Fish River basin, for example, are a recent phenomenon precipitated by elevated human population densities resulting from social engineering during a previous political dispensation (Ainslie 2002). This has resulted in an ecological "flip" due to the invasion of unpalatable shrubs (notably *Euryops* spp. and *Pteronia incana,* or blue bush), which outcompete other plants for moisture and thereby reduce forage production. The appropriate response is to rest invaded areas from grazing, thereby enabling more frequent fire regimes, and to reseed the area with shrubs and grass. But local people have never experienced these invasions until recently and have not evolved local knowledge to cope with them. The same applies to alien invaders, although in that case the coping strategy is to "switch" to invasive aliens as sources of fuel and building materials.

Local knowledge sometimes evolves inappropriately as a result of powerful external influences that override sensible local adaptations. In Richtersveld National Park, for example, Nama pastoralists believe that donkeys may not be harmed because of their biblical significance and that killing a feral donkey will lead to prolonged drought (Hendricks 2003). Local people have no use for feral donkeys, which compete with their goats and sheep as well as harm biodiversity and productivity, but the custom is religiously applied.

Local knowledge is often too fine grained and context specific to detect larger scale and slow change, and it does not respond to events and processes that do not have direct local repercussions. For example, local collectors of rare succulents in Lesotho and Richtersveld are unaware of the global conservation significance of the plants they illegally trade (H. Hendricks, personal communication).

Local knowledge also rarely responds to slow processes, such as gradual soil erosion, changes in the composition of palatable rangelands, siltation of water bodies, invasive plants, encroachments of mines on rangelands, and slow changes in groundwater quality due to salinization and cattle dips. Often local people's explanations for the causes of these slow changes are flawed, especially when they make spurious links between cause and effect. People at Machibi village in the Eastern Cape, for example, observed an increase in spider webs on unpalatable invasive shrubs. This was mainly because the webs, which were always there, became more visible in the structurally altered shrubland. People started believing that a linked drop in livestock fecundity was caused by spiders, rather than by the reduced productivity and palatability of the vegetation (C. Fabricius, personal observation).

Concerns and Challenges When Collecting Local Knowledge

Analysts have warned that local knowledge may not be relevant outside of the local context (du Toit, Walker, and Campbell 2004), and concern exists about the ability and impact of scaling local knowledge up to broader spatial scales (Lovell, Mandondo, and Moriarty 2002). Other analysts warn of a downplaying of environmental problems when local knowledge is overemphasized in line with "political correctness," and they are concerned about politicians using flawed local knowledge as a reason for ignoring environmental challenges (Burningham and Cooper 1999).

Some analysts also argue that integration with more dominant formal knowledge systems can marginalize local knowledge systems. By enabling the extension of the social and conceptual networks of scientific assessment (Latour 1987; Nadasdy 1999), integration can lead to the concentration of power in the hands of Western science, rather than the intended outcome of empowering local people. However, efforts to integrate or bridge different knowledge systems can help translate local knowledge into a form understandable and usable by scientists and formally trained resource managers (Nadasdy 1999).

Techniques Used to Collect and Integrate Local Knowledge

A wide range of participatory research techniques was used to collect and integrate local knowledge into the SAfMA process (Babbie et al. 2001). Among the techniques used to collect local knowledge were focus group workshops and interviews (Borrini-Feyerabend 1997), semistructured interviews with key informants (Pretty et al. 1995), a range of participatory rural appraisal (PRA) techniques (Chambers 1994; Borrini-Feyerabend 1997; Campbell 2002), participatory mapping (Alcorn 2000), and forum theatre. The range of PRA (also called participatory learning and action) techniques included matrixes, freehand and GIS mapping, pie charts, trend lines, timelines, ranking, Venn diagrams, problem trees, pyramids, role-playing, and seasonal calendars (Borrini-Feyerabend 1997; Jordan and Shrestha 1998; Jordan 1998; Department for International Development 1999; Motteux 2001).

Problem trees were particularly useful for identifying proximate and ultimate causes of ecosystem and social change. Mapping was an essential tool to define spatial change, while trend lines proved invaluable for recording local perceptions of change in key goods and services during predefined eras. Most valuably, these participatory techniques broke down barriers between scientists and villagers and enabled illiterate people to confidently participate in the process without being overwhelmed by grammatical and linguistic barriers.

However, these techniques proved useful only in collecting information. A larger challenge was posed by the need to integrate this information into the assessment findings. This integration was achieved in a number of ways. For example, data thus collected was converted into digitally enhanced charts, graphs, and reports by the specific researchers involved, thereby making tacit knowledge accessible to other scientists. However, to prevent an extractive process with a one-way transfer of knowledge (i.e., solely from local people to scientists), scientific knowledge was equally translated into a form that local participants could relate to. Story lines and drama, for example, were used to translate to local participants such complex issues as future scenarios developed at the national level. Reactions were then recorded and delivered to scientists working at coarser spatial scales. Forum theatre was particularly useful for converting complicated scientific scenarios of the future into dramatic presentations, to which local communities could relate (Burt and Copteros 2004).

Approaches to Validating Knowledge

Combining formal and local knowledge can produce a great deal of uncertainty. Thus it is essential to validate both formal and informal knowledge. Validation can be achieved through the cross-validation of both formal and informal knowledge. In other words, local experts validate scientific knowledge, and scientists validate informal knowledge. For example, to improve confidence in the data generated using the techniques outlined earlier, qualitative findings were validated through social and biophysical surveys, historical sources, and GIS and time-series mapping. Validation of scientists' interpretation of local knowledge took place through formal feedback meetings, where community members could challenge the validity of information. These feedback meetings were especially useful where local working group members, rather than scientists, provided the feedback. The most useful feedback meetings were those where scientists provided feedback by using modern technology—such as video, printed posters, and digital slideshows—followed by local people responding in their own language, using charts, hand-drawn maps, and verbal presentations.

Incorporating Formal but Tacit Knowledge

Formal knowledge can also be tacit, and formally trained scientists and managers have accumulated a large body of knowledge that is undocumented. "Expert opinion"–based processes are common enough in scientific assessments. For instance, uncertainty statements, a key feature of the Intergovernmental Panel on Climate Change (IPCC) Third Assessment Report, are virtually impossible to derive given current information sources and technology by formal statistical procedures. Almost all the IPCC uncertainty ranges are based on expert opinions but are nevertheless extremely valuable. An attempt is made to calibrate them and make them internally consistent by defining a shared vocabulary (Moss and Schneider 2000). Some formal processes, such as the "Delphi Method," exist for formalizing and making explicit such tacit knowledge in a transparent way. These processes are not without critics, because they may give a veneer of quantification and precision to what remains a value-ridden process.

SAfMA, at the regional scale, faced a problem in synthesizing the vast amount of data relating to biodiversity. Biggs, Scholes, and Reyers (2004) defined a "biodiversity intactness index" as a synthesizing framework for the information and then conducted sixteen independent (three- to five-hour) interviews with technical experts to solicit the information. The process was greatly aided by first carefully defining the purpose, the metric, a reference point (large protected areas), and the nature of the land use activities. The broad taxa were further subdivided into functional groups (i.e., groups of organisms that respond in similar ways to particular land transformations, such as "seed-eating birds" or "large mammal herbivores") in collaboration with the experts, and the total study region was divided into ecosystem types. The expert opinions were tested against the small body of independently gathered field data that exists (Scholes and Biggs 2005). The mean and range of the expert estimates of the effect of different land use practices on biotic populations in each ecosystem type were then used in calculating an aggregate impact, which can be thought of as the abundance of wild populations relative to their abundance in an untransformed state. The convergence in estimates between experts was remarkable, allowing the uncertainty range on the aggregate index to be estimated as ±7 percent around a mean of 84 percent.

Adding Value through the Assessment Process

If assessments work on existing data, as they claim to do, where does the added value come from that could justify the expense of undertaking the assessment? Feedback from end users—that is, local communities and government decision makers—suggests that well-conducted assessments are valuable to existing and future resource managers. The source of this value is the assessment process itself. Assessment moves data up the value chain, to information, then to knowledge, and in some cases, perhaps even to wisdom. Assessment achieves this movement through six basic processes: collation, evaluation, summarization, synthesis, dialectic, and communication.

Collation

Collation consists of making relevant information easily available. It is the most basic function of an assessment. The information is typically obtained from diverse, and often hard-to-access, sources, such as unpublished reports or "gray literature." For many policy makers in Africa, even the technically "open" literature, such as international scientific journals and books, is either inaccessible or incomprehensible. Policy makers everywhere are typically overworked and overwhelmed by information, so collated, well-organized, source-attributed information on a particular topic that is available all in one place is a significant benefit.

SAfMA contains many examples of this kind of activity. For example, the Zambezi Basin study brought together rainfall, evapotranspiration, and river flow data for all the subcatchments by combining climate databases with model outputs and GIS analysis (Desanker and Kwesha 2004). Another example is the use of GIS to capture fine-scaled local interpretations of land use change and changes in forest quality in the SAfMA local studies. This local knowledge about spatial changes was captured and made available to the assessment team working at coarser spatial resolutions.

Evaluation

Evalutation involves comparing, checking, and applying informed judgment to information. In this respect, an assessment differs fundamentally from a review. Scientific reviewers are expected to be "neutral," simply presenting all the sources of information while hesitating to provide an opinion. Members of their target audience are assumed to be in a position to draw their own opinions. Assessments, on the other hand, *are* expected to express an opinion on the validity and meaning of data, especially if competing or conflicting data sources are involved. If they fail to do so, the decision makers who are the assessment audience are forced to reach their own conclusions but often are not equipped to do so. This does not, however, violate the assessment stricture "to be policy relevant, but not policy prescriptive," and it stops short of making a normative statement about what *should* happen as a result. It should also include a statement of uncertainty, which can be formal (e.g., "the protein supply is 45 ± 5 g/person/day") or informal ("it can be concluded with high certainty that . . . ").

Evaluation is central to assessments, since their purpose is to act as a translator between the domains of technical knowledge and decision making. It is also the area where most classically trained scientists feel least comfortable; they like to be near certain before venturing an opinion. An example of this kind of process in SAfMA is the comparison of four different forest cover products at the regional scale, leading to the opinion that there is 4.5 ± 0.5 million square kilometers of forest in southern Africa (Scholes and Biggs 2004). Another example, one involving local knowledge, was the comparison of locally developed land use change maps with historical aerial photographs of the areas in question. This process of evaluation enabled the assessment team to make informed recommendations regarding land use.

Summarization

Summarization includes all approaches that help reduce the complexity and detail of data. This process operates differently, of course, when dealing with formal knowledge and local knowledge. In terms of formal knowledge, even in data-poor areas there are usually more data on hand than a decision maker can usefully assimilate. The volume needs to be reduced until each decision is informed by only one to five variables. Statistical summaries (means, medians, modes, standard deviations, and ranges) all fall into this category. Great care must be taken to perform the statistical summarization appropriately. For instance, there are important scaling considerations when accumulating averages from different-sized populations.

Indices and indicators also fall into this category. Indices are mathematical compilations of different types of data, forming a composite measure. Indicators are typically proxy data that suggest a trend in some other, more fundamental assessment variable. Indicators are a feature of state-of-the-environment reporting but run the risk of becoming so numerous that they fail to achieve the objective of simplification. An example of summarization in SAfMA is the biodiversity intactness index (Biggs, Scholes, and Reyers 2004), which combines thousands of observations at species level, with land cover and ecosystem maps, into a single score for biodiversity performance, with a confidence interval. The index can be progressively "unpacked" at different scales or for different taxa or land cover types.

In terms of informal knowledge, summarization is a more difficult task since it involves processed information rather than empirical data. It is also somewhat challenging to apply the inherently scientific approach of "summarization" to local knowledge since the knowledge systems faced often do not lend themselves to this process and value could be removed by so doing. Nevertheless, being part of an ecosystem assessment requires that information be summarized. This was achieved in the SAfMA in various ways, from using GIS technologies to capture spatial information to creating locally appropriate scenarios to summarize key drivers and trends within villages (see Burt and Copteros 2004). Feedback from local decision makers and resource users indicated that this information was enormously useful.

Synthesis

Synthesis consists of combining primary information in ways that provide novel insights. The simplest syntheses may be ratios. For instance, when yield data are divided by population data, the result is the average food supply per person. If this is then compared with a threshold (e.g., 2,000 cal/person/day), the result is information on food security that is not present in any one of the input variables alone but is a result of their combination through synthesis. Synthesis can also take place through applying much more complex models. An example from SAfMA is the regional-scale analysis of the grazing service—that is, the service provided by the ecosystem of grazing land for livestock. Data from subnational livestock databases were converted, through metabolic models, into forage demand values. Climate, soil, topography, and vegetation databases were the input to grass production models that calculated forage supply. The difference between supply and demand provided a synthesized, spatial assessment of the pressure on the service that could be related to independently derived satellite observations on land degradation (Scholes and Biggs 2004). Synthesis represents perhaps the most intellectually challenging aspect of assessment, but it is also the process that can add the greatest value.

Dialectic

A valuable assessment process is the dialogue and debate that occur when investigators with different analytical models apply themselves to the same problem. One example is the interaction between social scientists and biophysical scientists. Another is between researchers looking at the same issue at different scales. A third is the interaction between "Western" worldviews and "African" worldviews. Finally, even within one discipline (e.g., ecology, economics, or political science), different schools of thought usually exist. The assessment can be greatly enriched if these "conflicts" are not excluded or papered over but, instead, are actively encouraged as a source of constructive dialogue and critique. For example, SAfMA included researchers whose training, disposition, and experience caused them to favor aggregated, large-scale, generalized approaches to assessment, and others who for the same reasons favored disaggregated, place-based, specific approaches. We ended up using both—in some cases, as different lenses through which to view the same problem; in other cases, as approaches appropriate to different questions. If convergence can be achieved, then confidence in the robustness and wide acceptability of the finding is increased. Failure to converge, on the other hand, does not mean a failed process. It clearly establishes the uncertainty range of the issue.

Successful use of dialectic requires a high level of self-confidence and mutual trust among the participants. SAfMA was characterized by much dialectical debate, which quite unnerved new observers. The different approaches to scenario construction applied by the different subprojects are an example (compare Lynam et al. 2004, Scholes and Biggs 2004, Bohensky et al. 2004, and Burt and Copteros 2004). The coherence of the entire enterprise was built on the a priori agreement to use the Millennium Ecosystem Assessment's conceptual framework as the meeting point (MA 2003).

Communication

Communication transfers knowledge from the specialist and technical domain into a policy domain. It involves as much listening as speaking, remembering that communication is the message received, not the message transmitted. Assessment can be thought of as a translation device. It needs to render a signal intelligible and to deliver it where needed. The jargon-ridden, extremely detailed scientific discourse often needs simplifying (think of this as taking out the noise and leaving the main signal), but it should not be distorted in the process. The classical medium is the written report, because of its archival value and ease of use, but this format is increasingly being supplemented by electronic dissemination (Web pages, CD-ROMs), video productions, radio broadcasts, posters, and brochures. However, in the SAfMA, all of these devices proved themselves inadequate at the local level, so other methods were sought, such as visual displays, storytelling, theatre, and PRA (see Burt and Copteros 2004; Cundill 2005).

Face-to-face communication with the chosen target audience is an invaluable complement to the report in all instances. Assessment reports typically include a lot of graphical communication devices, such as maps, graphs, diagrams, photographs, and tables. Assessments often underestimate the time and resources needed for this process, without which the effort put into the preceding processes is fruitless. Ideally, communication should involve stakeholder involvement from the start. Although this is one of the guiding principles of integrated assessments such as the SafMA, full stakeholder involvement is difficult to achieve in practice unless enough time and resources are allocated for it. As a rough guideline, about a fifth of the total resources need to be dedicated to communication.

We suggest that the level of each process above can be used as a yardstick for "assessing assessments." An assessment that applies them all to a high degree is likely to yield a worthwhile outcome.

Acknowledgments

This work resulted from the Southern African Millennium Ecosystem Assessment, which was largely funded from the United Nations Environment Programme, using resources made available by the Government of Norway.

The contribution of the entire SAfMA team, and of a large number of scientific experts and holders of local and indigenous technical knowledge, to the project and to the ideas in this paper are acknowledged.

References

Ainslie, A. 2002. Cattle ownership and production in the communal areas of the Eastern Cape, South Africa. Research report no. 10. Cape Town: Programme for Land and Agrarian Studies, University of the Western Cape.

Alcorn, J. B. 2000. Keys to mapping's good magic. International Institute for Environment and Development. *PLA Notes* 39. http://www.iied.org/sarl/pla_notes/pla_backissues.

Babbie, E., J. Mouton, P. Vorster, and B. Prozesky. 2001. *The practice of social research.* Oxford: Oxford University Press.

Berkes, F., J. Colding, and C. Folke. 2000. Rediscovery of traditional ecological knowledge as adaptive management. *Ecological Applications* 10:1251–62.

———, eds. 2003. *Navigating social-ecological systems: Building resilience for complexity and change.* Cambridge: Cambridge University Press.

Berkes, F., and C. Folke, eds. 1998. *Linking social and ecological systems: Management practices and social mechanisms for building resilience.* Cambridge: Cambridge University Press.

Biggs, R., R. J. Scholes, and B. Reyers. 2004. Assessing biodiversity at multiple scales. Online proceedings of the Bridging Scales and Epistemologies Conference, Alexandria, March 17–20, 2004. Penang, Malaysia: Millennium Ecosystem Assessment. http://www.millenniumassessment.org/en/about.meetings.bridging.proceedings.aspx.

Bohensky, E., B. Reyers, A. van Jaarsveld, and C. Fabricius. 2004. *Ecosystem services in the Gariep Basin: Southern African Millennium Assessment Report.* Stellenbosch, South Africa: University of Stellenbosch.

Borrini-Feyerabend, G., comp. 1997. *Beyond fences: Seeking social sustainability in conservation.* Gland, Switzerland: IUCN (World Conservation Union)/Kasparek Verlag.

Burningham, K., and G. Cooper. 1999. Being constructive: Social constructivism and the environment. *Sociology* 33:297–316.

Burt, J., and A. Copteros. 2004. Dramatic futures: A pilot project of theatre for transformation and future scenarios. Unpublished report for the Environmental Education Department, Rhodes University, South Africa.

Campbell, B. 2002. A critical appraisal of participatory methods in development research. *Social Research Methodology* 5:19–29.

Chambers, R. 1994. Participatory rural appraisal (PRA): Analysis and experience. *World Development* 22:1253–68.

Cundill, G. 2005. Institutional change and ecosystem dynamics in the communal areas around Mt. Coke State Forest, Eastern Cape, South Africa. Master's thesis, Rhodes University, Grahamstown, South Africa.

Department for International Development. 1999. Sustainable livelihoods guidance sheets. http://www.livelihoods.org (accessed October 1, 2003).

Desanker, P. V., and D. Kwesha. 2004. *Ecosystem services in the Zambezi River basin.* Harare: Zimbabwe Forestry Commission.

Du Toit, J., B. Walker, and B. Campbell. 2004. Conserving tropical nature: Current challenges for ecologists. *Trends in Ecology and Evolution* 19:12–17.

Fabricius, C., and G. Cundill. Forthcoming. Restoring natural capital in communal areas. In *Restoring natural capital,* ed. J. Aronson and S. Milton. Washington, DC: Island Press.

Fabricius, C., B. Matsiliza, and J. Buckle. 2003. *Community-based natural resource management in Emalahleni and Mbashe: Eastern Cape Planning Process.* Pretoria: GTZ Transform.

Folke, C., F. Berkes, and J. Colding. 1998. Ecological practices and social mechanisms for building resilience and sustainability. In *Linking social and ecological systems,* ed. F. Berkes and C. Folke, 414–36. Cambridge: Cambridge University Press.

Hendricks, H. 2003. *Southern African Millennium Assessment: Lower Gariep Basin scoping report.* Kimberley: South African National Parks.

Jordan, G. H. 1998. A public participation GIS for community forestry user groups in Nepal. Paper presented at the Public Participation GIS specialist meeting, October 14–18, Santa Barbara, California. http://www.ncgia.ucsb.edu/varenius/ppgis/papers/jordan.pdf.

Jordan, G. H., and B. Shrestha. 1998. *Integrating geomatics and participatory techniques for community forest management: Case studies from the Yarsha Khola watershed, Dolakha District, Nepal.* Kathmandu: International Centre for Integrated Mountain Development.

Kay, J., H. Regier, M. Boyle, and G. Francis. 1999. An ecosystem approach for sustainability: Addressing the challenge of complexity. *Futures* 31:721–42.

Latour, B. 1987. *Science in action.* Cambridge, MA: Harvard University Press.

Lovell, C., A. Mandondo, and P. Moriarty. 2002. The question of scale in integrated natural resource management. *Conservation Ecology* 5 (2): 25. http://www.ecologyandsociety.org/vol5/iss2/art25/.

Lynam, T., A. Sitoe, B. Reichelt, R. Owen, R. Zolho, R. Cunliffe, and I. Bwerinofa. 2004. *Human wellbeing and ecosystem services: An assessment of the linkages in the Gorongosa-Marromeu region of Sofala Province, Mozambique to 2015.* Harare: Institute of Environment Studies, University of Zimbabwe.

Madzwamuse, M., and C. Fabricius. 2004. Local ecological knowledge and the Basarwa in the Okavango Delta: The case of Xaxaba, Ngamiland District. In *Rights, resources and rural development: Community-based natural resource management in southern Africa,* ed. C. Fabricius and E. Koch, 160–73. London: Earthscan.

Mauro, F., and D. Hardison. 2000. Traditional knowledge of indigenous and local communities: International debate and policy initiatives. *Ecological Applications* 10:1263–69.

Millennium Ecosystem Assessment (MA). 2003. *Ecosystems and human well-being: A framework for assessment.* Washington, DC: Island Press.

Moss, R. H., and S. H. Schneider. 2000. Uncertainties in the IPCC TAR: Recommendations to lead authors for more consistent assessment and reporting. In *Guidance papers on the cross-cutting issues of the Third Assessment Report of the IPCC,* ed. R. Pachuari, T. Taniguchi, and K. Tanaka, 31–51. Geneva: World Meteorological Organisation.

Motteux, N. 2001. *The development and co-ordination of catchment fora through the empowerment of rural communities.* Water Research Commission Report No. 1014/1/01. Pretoria: Water Research Commission.

Nadasdy, P. 1999. The politics of TEK: Power and the "integration" of knowledge. *Arctic Anthropology* 36:1–18.

Pretty, J., I. Guijt, I. Scoones, and J. Thompson. 1995. *A trainer's guide for participatory learning and action.* Sustainable Agriculture Programme. London: IIED Participatory Methodology Series, International Institute for Environment and Development.

Scholes, R. J., and R. Biggs. 2004. *Ecosystem services in Southern Africa: A regional assessment.* Pretoria, South Africa: Council for Scientific and Industrial Research.

———. 2005. A biodiversity intactness index. *Nature* 434:45–49.

Shackleton C., G. Guthrie, J. Keirungi, and J. Stewart. 2003. Fuelwood availability and use in the Richtersveld National Park, South Africa. *Koedoe* 46:1–8.

CASE STUDIES

CHAPTER 10

Keep It Simple and Be Relevant

The First Ten Years of the Arctic Borderlands Ecological Knowledge Co-op

JOAN EAMER

Interest in traditional ecological knowledge among resource managers and scientists in northern Canada and Alaska is growing (Urquhart 1998; Eamer 2000). This increased interest has led to more and more requirements for its use in many management, planning, and assessment processes (Berkes 1998; Canada Department of Justice 1998, 2002, 2003). This has prompted initiatives to interview elders and hunters and document their knowledge (Gwich'in Elders 1997; McDonald, Arragutainaq, and Novalinga 1997; Sherry and the Vuntut Gwitchin First Nation 1999), to examine methods of incorporating traditional ecological knowledge into resource management decision making (Huntington 2000; Usher 2000), and to examine and critique ways in which traditional and science-based knowledge are compared or synthesized (Krupnik and Jolly 2002; Nadasdy 2003).

This chapter describes and discusses the Arctic Borderlands Ecological Knowledge Co-op, a program that focuses on ecological monitoring from science-based and local knowledge sources. The Borderlands Co-op has operated for ten years in the western North American Arctic. During that time, it has faced challenges, adapted, developed its programs, and expanded geographically. Because of its longevity and its broad base of support, other organizations that are starting or expanding community-based and cumulative impacts monitoring initiatives in the Arctic look to the Borderlands Co-op for advice and assistance.

In the Canadian part of the Arctic Borderlands region, several comanagement regimes are operating, with direct participation of indigenous representatives in making decisions and advising local, regional, territorial, and national governments on many aspects of resource management (e.g., Bailey et al. 1995). The Borderlands Co-op builds on and collaborates with these research initiatives and management regimes but has no management authority itself. Key elements of the program are cooperative decision making in all aspects of the program's development and organization; involvement at the community level in direction and implementation of the program; and ongoing communication and discussion about the use of multiple information sources in ecological monitoring.

The Borderlands Co-op is, above all else, a collaboration. The term *we* in this chapter should be interpreted as "we, the people who are involved with and working to maintain and improve this program." This includes people representing community, comanagement and government councils and agencies, and researchers (see the acknowledgments section at the end of this chapter). Information on the Borderlands Co-op and the people and organizations involved is available on the program's Web site (http://www.taiga.net/coop). A discussion on the contributions of communities to coproduction of knowledge through the Arctic Borderlands Ecological Knowledge Co-op is available in a paper coauthored by Gary Kofinas of the University of Alaska Fairbanks and by the four initial participating communities (Kofinas et al. 2002).

The Borderlands Co-op focuses on strengthening the role of local indigenous knowledge in environmental assessment, planning, and management, and in exploring ways to bring local and science-based knowledge together to improve understanding of ecological conditions and trends.

Traditional knowledge studies in the region have documented a wealth of knowledge of place, way of life, culture, and spirituality (e.g., Nagy 1994; Gwich'in Elders 1997). Some studies have been conducted to document elders' and hunters' knowledge about animals of concern to management (e.g., Byers and Roberts 1995; Smith 2004). The Borderlands Co-op's community-based monitoring program differs from these studies. The program is not based on in-depth traditional knowledge interviews with elders. Instead, structured interviews, conducted by local residents of each participating community, focus on what the community's most active hunters, fishers, and berry pickers of all ages have observed over the preceding year. Interviewers also ask for interpretations of what people have seen, which are based on personal experiences and traditional knowledge. We use the term *local knowledge* to encompass this blending of observation and interpretation. The program documents and reports, annually, observations of indigenous people about the land, how it is changing, and how conditions and changes affect their lives. This work complements in-depth traditional knowledge studies, much as science-based monitoring complements science-based research.

The program operates in a diverse, multijurisdictional setting, bridging both geographic scales and organizational levels. Participants include community residents and representatives of boards, committees, government agencies, planning and assessment processes, and research projects—each with its own defined jurisdiction, and none covering the entire region.

The Borderlands Co-op itself works at several scales. Interview-based monitoring is conducted in communities in the Arctic Borderlands region. Results from the community-based monitoring program are summarized on the scale of each community and the land used by community residents for hunting, fishing, trapping, and berry picking. Information is also acquired from other monitoring and research programs at a range of scales, from local (climate station records) to regional (Porcupine caribou herd populations) to global (greenhouse gas levels in the atmosphere). This information is tracked, summarized, and presented in the context of its significance to the Arctic Borderlands region. Meetings of the Borderlands Co-op provide a forum for sharing and comparing information and for discussing the implications of global issues (such as climate change) to the region and the significance of local observations (such as observations on caribou distribution) to the region and its resources.

The Arctic Borderlands Region

The Arctic Borderlands Ecological Knowledge Co-op operates in the range of the Porcupine caribou herd (250,000 square kilometers) and adjacent marine and coastal areas, extending into the Mackenzie Delta (figure 10.1). This area is complex in terms of jurisdictions and is ecologically very diverse. The region contains tundra, taiga and coastal landscapes, mountains, large wetlands complexes, several major rivers, and one of the world's largest river deltas, the Mackenzie Delta. It contains internationally important wilderness and wildlife habitat. The Arctic Borderlands encompasses part of northern Alaska and, in Canada, parts of two territories: the Yukon and the Northwest Territories.

The human population is predominantly indigenous—Iñupiat (Alaska),

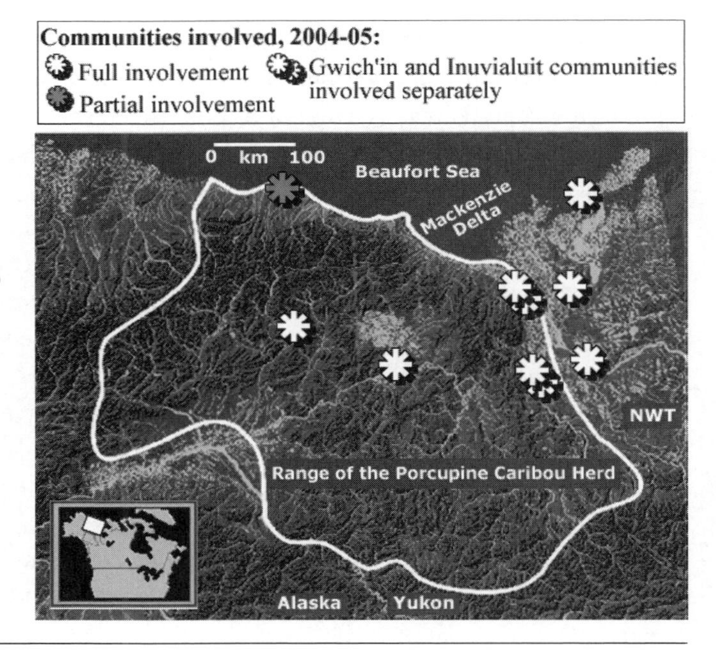

Figure 10.1 **Arctic Borderlands Ecological Knowledge Co-op region.** *(From base map from U.S. Fish and Wildlife Service, Fairbanks, Alaska.)*

Inuvialuit (Canada), and Gwich'in (Canada and Alaska)—and the area includes five major land claimant groups, each with its own governance and resource management structures. The communities range in size from fewer than two hundred people to about a thousand, with the exception of Inuvik, which has more than three thousand residents, of which about two thousand are indigenous. A total of ten communities, two of which are in Alaska, participated in the program in 2004–05.

Caribou have always been a key resource for people in the region. For most of the communities, the Porcupine caribou herd (named after the Porcupine River, a tributary of the Yukon River) is a major part of the diet and of the traditional culture. For the communities with coastal homelands, harvesting marine mammals is also important. Fishing, trapping, and berry picking are traditional activities for all of the communities. The economies of the communities are a mix of subsistence activities and wage economies. Oil and gas exploration and development are becoming increasingly important in some of the communities; indigenous and national, state, and territorial governments are important employers. Tourism currently provides limited job opportunities to local residents.

Although most of the Arctic Borderlands is sparsely populated and little developed, the region is not without its environmental stressors. The migratory Porcupine caribou herd's calving grounds are primarily in a narrow section of the coastal plain in Alaska (Griffith et al. 2002), a wilderness area with petroleum reserves; thus the herd has become the subject of a high-profile, bitter, and protracted dispute regarding its ongoing protection. Increased oil and gas exploration and preparations for pipeline development are taking place on the winter range of the herd in the Canadian side of the region. Levels of persistent organic pollutants and mercury (from atmospheric transport) in fish and marine mammals have raised concerns about the safety of traditional foods over the past fifteen years (Braune et al. 1999; Indian and Northern Affairs Canada 2003). The Arctic Borderlands is predicted by climate models to be among those regions that will experience the most severe impacts from climate change (Taylor and Taylor 1997; Zhang et al. 2000). Temperatures are measurably warming now, and the extent of permanent sea ice is decreasing. Changes in snow conditions in the Arctic Borderlands may now be contributing to the observed decline in population of the Porcupine caribou herd (Griffith et al. 1999).

Development of the Arctic Borderlands Ecological Knowledge Co-op

The Borderlands Co-op grew from a meeting of researchers, government managers and scientists, indigenous leaders, and community representatives in Dawson City, Yukon, in summer 1994. The purpose of the meeting was to come up with a plan to improve ecological monitoring in the range of the Porcupine caribou herd. Although the working relationships among the organizations represented at the meeting were fairly well established, it was clear that a rift existed between many scientists and community representatives in terms of the value and credibility of different types of information. All too often, the results of such a meeting are to respectfully acknowledge these differences and proceed with strengthening the science-based program, while perhaps increasing communications efforts but also leaving the communities frustrated and sidelined. At this meeting, people decided to tackle this issue head-on by developing a monitoring program that would strive to improve our collective understanding of ecological conditions and trends by using local observations, traditional ecological knowledge, science-based research and monitoring, and government records.

Meeting participants developed a set of guidelines for implementing this new program:

- Go slow.
- Keep it simple.
- Be relevant.
- Focus on the long term.
- Economize.

These guidelines have successfully stood the test of time and have been useful in implementing the monitoring program over the past ten years. Every year we review the guidelines to help keep us on track.

It was also decided at the founding meeting that this program would be developed and managed cooperatively, with major decisions being made by consensus at meetings, and with Environment Canada (a federal government department) leading but not "owning" the program. Environment Canada has maintained this lead role, providing staff time and core funding. Over the years this arrangement has evolved into a more formal model, with a not-for-profit society, set up and managed by the program's participants, administering the program. The goals of this not-for-profit society are as follows:

- To monitor and assess ecosystem changes in the range of the Porcupine caribou herd and adjacent coastal and marine areas
- To encourage use of both science-based studies and studies based on local and traditional knowledge in ecological monitoring and ecosystem management
- To improve communications and understanding among governments, indigenous and nonindigenous communities, and scientists with regard to ecosystem knowledge and management
- To foster capacity-building and training opportunities in northern communities in the context of the above-listed goals.

The gradual acceptance of the methods and results of the Borderlands Co-op cannot be separated from the organizational development. Control and ownership at the community and regional level are integral to the program.

In a 1996 workshop that was to become the first "annual gathering" of the

Borderlands Co-op, participants developed a list of about seventy potential indicators of ecological change for the region. Information was available for about half of these indicators from research and monitoring projects and programs and from other sources, such as transportation and census records. At this workshop, discussion also focused on how to document the communities' knowledge about ecological conditions and changes. A pilot project was started over the following year, based on interviews with people who were active hunters, trappers, berry pickers, and fishers.

Since then, a gathering has been held each year in one of the participating communities or in the regional centers of Whitehorse and Inuvik. These gatherings allow participants to discuss and make decisions about the Borderlands Co-op's programs. Each year an action item list is prepared, and each year the previous year's action item list is reviewed. Directors are elected, financing is discussed, reports are presented, indicators are reviewed, observations are compared, and the directions, goals, and operations of the program are argued over, fine-tuned, and reaffirmed. Decision making is by consensus. Most of the decisions taken at the gatherings are general ones regarding the directions, scope, and priorities of the program, with the details and follow-up being left to staff and directors. Key decisions, such as approval in principle of the information-sharing protocol, are made by consensus at a gathering and followed up with fine-tuning by the staff, a review, and a formal motion at a board teleconference.

The membership requirements have been kept flexible. As illustrated by table 10.1, the annual coming and going of individual participants presents challenges in maintaining the focus and continuity of programming. However, the relative stability in representation from the various boards, agencies, and processes provides evidence of support by these organizations.

When the not-for-profit society was formed in 1999, the consensus of the members was that the directors should make few decisions and that the main direction for the program should come annually from the broader membership. In subsequent gatherings, the members directed that the board should be more involved in operating the program, and there has been some evolution toward strengthening the role of the directors and formalizing structures and policies. Borderlands Co-op members are sensitive to the need to keep the participation in the program balanced between community and agency representatives and to keep all jurisdictions and land claim groups involved. The

Table 10.1
Borderlands Co-op membership

Scale of Representation or Primary Interest	Participant Affiliation	Number of Participants		Breakdown of Participation by Scale (average)
		7th Gathering	8th Gathering	
Community and traditional lands	Community monitor	4	4	36%
	Unaffiliated community member	2	3	
	Renewable Resource Council or Hunters and Trappers Committee	8	4	
	First Nation government	0	2	
Region (as defined by the planning or management process represented)	Regional wildlife or fisheries comanagement process	8	5	27%
	Land use planning or environmental assessment process	3	4	
Region (as defined by government agency jurisdiction)	Government agency (e.g., oceans, parks, refuges, environment and wildlife agencies)—federal and territorial, United States and Canada	8	11	25%
Various, depending on field of research	Researcher working in the Borderlands area	5	4	12%
Total participants		38	37	

Note: Participation at an annual gathering constitutes membership in the Borderlands Co-op. This table shows the breakdown, by scale, of participation at two annual gatherings. The Seventh Annual Gathering was in one of the participating communities (Fort McPherson), and the Eighth Annual Gathering was in a regional center (Whitehorse). In addition to these full-time participants, at the Fort McPherson Gathering interested local people attended portions of the gathering, and at the Whitehorse gathering interested government employees and college students dropped by the proceedings.

2004-05 board of fifteen directors includes representatives from each of the participating communities as well as people who were elected based on their strong interest and past involvement with the program, rather than on the basis of their affiliation.

Components of the Borderlands Co-op's Program

Three core features of the Borderlands Co-op Program involve the selection of indicators, community-based ecological monitoring, and mechanisms to ensure that research results are available to local communities in forms that they can use.

Indicators

The potential indicators identified at the First Annual Gathering in 1996 ranged from basic environmental measurements (such as temperature and the length of the ice-free period) to measurements of potential stresses (such as the number of airplane flights) and community and ecological measurements (such as the amount of time people spend on the land and the calving success of caribou). In developing these indicators, we have worked primarily with established data sets, in some cases requesting from the data holders additional data collection or manipulation to make the information more suitable for assessing conditions and trends. Most of the indicators are based on results of science-based monitoring (such as temperature records and animal population estimates) or on government records (such as community population census figures and airport flight records). The presentation of the data and the interpretive text accompanying each data set are developed or reviewed by the data holder.

Indicators follow a standard format that allows easy access to a wide range of information about the region (see box 10.1). Other examples of indicator titles include summer temperatures in the Arctic Borderlands, precipitation in Old Crow, snow depths at Eagle Plains,

Box 10.1
Anatomy of an Indicator

What is happening?
Usually a graphical presentation of the data, accompanied by a simple description.

Why is it happening?
Concise explanation of the factors affecting the conditions and trends observed.

Why is it important?
Significance of this indicator, in ecological and human terms. This section can also point out the relevance to policy and management and describe the actions being taken.

Technical notes
Information about the data set, including methods, frequency of measurement, references, and contact information.

Peel River Ferry operational period, Porcupine River break-up dates and ice-free period, early plant growth in caribou calving areas, salmon in the Porcupine River system, beluga abundance, caribou calving habitat use, mercury levels in marine mammals, airplane flights by community, community populations, development permits issued, carbon dioxide emissions, fur prices, marine oil spills, numbers of park visitors, and Dempster Highway traffic.

Each year at the annual gathering, the indicator set is reviewed and discussed by the general membership. The participants provide guidance regarding what indicators are most useful in assessing and communicating conditions and trends. For example, discussions at the 2003 and 2004 annual gatherings focused on what indicators could be developed that would help assess impacts related to recent and proposed oil and gas development in parts of the Arctic Borderlands.

The indicator set largely reflects what information is available; in 2004, we began a strategic assessment to select key indicators and identify gaps. Developed indicators are all available on the Borderlands Co-op Web site (http://www.taiga.net/coop) and are periodically printed and distributed to Borderlands Co-op participants.

Community-based Ecological Monitoring

Interviews with local experts are conducted annually by community monitors who are selected jointly by the Borderlands Co-op and each local participating organization (for example, the Hunters and Trappers Committee). A training and planning session is held each year with the community monitors to review the program and contract duties and to practice interview techniques. The first task for each community monitor is to develop (in consultation with the local organization) a list of knowledgeable, experienced people who have been active on the land over the past year. This list represents the community's selection of their local experts. The target is to interview twenty local experts in each community each year.

Prior to each interview, the community monitor reviews the basics of the program and discusses how the information will be used. An "informed consent" form with this information is signed, and a copy is left with the local expert. Interviews are anonymous (specific responses are not connected with names). In 2004, to provide an opportunity for better recognizing the local experts, interviewers asked people whether they wished to be recognized by name or photograph in the reports and posters. Each local expert receives an honorarium in the form of a coupon for gasoline at the local store. Gas prices are high in the region, and purchasing gas for snowmobiles and vehicles is often a factor limiting people's ability to get out on the land.

The interviews are conducted using an interview form developed by Gary Kofinas that is revised annually by Borderlands Co-op participants. The form has a mix of closed and open-ended questions. Here is an example of a "closed" question:

- How did the lakes freeze up this year?
  - A quick freeze-up
  - A slow freeze-up
  - Or just an average year?

And an example of an open-ended question:

- What kind of a year was it for cranberries?

Tape recorders are used only as an optional aid for note-taking for an open question that prompts people to discuss their main observations and concerns about environmental conditions and changes. If the person interviewed prefers not to be recorded, the interviewer takes notes on this question instead. A map is used for each interview to mark the areas being discussed. Questions are reviewed and adapted each year with the help of the community monitors and must be tailored to some extent to each community to reflect the differences in traditional areas and use patterns. The end product is always a compromise among several often-conflicting goals:

- Keep the questions simple, and keep the interview interesting and not too long.
- Make the interview form easy for inexperienced interviewers to use.
- Be comprehensive.
- Document information in a way that can be compared across areas and years.
- Ask questions in ways that are relevant to the people interviewed and that draw out observations and interpretations that reflect their traditional knowledge.
- Cover topics that will elicit observations from male and female experts of a range of ages.
- Adapt to needs for specific information for understanding issues that arise.
- Be consistent from year to year.

Figure 10.2

Excerpt from the 2003 Community Monitoring Report. *(From Allen et al. 2003.)*

Berries

Old Crow

- The berry blossoms started growing because of hot weather and early showers last spring. Then it rained, got damp, then it snowed. This killed and froze the berry blossoms resulting in hardly any berries last summer
- The few that grew were very small and had an unpleasant flavor. This includes all berries.
- People did not get enough to meet their needs.
- This also created problems for the animals because there were no berries to feed on. As a result the animals turned to grass and roots along the rivers to eat.

Fort McPherson

- This year was a very bad year for berries.
- Elders reported this resulted from extreme temperature changes this summer.
- There was an abundance of cranberries at Rat River. These were under shrubs, willows, and trees.

Observations about fish, berries, caribou, other animals, weather, and environmental conditions are documented. Many of the questions draw out observations about changes and interactions among environmental, economic, and community conditions, and the effects of these on people's ability to hunt, trap, fish, and collect berries.

Each community monitor prepares his or her own summary report on the interview results and presents it at the annual gathering. The community monitors' reports, along with added observations from the annual gathering, are reviewed by the local organizations and then compiled into an annual community report coauthored by all of the community monitors and widely distributed (e.g., Allen et al. 2003; Tetlichi et al. 2004). A copy of the report is mailed to each person who was interviewed in each community. This annual reporting by the community monitors to all contributors is crucial to the profile and success of the program. It allows people to see how their information is being used in developing a regional picture, and it reinforces community ownership of the results. Figure 10.2 shows an excerpt from a community summary report.

The current information management system has taken years to develop,

Figure 10.3
Database interface for creating reports.

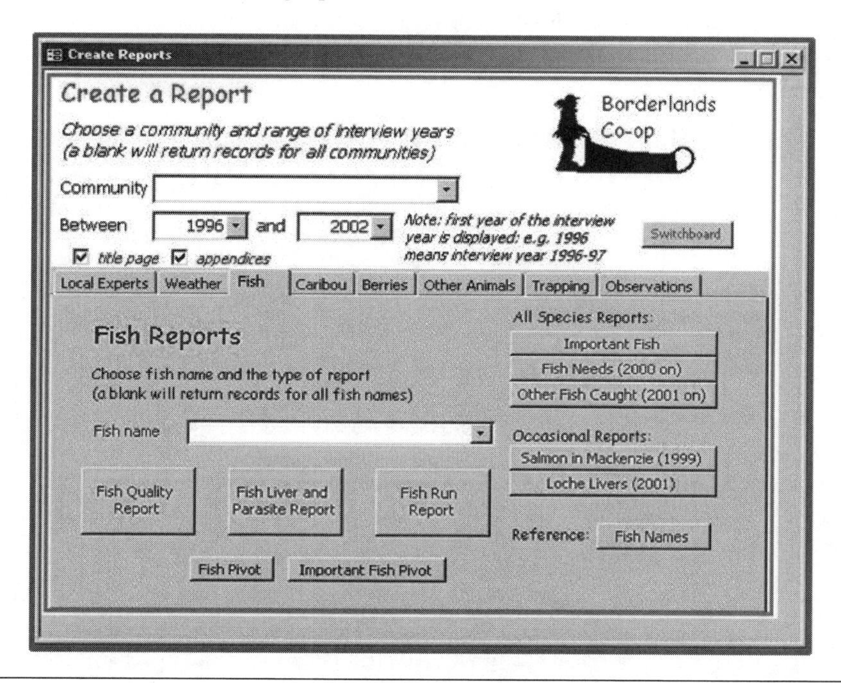

and work remains to be done on summarizing and interpreting the results. Information management is complicated by the broad scope of the interviews, the range of question types, and the variations in questions from year to year. The main characteristics of the current system include the following:

- The results from the interview forms and the community monitors' summary reports are stored in a customized relational database. This database is not publicly available; access to these "raw" results is based on the Borderlands Co-op's information-sharing protocol.
- The database has an interface that allows a user without database expertise to produce customized queries and reports by topic area, year, community, or keyword, or based on a word search (figure 10.3). Examples of types of data reports that can be produced follow:
    - Observations on fish quality for Fort McPherson, arranged by year and fish species

Figure 10.4
Example of a chart summarizing responses to a closed question. This chart is from a poster report for Whitefish in Aklavik. Other information that helps interpret the chart, including the number of people interviewed each year, is included on the poster.

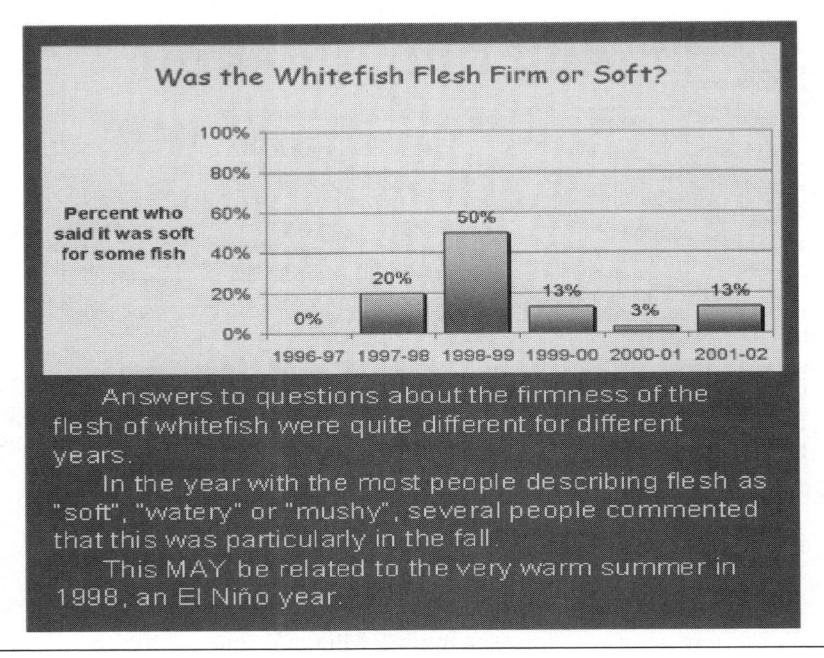

- Observations on any topic that references the place name "Peel River"
- Observations about the status, changes, and effects of different types of human activity for all communities in a selected year
- A report quantifying, by community and year, how many of the people interviewed met their needs for caribou in each season.

- Reports produced through the database contain documentation to link responses to specific questions and to assist the user in interpreting results.
- Maps that accompany each interview are digitized and linked to the database by each map reference code (for example, for each sighting of muskoxen or each marked location of increased stream bank erosion). Analysis of the spatially referenced results started in 2003 and is ongoing.
- Summaries are prepared on a topic basis for display and distribution. This work is currently done by Environment Canada staff and by contractors.

We are working toward a system in which resource managers in each community or comanagement region will have access to the database and will produce summaries to meet specific needs. The primary format used for summaries is large posters. Simple bar and pie charts are used where appropriate to show results from closed questions (figure 10.4). The main part of these posters consists of tables and text boxes summarizing and providing samples of the types of comments recorded in the interviews. The poster layout can be selected to allow comparisons across years, to scale up from the community level to the regional picture, and to examine topics in depth. The next step for us is to produce these materials in other formats (paper reports and Web versions).

- An information-sharing protocol, finalized in 2004, provides guidance for accessing and referencing results at the data and summary levels. This protocol reflects the desire of the Borderlands Co-op members to share information and to respect and recognize the local experts in each community.

Making Use of Research Results

One long-standing complaint from communities is that researchers come into the region, work for a bit, and then leave, but the communities do not receive the results of the research. Increasingly, researchers are reporting back to the communities, but it remains difficult for all parties to keep track of and find relevant information from past studies. Because of the importance of the Arctic Borderlands to wildlife, and because of the history of major petroleum-related development proposals, much research has been conducted in the region. To address needs for better access to and better understanding of research results, the Borderlands Co-op took the following steps:

- Worked with the Wildlife Management Advisory Council (North Slope) to develop an online database of information sources for the region and a literature review of coastal zone science and management. This database is accessed by resource management offices in the communities and regions.
- Produced a summary of what is known about contaminants from atmospheric transport in the region. This summary was presented at public meetings and distributed as presentation overheads and in a print version. (These products can be viewed at http://www.taiga.net/coop/reference.)

While this component of the Borderlands Co-op's program was stressed by participants at meetings in the first few years, it has not been a high priority in later years. As many of the agencies and organizations involved with research in the Arctic Borderlands now report back directly to the communities through meetings and reports, the need for communicating results through a separate process may be less than it was when the program began.

Putting It Together

Each annual gathering starts with an overview of the Borderlands Co-op's programs and a discussion of the relationships among the different program components. The following examples illustrate ways these program components have been used.

Providing Direction for Research and Making It Relevant

Local experts from the community of Old Crow, Yukon, observed that the lakes in Old Crow Flats were drying up. Scientists followed up on (and confirmed) these observations with remote-sensing studies and ground-truthing. Further assessment work will track this trend to see whether it continues and will look at the ecological implications (Jim Hawkings, Environment Canada, personal communication).

Following Up on Community Concerns

In the first three years of the community-based monitoring program, local experts in three communities identified an unusual number of diseased-looking livers from one species of fish (*Lota lota*, burbot or loche). There was concern that these fish might be contaminated and unsafe for human consumption. The Borderlands Co-op, through a partnership with a government department, followed up with a testing and analysis program. Experienced local fishers submitted "good" and "bad" livers for analysis. It was determined that contamination was not the source of the problem (Gary Stern, Fisheries and Oceans Canada, personal communication). This was communicated through the annual gathering and community meetings. In recent years, the incidence of reporting of diseased livers has dropped, and concern is rarely expressed. The community-based monitoring program continues to track this issue.

Assessing Conditions and Changes in a Region to Support Management Decisions

The Borderlands Co-op's indicator series, community-based monitoring program, and Web-based database of information sources are incorporated into the implementation section of the Wildlife Conservation and Management Plan of the Wildlife Management Advisory Council (North Slope), a comanagement council set up under the Inuvialuit Final Agreement land claim settlement. The Borderlands Co-op is a member of the implementation team for this plan and receives funding to provide information to the council in support of their assessment and management activities.

Specific action items undertaken by the Borderlands Co-op include assessing ecosystem health through ongoing monitoring and synthesis of information, especially related to climate change; tracking and reporting on the health of harvested fish and wildlife and on unusual sightings; maintaining the database of information sources to provide access to past research results; and producing educational materials on the effect of local pollution on wildlife and the environment (Wildlife Management Advisory Council [North Slope] 2003). In 2005 the council began work on a summary of what has been learned about the Yukon North Slope through the Borderlands Co-op's community-based monitoring and indicators.

Improving Understanding of Conditions and Trends of Ecosystems

The Porcupine caribou herd has been the subject of extensive research and monitoring over the past twenty-five years (Griffith et al. 2002; Russell, Kofinas, and Griffith 2000). The communities who are users of the herd hold knowledge based on centuries of observations. Caribou hunters observe and interpret the conditions they encounter each year while going about their activities on the land (Kofinas et al. 2004). These sources of information and interpretation are often at different temporal and spatial scales and inform one another. Examples include the following:

- Science-based methods provide estimates of herd size and calf survival and information on how snow conditions affect these (Griffith et al. 2002); local observations and traditional knowledge provide understanding of how caribou movements and feeding patterns are influenced by snow conditions and how these conditions affect the body condition of the caribou (Kofinas et al. 2004).

- Science provides regional trend information on climate variables (Whitfield and Russell 2004); local knowledge provides information on trends and quality of snow and forage in some key habitat areas.
- Harvest study records provide (often poor) records of total harvest (Hanley and Russell 2000); the community-based monitoring program provides information on whether each community has met its seasonal needs for caribou.

Conclusions: Some Lessons Learned

The development of this program has not been a steady progression. There have been difficulties obtaining support, financing, agreement on direction, and acceptance of the results. Nonetheless, the years have seen a steady growth in support and success of the Borderlands Co-op. Here is some of what we have learned:

- Keeping things simple and relevant to local concerns and needs, though not always easy, is crucial to the success of community-based programs.
- Developing a core set of people dedicated to the program is crucial. We have been fortunate to have strong supporters who are community leaders, elders, government managers, and academic scientists.
- Frequent reporting on the program and the results is very important. To reach all participants and interested parties, we use multiple means of communicating—newsletters, inexpensive photocopied reports, results posters, a Web site, and presentations at meetings.
- The organization of the program cannot be separated from its methods and results. Establishing a balance of power and ownership that communities, agencies, and councils are comfortable with is essential. For us, this is constantly evolving—as the profile of the program has risen, the need to structure and define the management of the program has grown.
- The community-based monitoring program presents significant challenges for data management and results interpretation. We did not sufficiently address this at the start, but we now have a system that allows us to access the results efficiently and to develop useful summaries that recognize the constraints and limitations imposed by the methods.

- One rather simple way of tackling the issue of multiple scales is to experiment with ways of displaying the information. We have found it helps to structure posters in a way that allows comparisons among communities and over time. This promotes discussion about impacts, issues, and trends at the regional scale.
- In this type of program, being independent from the management regimes has strong advantages. People are more open and relaxed about providing information and discussing its implications.
- Attention must be given to balancing the need for consistency and quality control with the need for local participation and ownership. At the outset, we recognized that involvement and control at the community level were essential for this program—although this has meant some inconsistencies in the documenting of local knowledge (with annual review of the methods, separate interviewers in each community, and often new people each year). This is part of the program and needs to be acknowledged when summarizing and interpreting results.
- The tension between science and traditional knowledge remains a part of the program. Results do not always agree; people remain entrenched in their views and traditions. This difficulty needs to be revisited periodically and examined openly.

Acknowledgments

I would like to acknowledge the many organizations and individuals who have worked on all aspects of the program since 1994. In particular, I acknowledge the long-term dedication of Gary Kofinas, Billy Day, Randall Tetlichi, Carol Arey, Annie B. Gordon, Aileen Horler, Deana Lemke, Robert Charlie, James Andre, Carol Andre, Mildred Allen, Joanne Gustafson, Barney Smith, Mike Gill, Scott Gilbert, the Wildlife Management Advisory Council (North Slope), the Gwich'in Renewable Resources Board, the Arctic National Wildlife Refuge, the Canadian Government (Departments of Environment, Parks, Fisheries and Oceans, and Indian and Northern Affairs), the Yukon and Northwest Territories Governments, and the University of Alaska, as well as others contributing funding, in-kind support, information, and interpretation assistance. Most importantly, this program could not exist without the support and input from the local resource councils and committees, the community monitors, board members, elders, and local experts from the communities of the Arctic Borderlands.

References

Allen, M., M. Andre, J. Gordon, D. Greenland, and R. Tetlichi. 2003. *Community reports 2002/03*. Whitehorse, Yukon: Arctic Borderlands Ecological Knowledge Co-op.

Bailey, J. L., N. B. Snow, A. Carpenter, and L. Carpenter. 1995. Cooperative wildlife management under the Western Arctic Inuvialuit Land Claim. In *Integrating people and wildlife for a sustainable future,* ed. J. A. Bissonette and K. R. Krausman, 11–15. Proceedings of the First International Wildlife Management Congress. Bethesda, MD: The Wildlife Society.

Berkes, F. 1998. Indigenous knowledge and resource management systems in the Canadian subarctic. In *Linking social and ecological systems: Management practices and social mechanisms for building resilience,* ed. F. Berkes and C. Folke, 98–128. Cambridge: Cambridge University Press.

Braune, B., R. Currie, B. de March, M. Dodd, W. Duschenko, J. Eamer, B. Elkin, et al. 1999. Spatial and temporal trends of contaminants in Canadian Arctic freshwater and terrestrial ecosystems: A review. *Science of the Total Environment* 230 (1–3): 145–207.

Byers, T., and L. W. Roberts. 1995. Harpoons and ulus: Collective wisdom and traditions of Inuvialuit regarding the beluga ("qilalugaq") in the Mackenzie River estuary. Report. Inuvik, NT: Byers Environmental Studies and Sociometrix Inc. for Indian and Northern Affairs Canada and the Fisheries Joint Management Committee.

Canada Department of Justice. 1998. *Mackenzie Valley Resource Management Act.* http://laws.justice.gc.ca/en/M-0.2 (accessed March 1, 2005).

———. 2002. *Species at Risk Act.* http://laws.justice.gc.ca/en/S-15.3 (accessed March 1, 2005).

———. 2003. *Yukon Environmental and Socio-Economic Assessment Act.* http://laws.justice.gc.ca/en/Y-2.2 (accessed March 1, 2005).

Eamer, J., ed. 2000. *Crossing borders: Science and community.* Program and abstracts of the 51st Arctic Science Conference, Whitehorse, Yukon, September 21–24, 2000. Whitehorse, Yukon: American Association for the Advancement of Science and Yukon Science Institute.

Griffith, B., D. C. Douglas, D. E. Russell, R. G. White, T. R. McCabe, and K. R. Whitten. 1999. Effects of recent climate warming on caribou habitat and calf survival. Paper presented at the Workshop on Biota and Climate Change, Ecological Society of America annual meeting, Spokane, Washington, August 11, 1999.

Griffith, B., D. C. Douglas, N. E. Walsh, D. D. Young, T. R. McCabe, D. E. Russell, R. G. White, R. D. Cameron, and K. R. Whitten. 2002. The Porcupine caribou herd. In *Arctic Refuge Coastal Plain terrestrial wildlife research summaries,* ed. D. C. Douglas, P. E. Reynolds, and E. B. Rhode (eds.), 8–37. Biological Science Report USGS/BRD BSR-2002-0001. Reston, VA: U.S. Geological Survey, Biological Resources Division.

Gwich'in Elders. 1997. *Nành' Kak Geenjit Gwich'in Ginjik* [Gwich'in words about the land]. Inuvik: Gwich'in Renewable Resources Board.

Hanley, T., and D. E. Russell. 2000. Ecological role of hunting in population dynamics and its implications of co-management of caribou. Special issue, *Rangifer* 12:71–78.

Huntington, H. P. 2000. Using traditional ecological knowledge in science: Methods and applications. *Ecological Applications* 10:1270–74.

Indian and Northern Affairs Canada. 2003. *Canadian Arctic Contaminants Assessment Report II, Highlights.* Ottawa, ON: Government of Canada.

Kofinas, G., Aklavik, Arctic Village, Old Crow, and Ft. McPherson. 2002. Community contributions to ecological monitoring: Knowledge co-production in the U.S.-Canada Arctic Borderlands. In *The earth is faster now: Indigenous observations of Arctic environmental change,* ed. I. Krupnik and D. Jolly, 54–91. Fairbanks, AK: Arctic Research Consortium of the United States.

Kofinas, G., P. Lyver, D. E. Russell, R. White, A. Nelson, and N. Flanders. 2004. Towards a protocol for monitoring of caribou body condition. Special issue, *Rangifer* 14:43–52.

Krupnik, I., and D. Jolly, eds. 2002. *The earth is faster now: Indigenous observations of Arctic environmental change.* Fairbanks, AK: Arctic Research Consortium of the United States.

McDonald, M., L. Arragutainaq, and Z. Novalinga. 1997. *Voices from the bay: Traditional ecological knowledge of Inuit and Cree in the Hudson Bay bioregion.* Ottawa, ON: Canadian Arctic Resources Committee and Environmental Committee of the Municipality of Sanikiluaq.

Nadasdy, P. 2003. Reevaluating the co-management success story. *Arctic* 56:367–80.

Nagy, M. I. 1994. *Yukon North Slope Inuvialuit oral history.* Occasional Papers in Yukon History No. 1. Whitehorse: Heritage Branch, Government of the Yukon.

Russell, D., G. Kofinas, and B. Griffith. 2000. Need and opportunity for a North American caribou knowledge cooperative. *Polar Research* 19 (1): 117–30.

Sherry, E., and the Vuntut Gwitchin First Nation. 1999. *The land still speaks.* Old Crow, Yukon: Vuntut Gwitchin First Nation.

Smith, B. 2004. *Aklavik Inuvialuit describe the status of certain birds and animals.* Whitehorse: Department of Environment, Government of the Yukon.

Taylor, E., and B. Taylor, eds. 1997. *Responding to global climate change in British Columbia and Yukon.* Vol. 1 of *Canada Country Study: Climate impacts and adaptation.* Vancouver: Environment Canada and British Columbia Ministry of Environment, Lands and Parks.

Tetlichi, R., M. Andre, A. M. MacLeod, A. B. Gordon, C. A. Gruben, M. Sharpe, B. Greenland, M. Allen, and E. Pascal. 2004. *Community reports* 2003–04. Whitehorse, Yukon: Arctic Borderlands Ecological Knowledge Co-op.

Urquhart, D., ed. 1998. Two eyes: One vision. Conference summary, Whitehorse, Yukon, April 1–3, 1998.

Usher, P. 2000. Traditional ecological knowledge in environmental assessment and management. *Arctic* 53:183–93.

Whitfield, P., and D. Russell. 2004. Recent changes in seasonal variations of temperature and precipitation within the range of northern caribou populations. Paper presented at the 10th North American Caribou Workshop, May 4–6, Girdwood, Alaska.

Wildlife Management Advisory Council (North Slope) [WMAC(NS)]. 2003. *Yukon North Slope: The land and the legacy (Taimanga Nunapta Pitqusia)*. Vol. 2, *Goals and actions*. Whitehorse, Yukon: WMAC(NS). http://www.taiga.net/wmac/consandmanagementplan_volume2/ (accessed March 1, 2005).

Zhang, X., L. A. Vincent, W. D. Hogg, and A. Niitsoo. 2000. Temperature and precipitation trends in Canada during the twentieth century. *Atmosphere-Ocean* 38 (3): 395–429.

CHAPTER 11

Cosmovisions and Environmental Governance

The Case of In Situ Conservation of Native Cultivated Plants and Their Wild Relatives in Peru

Jorge Ishizawa

Article 8(j) of the Convention on Biological Diversity (CBD) implicitly recognizes that valuable understanding for the sustainable use and regeneration of natural systems resides in practices of societies rooted in local cultures and ecosystems. In compliance with the CBD, the Global Environmental Facility (GEF) has provided funds for establishing project interventions for in situ conservation of the diversity of native plants and their wild relatives in centers of origin of agriculture.

This chapter examines the experience of one such project, called the In Situ Project, in the central Andes of Peru (2001–05) to explore the relationship among knowledge systems, the scaling up of project interventions, and environmental governance. The project's stated objective is to conserve agrobiodiversity in the cultivated fields (*chacras*) of campesino farmers in fifty-two locations in Peru.

The project addresses six areas of intervention: (1) the *chacra* and its surrounding areas, (2) the social organization of in situ conservation, (3) raising awareness of the importance of maintaining the diversity of native plants and wild relatives, (4) policies and legislation to promote in situ conservation, (5) markets for agrobiodiversity, and (6) an information system for monitoring agrobiodiversity.

The execution of the first three components has been contracted out to six implementing agencies, including two government research organizations and four nongovernmental organizations. Among the latter is Proyecto Andino de Tecnologias Campesinas (PRATEC), the Andean Project for Peasant Technologies.

PRATEC participates in the project by coordinating ten local community-based organizations (CBOs) in four different regions in Peru: the Altiplano region, the Central Southern highlands, the upper Amazon region of San Martín, and the northern department of Cajamarca.[1] PRATEC assists and coordinates fieldwork conducted by these CBOs in a range of ecosystems and communities across the country. It also participates in an interinstitutional technical steering committee along with the other implementing agencies involved in the execution of the project. This provides PRATEC the vantage point to reflect on the vicissitudes of implementing interventions on in situ conservation.

The value of "traditional knowledge" is also explicitly recognized in the Convention to Combat Desertification. However, traditional knowledge is generally expressed in the terms and protocols of technoscience. Even the traditions of *Farmer First* (Chambers, Pacey, and Thrupp 1989) and *Beyond Farmer First* (Scoones and Thompson 1994), influential works calling attention to the need to value and recognize farmers' knowledge, are ultimately centered on the technical outsider. We contend that instead of attempting translation, vernacular wisdom should be considered in its own right, and bridges between scientists and indigenous holders of equally valid paths to knowledge should be sought. The concept of translation refers "in its linguistic and material connotations . . . to all the displacements through other actors whose mediation is indispensable for any action to occur" (Latour 1999a, 311). In general, the knowledge domains do not overlap; they must be displaced in order to be meaningful. In effect, scientific knowledge is thus often constructed within a confining explanatory framework, defined by expert consensus about what constitutes a scientific "fact" (Latour 1987).

"Respect" for indigenous and local cultures should be understood as going beyond the recognition of their existence as privileged informants for technical outsiders. This entails also going beyond means to value and make visible this local knowledge and to consider the cosmovision of the indigenous and local peoples in its own terms as entirely equivalent to any other as valid modes of being-in-the-world.

For PRATEC, the challenge of the In Situ Project relates to the position of the central Andes as a global center of origin of agriculture where the domestication of plants dates back at least eight thousand years (National Research Council 1989, 163). The extraordinary interspecific and intraspecific diversity of plants and animals is a distinctive characteristic of the Andean campesino agriculture today and has been nurtured for millennia by campesino communities. Logically, these communities should be acknowledged as the real experts in conserving agrobiodiversity. Instead, the strategy privileged in the project document was to translate and reformat the campesinos' knowledge into the binding framework of a technoscientific approach. Most of the fieldwork was to be devoted to gathering data on campesino practices and knowledge, thus approaching conservationist farmers as informants. PRATEC argues that this approach is confining, restrictive, and ultimately distorting to this knowledge form.

Fortunately, the In Situ Project left room for diversity in institutional approaches to collaboration with the campesino communities. Taking advantage of this policy, PRATEC and its ten associated CBOs adopted in their participation an incremental approach that builds on what the campesinos already do for regenerating the diversity and variability of plants and animals based on their own cosmovision, knowledge, and practices.[2]

A summary of the Andean campesino cosmovision is presented below. The epistemological questions raised by technical interventions in order to conform to this cosmovision are taken up next, followed by an account of the PRATEC approach of cultural affirmation. The connections with the issue of environmental governance are then briefly explored. The chapter concludes by suggesting that an effort to take alternative cosmovisions at face value would advance the international conventions' purpose of achieving the planet's well-being.

Andean Campesino Cosmovision and Cultural Affirmation

The In Situ Project's objective is to conserve agrobiodiversity in the campesinos' *chacras*; thus the project demands that its implementing institutions go beyond the management and monitoring of biodiversity, to which the technoscientific approach is confined, toward an effective intervention that promotes conservation. Before the project's inception, PRATEC had found that, for the Andean campesinos, the in situ conservation of plants and animals is tantamount to their ancestral nurturance of life as it is lived in the Andes. In other words, in situ conservation of the diversity of native cultivated plants and their wild relatives is equivalent to Andean Amazonian campesino agriculture. Hence, PRATEC's approach of cultural affirmation has consisted of the strengthening of agriculture carried out by the traditional nurturers of that diversity.

As shown in figure 11.1, the strengthening of the campesino agriculture in

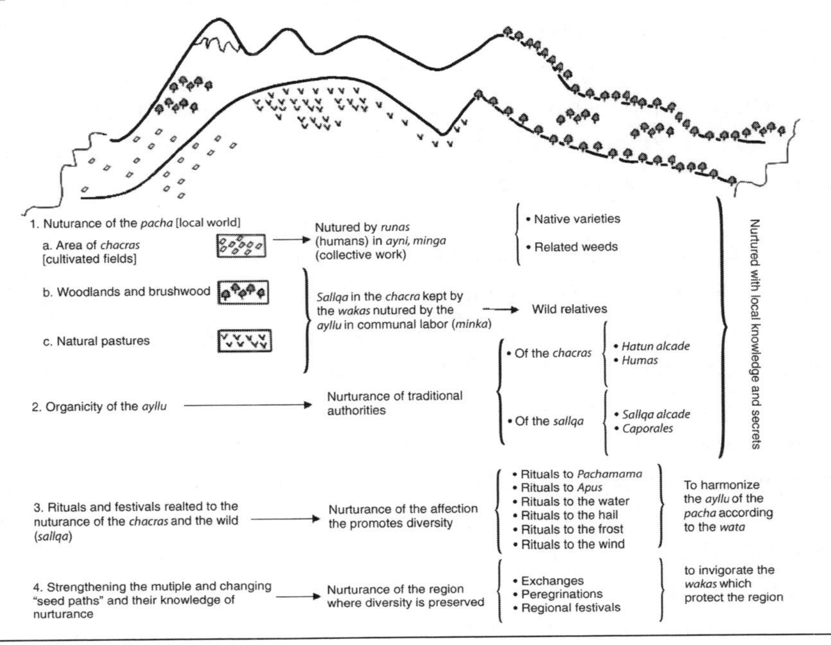

Figure 11.1
Dynamics of seed regeneration in the Central Andes.

the central Andes builds on the overall process of seed regeneration carried on by the nurturing communities. It includes several areas of intervention:

- The local landscape, or *pacha,* including the area of the *chacras* (cultivated fields), the *montes* (woodlands and brushwood), and the area of natural pastures
- The organicity of the *ayllu*—that is, all the entities inhabiting a local landscape[3]
- The rituals and festivals related to the nurturance of *chacras* and the *sallqa* (the wild)
- The multiple and variable paths of the seeds and of knowledge of their nurturance.

The regeneration of the local landscape comprises the area of the *chacras,* the area of the *montes,* and the pasture areas. The *chacras* are nurtured by the *runas* (humans) in *ayni* or *mingas* (collective work). In the Andes, the *montes* and pasture areas are considered as being nurtured by the *wakas,* or deities.

Mutual nurturance, then, is the basic mode of being in the Andes. All nurture, and all are nurtured, in every moment. An expression found in both native languages (Quechua and Aymara) is "we nurture while being nurtured." Julia Pacoricona Aliaga, from Conima, Puno, clarifies this expression with reference to the potato plant:

> The potato is our mother because when it produces fruits it is feeding us, clothing us and giving us happiness, but we also nurture it. When they are small, we call them *wawas* (children) because we have to look after them, delouse (weed) them, clothe (hill soil) them, make them dance and feast them. This has always been done. My parents taught me to nurture them with affection and good will as we do with our children. (Terre des Hommes Germany 2001, 23)

The diversity in the *chacras* consists of native species and varieties, their wild relatives, and "related weeds." The wild relatives and related weeds are also found in the *montes* and pasture areas. The regeneration of the local landscape is undertaken with the *pacha*'s (or community's) own knowledge of nurturance embodied in the signs of climate, soil, water, and the "secrets" of nurturance.

The care of the *ayllu*'s organicity is encharged to the traditional authorities of the *chacra* and the *sallqa*.[4] These authorities are not bearers of power but are mother and father to the community in their tenure, which is centered on the nurturance of the whole *pacha,* or locality (*chacras,* pastures, and *montes*). The *chacra* has distinct authorities in charge of its care and in care of the communal rituals in the agricultural cycle and of the *sallqa:* the community herds, pastures, and montes.

In the Andean cosmovision, it is affection and respect for the plants and their seeds that conserve diversity in the *chacras, montes,* and pastures. Affection and respect are vividly expressed in the rituals and festivals related to nurturing the *chacras* and the *sallqa.* They include rituals to Pachamama (Mother Earth); rituals to the Apus or Achachilas (mountain deities) for the nurturance of animals; rituals to the water asking for rain; rituals to hail, frost, wind, and snow; and *avios* (rituals of dismissal). Rituals bring harmony to the *ayllu* in accordance with the agricultural cycle. The testimony of doña María Lázaro from the community of Vicos, district of Marcará in Ancash, in the northern highlands of Peru, is eloquent:

> This little potato of mine I greatly care for. I converse with my seeds. My seeds know me because I am constantly speaking to them. This is

> the reason why my potatoes never leave me. In the same way I care for my *chacra* every time I go to visit her. I always talk with her and I do likewise with all my seeds. In my house we are always together. I sleep with my seeds. I store my seeds in my *pucu* (small storehouse). There I accompany them at night. . . . This potato never disappears because she likes me a lot. (Asociación Urpichallay 1999, 24)

The motivation of the Andean campesinos to conserve is intrinsic. Conservation of seed diversity is the result of a way of life. The affection for the seeds makes their regeneration a part of the campesinos' lives.

The nurturance of the region where diversity is ritually conserved is done through strengthening the multiple and ever-changing seed paths whereby the campesinos exchange seeds. The activities include regional pilgrimages as well as regional festivals for the nurturance of the deities that protect the *pacha*.[5]

An Epistemology for Cultural Affirmation in the Andes

Implementing an approach that affirms Andean culture demands an epistemology that derives from the campesinos' cosmovision, from their lifeworld. PRATEC understands the term *lifeworld* as "the world of our immediately lived experience *as* we live it, prior to all our thoughts about it. It is that which is present to us in our everyday tasks and enjoyments—reality as it engages us before being analyzed by our theories and our science" (Abram 1996, 40).

Such epistemology is based on PRATEC's interpretation of testimonies of people who experience the Andean lifeworld and are able to lend it a voice. Nurturance, or attentive care, among all persons in the *pacha* is central to this lifeworld. Another important characteristic is that distinctions such as those required by Aristotelian logic are misplaced: entities share some attribute that makes them appear the same.[6]

On the origin of the diversity of maize land races, don Cipriano Armas, from the community of Recuayhuanca in Marcará, Ancash, gives his version of an explanation that we have found to be widely held in the Andean communities:

> My hilling I have to finish the same day, since if I leave it for the following day, my maize plants will not go together to take their bath in

> the sea. For this reason I have to finish the same day at any cost. Also, when you finish the following day, the part you have finished go to take their bath, but the other part are only ready the next day. Then they meet the plants that you hilled the day before on their way to the sea. When they reach the ocean, where they take their bath, they mix and return all mixed, of different colors. Then in the harvest you find different colors that you have not sowed, that is, it is not your original maize. (Asociación Urpichallay 1999, 28)

The epistemology that we, as external agents, bring recognizes that any interpretation we can make of such testimonies is only a working hypothesis. We can demand a coherent interpretation but must renounce the notion of a general discourse on Andean cosmovision or a unique access to it. The discourse remains limited to a specific area of application and stands or falls on its own coherence and on the consequences of the actions it suggests.

We recognize that our epistemology is external and alien to the Andean cosmovision. The epistemological enterprise is undertaken only as an exercise in an attempt to build passerelles between cosmovisions. Two distinctive characteristics must be noted in these efforts:

- There is inherently no possibility of completeness or uniqueness in the expression of Andean cosmovision. Such expression is local and circumstantial; it requires a personal voice.
- The fact that explanations similar to the one offered on the origin of the diversity of plants exist in different cultural and geographical settings is illustrative, but it is not invoked as criterion of validity. There is never a pretension of transcendent objectivity.

Technoscientific Knowledge and Modes of Intervention in In Situ Conservation

The case of in situ conservation of the diversity of native plants and their wild relatives in the central Andes is particularly interesting. Project execution has shown that substantive knowledge is present in the practices of the Andean peasant nurturers of agrobiodiversity. Indeed, they are now being recognized as longtime experts in domestication of plants and animals. In contrast, scientific interest in in situ conservation is fairly recent. Nevertheless, most of the projects now being implemented in agrobiodiversity-rich areas for their conservation in situ still adopt a technoscientific approach.

Maxted et al. (2002) provide a recent account of on-farm conservation of germplasm from a technoscientific point of view, understood as "the sustainable management of genetic diversity of locally developed crop varieties (land races), with associated wild and weedy species or forms, by farmers within traditional agricultural, horticultural or agrisilvicultural systems." They recognize that there is no scientific tradition in the subject area and that "the farmers ultimately undertake the conservation, not the scientists." Even though the "farmers are aware of the importance of land races and the need for broadly-based agricultural biodiversity . . . their principal goal is economic. Agricultural security for them and their family is paramount and not the more nebulous conservation of genetic diversity. Thus, the role of the conservationist [the technical outsider] is . . . to help promote and preserve the conditions in which the traditional farmer can maintain genetic diversity in land races and related crop weeds, within the traditional production systems employed" (Maxted et al. 2002, 34).

Based on the critical assumption of the farmers' economic motivation to conserve biodiversity, the methodology proposed consisted of setting up a process divided "into three phases: (1) project planning and establishment, (2) project management and monitoring, and (3) on-farm utilisation of diversity" (Maxted et al. 2002, 33–34).

Site selection and material incentives are key to in situ planning and establishment; both seek to ensure that selected farmers continue to cultivate and manage "the maximum possible range of genetic diversity" of the target crops and land races within their farming systems (34–37). Formulating project activities demands research on why the land races exist at the site and whether they will continue to consider the influence of modern varieties, culture, and various socioeconomic factors, including availability of land, labor, and capital; macroeconomics; and extension workers (39).

On-farm project management and monitoring starts with a baseline study documenting levels and patterns of genetic diversity, local management practices of diversity through the agricultural cycle, and the physical and biotic environment. On this basis, monitoring will be attentive to genetic erosion (Maxted et al. 2002, 41–42). Only utilization will promote conservation, especially the traditional use by the farmers for their livelihood and the use of the germplasm by breeders (43–44).

Thus stated, this technoscientific approach to on-farm conservation does not take into proper account that the real proven experts are not the "professional conservationists"—that is, scientists and technical personnel—but are the campesino nurturers themselves, who have conserved for millennia without professional help. The "professional conservationists" do not conserve. None of the activities undertaken by the project management team is directly concerned with conserving agrobiodiversity. Most of them are part of a research effort to monitor the levels of agrobiodiversity in the campesino *chacras*. Knowledge obtained through research is to be applied to conservation management and agrobiodiversity monitoring. No direct intervention for conservation purposes or for promoting conservation is contemplated.

Despite the claims of its proponents, the major blind spot of the technoscientific approach remains sustainability of conservation activities, which hinges on the motivation of the campesino nurturers to conserve biodiversity. This blind spot is basically a cultural one. The assumption of the practitioners of the technoscientific approach is that motivation is fundamentally economic or that it can be turned into such through "some form of incentives to encourage the farmer to continue cultivation of the land races." This assumption has yet to be substantiated. To our knowledge, no research has been undertaken to test its plausibility. It has been PRATEC's experience in the In Situ Project—contrary to what Maxted et al. (2002) state—that household livelihood effectively turns around the "conservation of genetic resources" (and diversity, generally) and hence, "conservation" is, however indirectly, the primary focus of their concerns when undertaking agricultural activities.

The following sections present some conceptual elements of an approach to in situ conservation of native plants based on the Andean campesino cosmovision.

The Concept of Contact Zone

The "contact zone" is the meeting space shared by the project personnel and the local communities. In the In Situ Project's contact zone, importantly, peoples entertaining different cosmovisions encounter one another. It is thus a culturally and intellectually diverse space, in which the quality of the contact determines the success of the cooperative intervention. The concept is adopted here to allow an analysis beyond the populist view of the *Farmer First* and *Beyond Farmer First* perspectives on agricultural research and extension practices (Scoones and Thompson 1994, 16–32). These perspectives assumed an agenda of active farmer participation, empowerment, and poverty alleviation. In the *Beyond Farmer First* tradition, the concepts of "interface" and "encounters" are used in an actor-oriented perspective (Long and Villarreal 1994, 41–52). They make it possible to analyze relationships between actors holding differing interests and placed in an asymmetric power relationship based on a differential access to privileged knowledge (Foucault 1980, 78–108).

In contrast, here the emphasis is on a collaborative perspective of participants whose assumption of their basic equivalence puts the focus on the relationships rather than on the actors themselves. A major reason for this approach is to consider the situation in which different actors across the "interface" are part of the same community—for instance, when the technical "outsiders" are in the process of returning to their own communities. In this case, the question is not what negotiations take place at the interface but how to dissolve the interface altogether. Hence, a different label is needed for an apparently similar concept.

Accompanying Agrobiodiversity Conservation

How can the process of seed regeneration be strengthened through project interventions? In an in situ conservation project in the central Andes, two possibilities are open at the "contact zone." First, the project personnel may accept their role as external agents and keep to their management and monitoring tasks, inducing the campesino conservationists to continue conserving by offering, through project activities, appropriate incentives (such as markets and promotional policies) or by removing barriers. The other possibility is for the project personnel to demonstrate their belief that the campesinos' ways of agrobiodiversity conservation are basically sound since they have worked for millennia. PRATEC's approach proceeds from the latter possibility, using an informal partnership in which the project personnel accompany the continued regeneration of biodiversity undertaken by the Andean Amazonian peasants and the entities that make up the Andean *pacha* as depicted in the dynamics of seed regeneration (see figure 11.1).[7]

The technical personnel's motivation for accompanying the campesino communities is important. Here it is assumed that the professionals believe that the Andean campesino ritual agriculture is a mode of life appropriate for the specific conditions of the central Andes. This is their basis for accompaniment, by which they affirm the Andean campesino mode of life.

The Concept of Incremental Interventions

Another central concept in the execution of the In Situ Project is the notion of incrementality. All interventions are incremental in that they add, in depth and extension, to what the communities involved are already doing or are, in principle, willing to do by themselves. External intervention is restricted to helping enlarge the living web of cooperative relationships already in place. The concept is closely associated with the notion of contact zone. In effect, it is postulated as a working hypothesis that both concepts will aid understanding of the partners' role in furthering the joint action while maintaining focus on the lifeworld, wisdom, and norms for governance of the local and indigenous communities. If, in the *Farmers First* and *Beyond Farmers First* perspectives, the technical outsiders still had something to bring to the encounter at the interface, in the case of in situ agrobiodiversity conservation, they come empty-handed in terms of expertise. The expertise already exists in the communities of nurturers themselves. Thus, the populist approach of *Farmers First* and *Beyond Farmers First* falls short of dealing with this case.

What are the terms in which the project's technical personnel meet with the campesino nurturers of biodiversity in the contact zone of an in situ conservation project? In PRATEC's approach, the technical participants recognize the campesino's expertise and come to the encounter with eyes, ears, and heart wide open to learn from a millenary wisdom in its own terms. The initial approach includes such technical activities as the inventory, recovery, and collection of local and regional germplasm and the testing of new germplasm for its gradual incorporation in the *chacras*. These practices involve local knowledge that is documented for later publication and dissemination in technological booklets. The project's technical personnel also accompany diverse activities of nurturance, such as sowing in communal and collective lands as germplasm *chacras* and the exchange of seeds and knowledge in communal and intercommunal meetings.[8]

What the nurturers of biodiversity bring to the encounter is their general concern about the loss of respect they feel affects their mutual relationships with seeds, deities, nature, and other people. They feel that their rituals—the show of respect and affection for their deities (mountains, lakes, and Pachamama, or Mother Earth) and for nature at large—are being forgotten. Hence, climatic variations have become unpredictable, harvests have declined, and life has turned precarious in general. There is no word in Quechua to designate respect. The understanding of respect is obtained from concrete personal behavior.

In PRATEC's case, the project's technical personnel are organized in a network of local CBOs called Nuclei for Andean Cultural Affirmation (NACA). They base their action on the local understanding of the loss of respect and affection among all entities in the *pacha*. The NACAs accompany the communities in remembering the ways in which their ancestors learned respect. Traditionally, this has been achieved through participating in rituals and exercising a *cargo* (duty) in the system of traditional authorities. In the project activities, the NACAs contribute limited material inputs, such as fresh seeds from other regions and agricultural tools from urban origin. To document the project's progress, local team members register and systematize both the diversity of seeds and the *saberes* (traditional knowledge) and secrets of nurturance involved. They also help regenerate the ancestral ways of seed provision and exchange by accompanying community groups in visits to other communities following the seed paths.

As a coordinating or second-level implementing agency, PRATEC has been providing administrative support and technical backstopping to the NACAs. A close monitoring of the activities at the project's contact zone provides PRATEC privileged access to learning from the campesino lifeworld and the NACAs' lived experience. PRATEC conceives its role as an accompanist of the NACAs and thus as a second-order accompanist of the campesino communities. A major part of the accompaniment focuses on the NACAs' personnel providing training programs for the accompanists as well as workshops for the exchange of experiences.[9] The formation has evolved from one devoted to the training of accompanists to the communities to the training of cultural mediators. This is a major shift, since cultural mediation requires the accompanists to understand two different cultures in their roots.

Adopting the communities' diagnosis of loss of respect as the major threat to communal well-being demanded going well beyond the project's technical format to align activities around recovering respect in all its expressions. Activities included supporting the recovery of rituals associated with the regeneration of biodiversity, and promoting exchange visits by community members who wished to learn how other communities remember and strengthen rituals and celebrations and recover their traditional authorities.

The role of the accompanist at the project's contact zone can be characterized as cultural mediation. The cultural mediator is the intellectual hinge between cosmovisions—in this case, between the Andean cosmovision and the one implicit in the technoscientific approach. Two aspects of the mediation relate to traditional knowledge vis-à-vis science and its application to environmental governance—that is, the values and norms implicit in the agreement to care for the Earth's balance or to respect Earth law (Berry 2002; Stutzin 2002).

The accompanists have been educated for long years in a worldview that does not correspond with their original mode of living, of which they have a lived experience. Through their professional experience, they have corroborated the validity of their ancestors' knowledge and the customs regenerated by past generations as a basis for well-being. Juan Arturo Cutipa, a young and accomplished accompanist, member of the Asociación Chuyma Aru, tells in a book written by Loyda Sánchez (forthcoming from the Asociación) how he learned the traditional knowledge from his mother, doña Anastasia Flores Chambilla, from the community of Ccota, Puno:

> When I helped my parents in the fields on Saturday and Sundays I saw at harvest time that what they did was out of affection and reciprocity. I used to tell them that they were wasting money. Why are you so spendthrift with helpers? If they are not good at work stop hiring them. You can replace them with more efficient hands. My mother used to say: "This lady has no one to work on her *chacra*. Even if she works little, she talks with us and makes us laugh and thus further our work. Moreover if we do not share food with her it could even be a sin and God would chastise us. Who can give her something, if she does not have anyone to make *chacra* for her?" With my university student's eyes I had completely forgotten mutual aid, reciprocity, compassion, even respect which is most important in the field.

The role of cultural mediators is to aid the conversation between different cosmovisions by becoming a competent interlocutor. They must realize that they are subject to colonization—that is, to the unconscious submission to alien values and norms (Sartre 1967; Freire 1969). Colonization is dual. The training the accompanist received during a long period of schooling devalues the campesino mode of life as a stage in the history of humanity that is presently obsolete. Knowledge of the ancestors is looked on with contempt as a source of the poverty that outsiders perceive in the campesino lifeworld. Thus, the professional becomes dependent on external knowledge handed over without context, the pertinence of which, in a new milieu, is based on faith and the power of those who originated it, and not on factual verification. What makes colonization difficult to overcome personally for mediators is the apparent impossibility of renouncing the privileges of professional status and the cognitive authority that training bestows.

Scaling Up: The Incremental Approach

Our approach addresses an interesting question raised by a GEF consultant during project elaboration: how can a coherent program be made out of multiple local projects? To account for the diversity of circumstances in each project location, the local CBOs demanded autonomy. At the same time, coherent action was required. In PRATEC's approach, the contact zone between local communities and project personnel has primacy because activities jointly undertaken by the communities and the NACAs must be rooted in the community's lifeworld. We achieve coherence of our collective undertaking in different places and circumstances by different peoples and teams by adhering to a shared cosmovision of nurturance that is still present in the peoples of the central Andes. Coherent scaling up firmly rooted in specific places is thus made possible. The condition is nonetheless quality of the contact, which expresses itself in the respect and affection among participants in the collective action.

Scaling up the contact zone to the level of second-order coordinating institutions like PRATEC requires bridging the gap opened by the value–fact distinction implicit in the technoscientific approach to in situ conservation. This derives from that approach's basic assumption about the economic motivation of the farmers to conserve biodiversity. The distinction is a legacy from the founding fathers of modern science, who endeavored to create a space in which rational argument would prevail (Shapin and Schaffer 1985). Thus, in scientific activities, values were to be neatly distinguished from facts.

PRATEC's approach contains a second-order level of the contact zone at which PRATEC's action itself is located. This level allows for the reflection that produces some degree of generality needed to orient the collective action. Bringing in the values of respect and affection, which are central to the in situ conservation of agrobiodiversity to this reflection, requires the level of intellectual rigor found in the good and responsible practice of science, in which judgment and discernment must be exercised.

In a lecture titled "Moral Judgment and Political Action," included in the book *A Rumor of Angels* (Berger 1990), sociologist Peter Berger advances his understanding of what the social sciences can contribute to exercising judgment. He proposes four criteria. First is the *discipline of detachment*—that is, demonstrating the quality of social scientists to act, not as moralists, but in their "trained capacity to assess empirical evidence." Even though Berger restricts his injunctions to social scientists, we believe they can be applied to the practice of all scientists, especially when considering the consequences of the actions they recommend. He writes: "Part and parcel of [their] training is the discipline of detachment, that is, an ability to look at a situation clearly, to bracket off one's own feelings and convictions in the effort to understand what others feel and believe, to listen rather than to preach. Most important is [their] ability to look at *reality* even if what comes into view is very much different from what one would wish to be there" (148).

Berger continues:

> The second is the *clarification of normative and cognitive presuppositions*. In everyday life we constantly employ both kinds of presuppositions: Norms tell us what the world ought to be and how we ought to act; but these norms are supposed to maintain in a world that is real, and we hold a large number of assumptions, or cognitive presuppositions, as to what reality is. It is important to understand that norms have little if any meaning without the cognitive presuppositions that go with them. (Berger 1990, 149–50)

A major normative presupposition of the technoscientific approach to in situ conservation is that the motivation of the campesino nurturers of biodiversity for conserving is strictly economic. This is why a whole area of field research is devoted to clarifying farmers' decision-making criteria in selection procedures, farming practices, size of plant population, and seed source. This assumption is implicit in formulating in situ conservation projects, and consequently, project activities do not include field research to substantiate it.

Berger continues in his lecture:

> The third contribution is *the social location of actors and their interests*. . . . The sociologist is the character who, when confronted with any statement of belief or value, will invariably ask the prototypically mistrustful question, "Says who?" This question, disagreeable though it sounds, is of great importance in clarifying any situation in society and especially any situation within which one intends to act politically. (Berger 1990, 154)

This important aspect has to do with responsibility, both personal and corporative. The proponents of an in situ conservation project have their own interests and are socially located.

Berger concludes: "Finally, the fourth contribution—*the assessment of trade-offs*. . . . It is the easiest thing in the world to proclaim a good. The hard part is to think through ways by which this good can be realized without exorbitant costs and without consequences that negate the good" (Berger 1990, 159). This is probably the most neglected aspect of projects. The implicit costs must be considered along with the obvious benefits of well-meaning proposals.

Governance and Knowledge

By "environmental governance," we refer to the values and norms implicit in the idea of an Earth law that has the purpose of maintaining the Earth's balance. The Andean understanding of such Earth law involves mutual nurturance and respect among all entities in the *pacha,* or local world. Environmental science in the dominant technoscientific tradition has been used to produce pertinent knowledge for environmental governance. In this chapter, we have argued for the need to qualify this assertion and to explore more closely the relationship between science and "traditional knowledge."

The above account of Andean cosmovision sought to outline a perception of the world that differs radically from that of science. This cosmovision has been at the very root of a millenary form of approaching the conservation in situ of the diversity of plants (and animals) native to the Andean region, a form that has effectively conserved it. Our contention is that, viewed only from the cosmovisions, it is possible to approach the issues of environmental governance on equal footing.

Exclusively considering the knowledge from environmental science for the purpose of defining policies and adopting norms has at least two consequences. One is that laypersons are kept out of this process, and thus the popular (vernacular) knowledge they hold cannot influence decisions that may affect them—hence, the need to "translate" scientific knowledge into a format that people can understand if science is to provide space for democratic inclusion. However, this translation may not get the message through if the difference in cosmovisions is not carefully considered. The other consequence is that the definition leaves out all spiritual (or nonmaterial) connotations. The Andean case shows that the spiritual plays a crucial role in the campesinos' lifeworld. The limits of the technoscientific approach may preclude a deep understanding of issues of environmental governance that have spiritual roots.

In the In Situ Project, environmental governance was restricted to those aspects intending to "encourage the equitable sharing of the benefits arising from the utilization of [traditional] knowledge, innovations and practices" (article 8[j], Convention on Biological Diversity). These aspects only partially address the issues raised by the campesino communities on the loss of respect for their knowledge and cosmovision. This question expresses a concern that goes beyond the project and is central to environmental governance around the globe.

Conclusions

Bridges or Common Worlds?

"Bridging epistemologies" seems a viable idea if the underlying cosmovisions are considered and made explicit. The CBD has opened avenues for a fruitful collaboration between scientists and holders of vernacular wisdom. However, our experience warns against attempting one-sided translation in the implicit belief that traditional knowledge is just an input to the scientific enterprise. PRATEC's proposal is to undertake the challenge of considering the cosmovision implicit in Western technoscience and the cosmovision at the basis of vernacular knowledge as valid complementary modes of approaching the issue of environmental governance. Only with this explicit understanding can bridges be built between scientists, policy makers, and other actors, irrespective of the culture they embody.[10]

It is further proposed that the meeting ground between cosmovisions occur at the level of the contact zone, where problem identification from the grassroots can be agreed on and reformulated as a global concern. In effect, the loss of respect that affects biodiversity regeneration identified by the Andean traditional authorities can be recognized as the same basic problem at the root of the present ecological crisis. This common understanding can be the basis for interventions that incrementally contribute to problem solving.

However, bridges between technoscience and traditional knowledge may prove infeasible if conceived as entirely rational constructions. The attempt should be to build a good world in which many cosmovisions are welcomed, respected, and valued. This is always possible and desirable.

References

Abram, D. 1996. *The spell of the sensuous: Perception and language in a more-than-human world.* New York: Pantheon.

Asociación Urpichallay. 1999. *Así converso con mis semillas. La agrobiodiversidad en la cuenca del Marcará: Una perspectiva campesina* [I converse with my seeds. Agrobiodiversity in the Marcará watershed, a campesino perspective]. Marcará, Ancash, Peru: Asociación Urpichallay.

Berger, P. L. 1990. Moral judgment and political action. In *A rumor of angels: Modern society and the rediscovery of the supernatural,* 143–166. New York: Doubleday.

Berry, T. 2002. Rights of the earth. *Resurgence* 214 (September/October): 28–29.

Chambers, R., A. Pacey, and L-A. Thrupp. 1989. *Farmer first: Farmer innovation and agricultural research.* London: Intermediate Technology Publications.

Foucault, M. 1980. *Power/knowledge: Selected interview and other writings 1972/1977,* ed. Colin Gordon. New York: Pantheon Books.

Freire, P. 1969. *Pedagogy of the oppressed.* Cambridge, MA: Harvard University Press.

Latour, B. 1987. *Science in action: How to follow scientists and engineers through society.* Cambridge, MA: Harvard University Press.

———. 1999a. *Pandora's hope: Essays on the reality of science studies.* Cambridge, MA: Harvard University Press.

———. 1999b. *Politiques de la nature: Comment faire entrer les sciences en démocratie* [How to bring the sciences into democracy]. Paris: La Découverte.

Long, N., and M. Villarreal. 1994. The interweaving of knowledge and power in development interfaces. In *Beyond Farmer First: Rural people's knowledge, agricultural research and extension practice,* ed. I. Scoones and J. Thompson. London: Intermediate Technology Publications.

Maxted, N., L. Guarino, L. Myer, and E. A. Chiwona. 2002. Towards a methodology for on-farm conservation of plant genetic resources. *Genetic Resources and Crop Evolution* 49:31–46.

National Research Council. 1989. *Lost crops of the Incas: Little-known plants of the Andes with promise for worldwide cultivation.* Washington, DC: National Academy Press.

Sartre, J. P. 1967. Preface. In F. Fanon, *The wretched of the earth.* Harmondsworth, Middlesex, England: Penguin Books.

Scoones, I., and J. Thompson, eds. 1994. *Beyond Farmer First: Rural people's knowledge, agricultural research and extension practice.* London: Intermediate Technology Publications.

Shapin, S., and S. Schaffer. 1985. *Leviathan and the air-pump: Hobbes, Boyle and the experimental life.* Princeton, NJ: Princeton University Press.

Stutzin, G. 2002. Nature's rights. *Resurgence* 210 (January/February).

Terre des Hommes Germany. 2001. *Children and biodiversity in the Andes.* Lima, Peru: Terre des Hommes Germany.

CHAPTER 12

Harmonizing Traditional and Scientific Knowledge Systems in Rainfall Prediction and Utilization

RENGALAKSHMI RAJ

Historically, and still today in India, farmers have used traditional knowledge to understand weather and climate patterns in order to make decisions about crop and irrigation cycles. This knowledge is adapted to local conditions and needs and has been gained through many decades of experience passed on from previous generations. However, in recent years, farmers have perceived that the variation in rainfall is becoming increasingly erratic and difficult to predict. This has reduced their confidence in traditional knowledge and has led them to seek out scientific weather forecasts. These scientific forecasts are formulated at a much larger scale, diverging with local needs.

This chapter describes a project initiated by the M. S. Swaminathan Research Foundation (MSSRF) focused on harmonizing scientific and local knowledge of rainfall prediction for better use at the grassroots level in a few selected villages. The project activities are implemented through already-established computer-based "Village Knowledge Centers" (managed by a local farmer association), in which a central "hub" receives the scientific information and disseminates it to farmers in their local parlance. At the same time, the project has sought to link the traditional knowledge system on rainfall prediction with this scientific information through a multistakeholder participatory approach.

Degradation of soil, decreasing water resources, and changes in the climate are the three main obstacles to sustainable agricultural development in India.

Climate, including weather, is an important abiotic variable influencing crop production, especially in semiarid regions. The most effective way to deal with increased vulnerabilities caused by climatic variability is to integrate climate and weather concerns in agricultural planning and implementation processes. Scientific weather forecasts in India have until recently provided information on seasonal rainfall and weather only at a regional scale, which is generally too coarse to help decision making at the farm level. However, recent developments in weather and seasonal rain forecasting have increased the accuracy and reliability of the prediction of the Indian monsoon. General circulation models (GCMs) using sea surface temperature as boundary/initial conditions now enable researchers to predict seasonal rainfall (Blench 1999). Similarly, the capacity to generate and supply site-specific, medium-range weather forecasts has been enhanced in recent years. Despite these advances, access to the location-specific rainfall forecasts needed for proper decision making at the farm level is very limited.

Traditionally, farmers in India have made their own rainfall predictions, based on knowledge that has evolved through observation and experience over a considerable period of time. They base their predictions on a set of indicators, each of which has a different level of reliability. In response to the variability of the climate and weather as well as the uncertainty in their forecasts, they have evolved several coping strategies and mechanisms in rain-fed systems across the country.

Many international development agencies, government sectors, universities, and research institutions have begun to emphasize more strongly the value of indigenous knowledge in development. The dichotomy between indigenous knowledge and modern scientific knowledge is increasingly seen as a cause for underdevelopment; hence, work is now under way to bridge these two systems. Participatory research and farmer-back-to-farmer models of technology transfer (Amanor et al. 1993) are examples of attempts toward establishing such a bridge. Similarly, disciplines such as ethnobiology have sought to build bridges between indigenous knowledge and modern science. However, the challenge is how to bring together traditional knowledge and modern science without substituting one for the other and while respecting these two sets of values and building on their respective strengths.

In October 2002, the MSSRF, based in Chennai, India, initiated a project called "Establishing Decentralized Climate Forecasting System at the

Village Level" to enhance farmers' capacity to use locale-specific seasonal rainfall and weather forecasting. The project was undertaken in collaboration with Reddiyarchatram Seed Growers Association (RSGA), a farmers' association in Kannivadi in the Dindigul district of Tamil Nadu state, India. The main objectives of the project are to document the traditional knowledge on climate and weather forecasting, to examine the reliability of that knowledge, and to create a mechanism to provide access to scientific forecasting information that can supplement the traditional system. The project strives to link the two different knowledge systems through a multistakeholder participatory approach.

MSSRF has been working in this region since 1996 to develop biovillage models that address the twin problems of natural resource management and livelihood security for sustainable rural development. Participatory research, capacity building, and grassroots institution building are the three major dimensions of this work. Grassroots institutions help reduce the environmental and social transaction cost in development process. RSGA is one of the grassroots institutions developed in this region to promote sustainable agriculture and rural development. Small and marginal farmers constitute the majority of RSGA members, and one-third of the members are women. RSGA has organized need-based capacity-building programs with the support of Commonwealth of Learning, Canada. RSGA also manages a meteorological station as well as a Web page providing information on commodity prices and other economic market information, and it serves as the hub of Internet-based village knowledge centers (VKCs). The hub connects with four VKCs located in four different villages through a wireless local area network. This helps to serve the local villagers in four villages, especially those deprived of access to information important to their day-to-day life.

The VKCs are community-managed, demand-driven centers, each with three computers and an Internet connection. They dynamically supply information to meet different needs of the community, such as the product price in several nearby market centers, employment information, and government entitlements. The hub at RSGA receives the generic information and adds value by converting it to locale-specific information. Each VKC employs two women from the respective villages with secondary school education, who are trained in handling the computers to manage the unit. Access is ensured to all members of the community, irrespective of caste, class, gender, or age.

Methodology

The project has been undertaken in Reddiarchatram block, a semiarid region located in Dindigul district of Tamil Nadu state that contains twenty-four villages (administrative). More than 80 percent of the households depend on agriculture. Important planting seasons are June to July and October to November for both the irrigated and rainfed crops, in addition to the summer irrigated crop. The mean annual rainfall is 845.6 millimeters. Rainfall in the region varies greatly among seasons. Although the area benefits both from the northeast monsoon (October to December) and southwest monsoon (June to September), the maximum percentage (52.5 percent) of rainfall is received during the northeast monsoon, with only 25.8 percent of the annual rainfall received during the southwest monsoon. The area receives only 5.4 percent of the total annual rainfall during January and February, and nearly 16.3 percent during the summer seasons between March and May.

The total area under cultivation is 24,624 hectares, which includes both dry and irrigated lands. Approximately 29,600 households are involved in agriculture, and more than 50 percent of these are small and marginal farmers. Sorghum, small millets, grain legumes, cotton, and chickpea are the major annual crops cultivated under rain-fed conditions. Cotton, maize, flower crops, vegetables, gherkins, sugarcane, annual moringa, paddy, and onion are the most important annual crops grown in this region. The major source of irrigation is underground water through wells, followed by small tanks and reservoirs.

Four villages with functioning VKCs (one of which includes the hub center) were selected for the project. The four villages are located in the three distinct agroecosystems of the block: (1) Pudupatti and Kannivadi (with the hub center) villages, dominated by irrigated agriculture; (2) Samiyar patti village, which represents a subsistence cereal and legume production area with vast area under rain-fed cultivation of cotton and chickpea; and (3) Thonimalai, a hill zone where a coffee-based, multitier cropping system is being practiced.

The project was carried out between October 2002 and March 2004. In each village, the traditional knowledge concerning weather and climate forecasting was studied through the use of a conventional survey questionnaire; anthropological tools, such as participant observation; and participatory developmental tools, such as the Venn diagram and focus group discussions. Through questionnaires, the traditional weather and seasonal rainfall indicators and predictors were identified among 20 percent of the selected sample households in each of the villages. Anthropological tools (such as open-ended interviews) were used to study the metaphors, folklore, and proverbs that gave a better perspective on the traditional knowledge. A series of participatory rural appraisals was organized in representative villages in the block to obtain additional information on the social system, existing natural resources, agricultural practices and seasons, rainfall patterns, and the prevailing pattern and system of information flow. The needs, constraints, and coping strategies of farmers and agricultural laborers in response to variability of weather and climate were assessed through focus group discussions; this information was validated through triangulation with the information obtained through informal discussion with other knowledgeable men and women farmers.

MSSRF facilitated access for the VKC hub to scientific forecast information from the National Centre for Medium Range Weather Forecast (NCMRWF) for medium-range weather forecasts and Tamil Nadu Agricultural University (TNAU) for seasonal rainfall forecasts. The hub also manages a weather station, and the animators (staff) at the hub were trained in weather observatory management by TNAU. MSSRF also trained the animators to convert the generic information into farmer-friendly versions. The animators regularly recorded local weather parameters (maximum and minimum temperature, soil temperature at different depths, hours of sunlight, wind direction and velocity, evaporation rate, relative humidity) according to the norms of the Indian Meteorological Department and communicated that information to NCMRWF twice each week through electronic mail. In turn, NCMRWF provided weather forecasts of cloud cover, precipitation, temperature, wind direction, and wind velocity to the hub twice each week.

After the hub received the scientific forecasts, the staff converted the generic information into locale-specific, farmer-friendly language. For example, a forecast of a wind direction of one hundred degrees might be communicated to a particular village as *"lessana kathu sanimolaiyilurunthu addikkum,"* which means "mild wind blow from northwestern direction." The forecasts were then disseminated to farmers and agricultural laborers through the VKCs, bulletin boards, and local newspapers. After they received the information, the four VKCs communicated the forecasts to fifteen additional nearby villages through bulletin boards.

Similarly, linkages were established for the VKC hub to receive the seasonal rainfall forecast from TNAU. For example, in 2004, TNAU forecast that the region had a 40 percent probability of normal rainfall, a 40 percent probability of above-normal rainfall, and a 20 percent probability of below-normal rainfall for the northeast monsoon season. A focus group discussion was carried out in each of the four villages to communicate the forecast. In each village, the initial focus group discussion involved an explanation of how the scientific forecasts were generated and a discussion of the forecasts' attributes. The probabilistic nature of the seasonal rain forecast was explained to the farmers, and small games were organized to clearly explain the concept of probability. Then, using the climatological data, a "probability of exceedance" graph was generated to explain the relationship between rainfall amount (forecast) and probability.

Attempts are being made to communicate only the forecast information to the people instead of giving agro advisories, since this allows the farmers to make their own decisions. Under varied cropping pattern and rain-fed situations, farmers make follow-up decisions based on the event of rainfall and follow dynamic strategies instead of single strategy as most of the forecasters recommend. The discussions also focus on the areas of agreement or disagreement between weather indicators established in the traditional knowledge system and the available information supplied by the scientific forecasting system. The entire process of providing the scientific forecasts and interacting with farmers in participatory discussions is institutionalized through the VKCs.

Traditional Climate Knowledge and Forecasting

The term *traditional knowledge* as used in this chapter is defined as the "knowledge of a people of a particular area based on their interactions and experiences within that area, their traditions, and their incorporation of knowledge emanating from elsewhere into their production and economic systems" (de Boef et al. 1993). It differs from Western or scientific knowledge in the way it explains and establishes knowledge claims (Millennium Ecosystem Assessment 2003). Traditional knowledge is a cultural tradition that is constantly being developed and adjusted, and transmitted from generation to generation. It continues to play a major, even if largely unrecognized, role in the modern world. For example, the pharmaceutical industry still uses knowledge of traditional medicines to develop modern drugs. And interest in the use of indigenous knowledge has surged in such areas as agriculture and the conservation of genetic resources.

Understanding people's perceptions and knowledge of weather and climate is critical for effectively communicating scientific forecasts. Likewise, traditional weather and climate knowledge can provide significant value alongside the scientific forecasts. Traditional knowledge is learned and identified by farmers within a cultural context, and the knowledge base reflects the specific language, beliefs, and cultural processes. The local weather and climate are assessed, predicted, and interpreted by locally observed variables and experiences using combinations of plant, animal, insect, and meteorological and astronomical indicators. The different weather and seasonal rainfall indicators used to predict the occurrence of the rainfall are given in table 12.1.

Farmers use different kinds of traditional knowledge to predict rainfall based on their observation of such phenomena as wind movement, lightning, animal behaviors, bird movement, halos or rings around the moon, and the shape and position of the moon on the third to fifth days from the new moon. These types of information provide a framework that farmers use to explain relationships between particular events and changes in the climate and weather. Farmers combine different predictors and indicators to inform critical farm decisions and to decide on adaptive measures. Shaped by local conditions and needs, the knowledge is dynamic and nurtured by observation and experiences of both the men and women farmers. Also, the farmers in the region have developed certain beliefs with regard to rainfall. For example, farmers consult a local calendar to see whether *sani* (the god Saturn) takes an upper position, in which case the rainfall will increase for the coming season. Farmers also listen to local fortune-tellers. If there is a long dry season, in one ritual the community collects food from each household and then eats together; after the meal the people cry and then plead with the god to bring rain.

Men and women have different kinds of knowledge and use it for different purposes. Similarly, elder persons are more knowledgeable and able to use more indicators with greater understanding of their individual reliability. The older men and women used more than twelve indicators for weather forecasting, whereas the middle-aged persons (twenty-five to thirty-five years old) used only three or four indicators. Farmers as well as agricultural laborers have their own indicators that are based on their own needs and experiences. Also, farmers can list more indicators than the agricultural laborers.

The variations in indigenous knowledge in a community are based on age, gender, kinship affiliation, ideology, and literacy. Thus, social stratification influences the evolution and management of knowledge. Socialization and social heredity (the process of learning) take place within a particular

Table 12.1

Indicators/beliefs, reliability, related decisions, and user groups

Indicators	Decisions	Reliability (low, medium, high)	Users
Weather Indicators (twenty-four-hour forecast information)			
If lightning occurs from the east, west, and south, expect rain immediately	Indirectly helps to mobilize labor for weeding, day-to-day activities like shifting the cattle and other livestock and poultry to the shed, drying and collecting the dried products from the drying yard, and organizing fuels under shade; very rarely used for pesticide application	High, commonly used	Farmers and other people in the community
If lightning comes in an opposite direction (east to west), expect rain in one hour	Indirectly helps to mobilize labor for weeding, day-to-day activities like shifting the cattle and other livestock and poultry to the shed, drying and collecting the dried products from the drying yard, and organizing fuels under shade; very rarely used for pesticide application	Very high, commonly used	Many farmers
If lightning happens in the southeast and northwest directions, expect rain in the night	Used to make decisions on picking fruits and flowers that would generally occur the following morning	High	Commonly used by all farmers
Rings around sun	Used to decide irrigation, labor arrangement, and fertilizer application	High	Commonly used by all farmers
Increased number of mosquito bites	Supportive indicator	Low	Used by a few women agricultural laborers
Weather Indicators (two- to ten-day forecast information)			
In April and May, if the shade appears in the nearby hilltop (*thonimalai*), expect rain in another two to five days (*konamalai* for Kannivadi region; *gopinathan malai*/*palani malali* for Pudupatti village)	Used to initiate activities like arranging the dry fodder heap, land preparation (e.g., summer plowing and organizing/booking for country plough and tractor), and sowing of some vegetables under irrigated conditions	High	Majority of farmers

Table 12.1, continued

Indicators	Decisions	Reliability	Users
If there are small streaks in the clouds, expect rain in another two days; called "mazhai sarai"	Not used directly to make decisions, but farmers believe this indicator strengthens the reliability of other indicators	Low	Farmers and other people in the community
If circles are found around the moon, expect less rainfall; if small circles appear, expect less rainfall within two days	Land preparation: making arrangements for plowing, labor allocation, irrigation decisions, harvest decisions, and arrangements for post-harvest processing	High	Farmers
Northwesterly wind blows: expect rain in another two days; popularly called "mula kattru" (brings rain-bearing clouds)	Fodder and fuel arrangement; weeding, harvesting, sowing, irrigation, and postharvest decisions	High	Farmers and other people in the community
Northeast wind brings rain, and northwest wind (*sanimuzhai*) prevents the rain; south or northeast wind (female wind) helps to mobilize the clouds, and southwest wind (male wind) condenses clouds	Generally, people are more confident	Common and reliable	Both farmers and agricultural laborers
If frog in the well makes continuous sound	Arrangements for raising the motor in the well, making bookings for starting bore wells; decisions on irrigation and weeding	High	Farmers and other people in the community
If crab makes a bigger hole in the channel	Making arrangements for weeding and harvesting, organizing threshing floor and accessories, making bookings for implements like a plough, and arranging seeds for sowing	High	Farmers
During evening if the lower cloud appears red followed by a black cloud at the top, expect rain in another two days; if, during that time, the wind comes from the southeast (Saturn side), rain is unlikely	Decisions on irrigation, plowing, sowing, harvesting, and threshing	High	Farmers

Table 12.1, continued

Indicators	Decisions	Reliability	Users
If dragonflies and black sparrows fly in a group at three to four meters above ground level, expect rain the same day	Decisions on threshing floor, making arrangements for fuel and fodder; keeping the livestock under protection	High	Farmers
Black clouds with no stars bring rain; white clouds do not bring rain	Decisions on irrigation, postharvest operation, vegetable and flower plucking, drying of fodder and fuel	Medium	Farmers
Increase in body sweating	No decision is made based on this indicator, but it is used to support other indicators	High, nowadays decreasing	Farmers and other people in the community
Expect rain if dog jumps irregularly on the road at midday, poultry sit in a place for a long time, and sheep move in a group; popularly called "requesting a favor from god" (*mazhai varam ketkirathu*)	Used as a supportive indicator	Medium	Women and herders
In the cyclone period, if the clouds move in a group from east to west, expect rain in next two days	Irrigation, fertilizer application, harvesting, postharvest operations	High	Farmers
Seasonal Indicators (one- to three-month forecast information)			
South- and east-side winds indicate good rain during summer months; locally known as "thennal"	Plowing, sowing, and making arrangements for seeds and decisions on crops and cropping system	High	Farmers
Westerly wind brings heat and southwesterly wind bring coolness and clouds—rains during monsoon season	Sowing, planting, weeding, and threshing	High	Farmers
If there is more wind during July and August, expect good rain in October and November	Decision on cropping pattern and farm investment	High	Farmers

sociocultural realm, which is determined by class and caste (or caste in class). Gender is another important dimension of the social stratification. Knowledge is transferred to younger generations by the older ones through causal conversation, observations in the field, folk songs, metaphors, and so forth. In the ritual of "ceremonial plowing," all of the farmers in a village come together and initiate the first plowing. This traditional practice or ritual communicates to the entire community about the onset of rain. The elders use the same occasion to informally educate the younger generation about the traditional rain classification and appropriate cropping practices.

The indicators listed in table 12.1 reveal the qualitative nature of the indigenous knowledge concerning seasonal rainfall and weather. Weather predictions are used to make short-term decisions in both the irrigated and rain-fed systems. These predictions help the small and marginal farmers to plan various agronomic practices more effectively, especially at the time of sowing, weeding, spraying of chemicals, and harvesting and postharvest operations. Farmers also use seasonal rainfall predictions to prepare themselves for anomalies related to rainfall. For example, the predictions help determine the appropriate cropping pattern for the season. If the rainfall is normal, farmers plant high-value crops with high-yielding varieties (such as maize); however, if the rainfall is forecast to be below normal, they are more likely to plant short-duration, drought-resistant pulses and small millets. Farmers have been using different strategies to adapt and cope with uncertain weather and climate based on their experience and acquired knowledge from previous generations (box 12.1).

In the focus group discussions, farmers indicated that recent increases in the variability in rainfall have reduced their confidence in their own predictors, leading them to rely more on scientific forecasts. They indicated that variability has increased in terms of more water deficit years, late onset of rain and premature end of rains, and irregular distribution of rainfall in time and space. A climatological analysis of the interannual variability using twenty years of annual rainfall in this region indicated a coefficient of variation of about 36 percent; across the season, the variability in terms of coefficient of variation is high during the southwest monsoon season (71.6 percent) followed by the northeast monsoon season (52.2). Hence, the challenge and necessity are to provide reliable forecasting through appropriate methods based on the farmers' needs.

Box 12.1

Examples of Traditional Weather Prediction

Farmers use several meteorological indicators for seasonal rainfall prediction. For example:

- Westerly wind during *Adi* (June and July) brings rain in *iyappsi* (October and November).
- If there is no rain in the summer and there is wind in *Adi* (June and July), farmers prefer short-duration crops (such as cowpea), they reduce their farm investment, and some will invest in livestock, especially goats.

Farmers have evolved contingency cropping systems as a risk-averse strategy to reduce the potential for crop failure associated with climate fluctuations, especially for the rain-fed systems. It is common for farmers to start planting at the onset of rains. The following example shows how their crop selection changes according to the variation under a rain-fed agroecosystem.

- If rain sets in during June and July: lablab, sorghum, redgram, groundnut, vegetable cowpea
- If it is late by fifteen days: cowpea, fodder sorghum
- If it is late by another fifteen days: green gram and blackgram
- If it delays by yet another fifteen days: minor millets/short-duration sorghum

Other decisions involve mobilizing seed, fertilizer, and application; decisions on sowing (early or late); land and bed preparations; and midseason corrections, such as reducing population or providing irrigation.

Bridging Knowledge Systems

Scientific forecasts differ from traditional prediction in scale and, to some extent, in the type of factors used to predict weather patterns. Some of the weather predictors (e.g., wind speed and direction, temperature changes) used by farmers are similar to those used in the scientific forecast. However, while farmers have been using a combination of various biological, meteorological, and astronomical indicators to predict the rainfall, the scientific forecasts rely primarily on meteorological indicators, such as wind and sea surface temperatures. Traditional forecasts are highly locale specific, mostly at the village level within a radius of one to two square kilometers, and are derived from an intimate interaction with a microenvironment observed over a period of time. In contrast, the scientific forecasts encompass much larger geographic scales of fifty to three hundred square kilometers and depend on global meteorological parameters and their dynamics. Also, farmers perceive a high-rainfall year or season based on the time of onset and distribution instead of the total amount of rain received in a year.

The scientific forecast provides a probability distribution for the amount of seasonal rainfall as well as specific quantitative forecasts for medium-range rainfall amounts. The seasonal rainfall forecast does not provide information on the likely onset of rainfall and its distribution. The amount of rainfall and the timing of onset are the two most significant variables farmers use to make decisions on their initial agricultural activities.

On the other hand, traditional forecast knowledge is able to help the farmers in terms of the possible onset of rainfall using such indicators as direction and intensity of the wind during summer season, position of the moon on the third day, and traditional calendars (including other supportive indicators). Although the reliability of the traditional indicators varies, they do help farmers prepare for the timing and distribution of rain, while a scientific forecast may help them prepare for the amount. The different strengths of the two systems, when combined, provide farmers with more valuable information than either system can provide in isolation. The benefits of the two systems were elaborated during the group discussions with the farmers concerning seasonal rainfall forecasts. In this way, it was possible to establish a continuum between scientific and traditional forecast, which combines the scale and period of the onset of rainfall.

During the 2003 winter monsoon, the quantity of rainfall was predicted and communicated to the farmers in probabilistic mode two months in advance. The forecast indicated a 40 percent probability of normal rainfall (approximately 375 millimeters) between October and December and only a 20 percent probability of below-normal rainfall. At the same time, farmers also used traditional knowledge and observed wind pattern during the 2003 summer months (May through July) as a predictor. During May through June, the wind flow from west to east was very weak, but it increased during the subsequent month of August. According to the farmers, this delayed wind movement indicated that there would be a delayed onset of rain of about two to three weeks beyond the normal onset (i.e., the onset was expected to shift from the fourth week of September to the third week of October). The farmers also predicted a slightly below normal rainfall.

During that season, 230 millimeters of rainfall was received, which is 39 percent lower than the average rainfall; thus, the traditional forecast was more accurate than the scientific forecast. The onset of rainfall was delayed by more than three weeks. The farmers prepared themselves and practiced mixed crop-based, drought-tolerant, short-duration crops, such as sorghum and cowpea. Based on this experience, farmers indicated that they needed time to observe the effectiveness of scientific forecasts over seasons and years. It is unlikely that farmers will rely heavily on scientific forecasts until the forecasts prove their reliability.

Regarding the medium-range weather forecasts, the scientific forecast provides an expected quantity of rainfall, while farmers predict the possible period of rain using such meteorological indicators as lightning, cloud density, and wind movement. They also use supportive indicators derived from changes in the behavior of animals and insects. Thus, for medium-range forecasts, both the traditional and scientific knowledge systems help prepare the farmers for amount and timing of rainfall. This experience demonstrates the possibility that exists to harmonize the two knowledge systems in the context of the farmers' cognitive landscape, which is tuned to incorporate multiple sources of information.

Conclusion

This chapter documents the vast traditional knowledge that farmers hold concerning rainfall prediction and the level of their sophistication in understanding the reliability of these predictions. The traditional weather forecasting system in these communities provides information not available in the scientific forecasts. But scientific forecasts also have their benefits, and it is necessary to understand local peoples' perception of rainfall prediction in order to communicate scientific forecasts effectively, since weather-related information is learned and identified by farmers within a cultural context and their knowledge base follows specific language, beliefs, and social processes. Acknowledging the importance of the traditional knowledge base also aids social interaction and acceptance of scientific forecasts among the farmers. Thus, the two different knowledge systems need to be bridged.

Intensive dialogue between the scientific knowledge providers and user groups helps to define the strategies for bridging these two knowledge systems. The project described in this chapter shows that farmers were able to bridge the two different knowledge systems, which is not surprising since the farmers are used to operating in multiple cognitive frameworks. Participatory approaches were critical in developing a decentralized forecasting system at the village level that could effectively bridge the traditional and scientific knowledge base. On the other hand, access, availability of infrastructure, skill, and expertise are crucial to developing reliable region-specific scientific forecasts to serve the farming societies.

It is too early to judge what the final balance will be in the use of traditional and scientific knowledge in decision making by farmers in this region. But a system and process now exist within which farmers' understanding and confidence in scientific forecasts can be developed without undermining the benefits provided by traditional systems. There is a vast scope to link two different knowledge systems with the participation of local people.

References

Amanor, K., K. Wellard, W. de Boef, and A. Bebbington. 1993. Introduction. In *Cultivating knowledge: Genetic diversity, farmers' experimentation and crop research,* ed. W. de Boef, K. Amanor, K. Wellard, and A. Bebbington, 1–13. London: Intermediate Technology Publications.

Blench, R. 1999. Seasonal climate forecasting: Who can use it and how should it be disseminated? *Natural Resource Perspectives* 47 (November): 115. http://www.odi.org.uk/NRP/47.html (accessed April 9, 2006).

de Boef, W., K. Amanor, K. Wellard, and A. Bebbington. 1993. *Cultivating knowledge: Genetic diversity, farmer experimentation and crop research.* London: Intermediate Technology Publications.

Millennium Ecosystem Assessment. 2003. *Ecosystems and human well-being: A framework for assessment.* Washington, DC: Island Press.

CHAPTER 13

Managing People's Knowledge

An Indian Case Study of Building Bridges from Local to Global and from Oral to Scientific Knowledge

Yogesh Gokhale, Madhav Gadgil,
Anil Gupta, Riya Sinha, and K. P. (Prabha) Achar

Humans owe their domination of the living world to their occupation of what has been termed the "knowledge niche." They have been acquiring, organizing, and using knowledge for hundreds of thousands of years. Knowledge is power, and since time immemorial people have exercised control over who accesses what knowledge. Thus, dispensers of herbal medicine have often shared their knowledge only with a select group, such as their eldest sons. This is equivalent to the intellectual property rights (IPR) system of trade secrets. Knowledge began to be organized more systematically with the invention of agriculture, the growth of villages and towns, and the discovery of writing.

Similarly, the classical Indian system of medicine, Ayurveda, grew out of a collation of folk knowledge of herbal medicine. Indeed, Ayurvedic texts urged healers to absorb the knowledge of hunters, herders, and forest-dwelling people subsisting on tubers. At the same time, Ayurvedic practitioners attempted to establish a monopoly over this knowledge by forbidding the study of Sanskrit, the language of Ayurvedic texts, to castes lower in social hierarchy.

The growth of knowledge sped up with the elaboration of the scientific method beginning during the sixteenth century in Europe. The growth of scientific knowledge depended on broad access to scientific information and the ability of people from all sections of society to contribute to the scientific enterprise. At the same time, commercial interests wanted monopoly over applicable knowledge. These conflicting interests led to the development of a new system of IPRs called patents. Patents allowed for knowledge to be shared (and, indeed, demanded that it be made public) in the form of specifications accompanying a patent application. At the same time, it permitted monopoly rights to the patent holder over the commercial application of the patented knowledge for a limited period.

The patent system permits monopoly rights over specific, limited knowledge that is claimed to be novel, nonobvious, and applicable. This excludes anything already known in the so-called public domain. It also excludes anything that is not codified, such as orally transmitted knowledge. As a result, a pharmaceutical company may build on orally transmitted, community knowledge of medicinal uses of an herb and then establish its own monopoly rights over the application of the knowledge.

Such expropriation of knowledge that is outside the system of modern science and technology and that is not explicitly protected through establishment of intellectual property has, of course, been going on for a long time. A well-known Indian case in this context is the development of a drug to treat hypertension, reserpine, from *Rauwolfia serpentina* (Gupta 2000). Another recent case is the successful defeat of a U.S. patent application on a cream prepared from turmeric on grounds that such an application is not novel, being already documented in several classical Ayurvedic texts.

Combating Biopiracy

Recent years have witnessed a growing perception that such expropriation of knowledge, sometimes labeled "biopiracy," is unfair. A measure to recognize and reward community knowledge of sustainable uses of biodiversity was incorporated as Article 8(j) of the international Convention on Biological Diversity (CBD), in force since 1993. This is a very significant measure for a country like India, which is rich in knowledge of biodiversity both in the form of such classical medical systems as Ayurveda, Sidha, and Yunani, and in folk knowledge of uses as pharmaceuticals, neutraceuticals, dyes, pesticides, and others. Attempts to document such knowledge in the parlance of modern science began with early contacts of Europeans with India. *Hortus Malabaricus,* composed between 1678 and 1703 by van Rheede, a Dutch resident of Kochi, documented the knowledge of four local physicians (Manilal 1980). In the nineteenth century, the British organized the systematic explorations and documentation of uses of Indian biodiversity, culminating in the "Wealth of India" series. Many other texts and research papers on these topics continue to be published, by and large without assigning any credit to knowledge providers. At the same time, the documentation serves to bring this knowledge into the public domain. Technopreneurs can then add a small element of a novel application to such knowledge and establish their IPRs over it.

Rewarding People's Knowledge

It is against this background that an initiative called the Honeybee Network was launched in the late 1980s. The Honeybee Network aims to document people's knowledge by assigning full credit to knowledge providers and then to disseminate it widely (Gupta 1999). Its emphasis is on sharing, not on assertion of any intellectual property rights over the knowledge. However, in the 1990s, it began to examine issues relating to IPRs, eventually leading to the establishment of a new initiative called the National Innovation Foundation (NIF). In the 1990s too, after the ratification of the CBD but before passage of India's Biological Diversity Act, the Tropical Botanical Garden and Research Institute (TBGRI) in Thiruvanathapauram in Kerala volunteered to share benefits with providers of orally communicated, community knowledge of Kani tribals. Two members of the Kani tribe had informed TBGRI scientists of certain therapeutic properties of the forest-floor herb *Trichopus zeylanicus*. Using this as a starting point, TBGRI developed a commercial product for which a pharmaceutical company paid Rs. 10 lakh (about US$23,000) to the institute. TBGRI deposited half this amount in a trust established for the purpose of benefit sharing with the Kanis of Kerala (National Innovation Foundation 2002).

This experiment, which at that time had no legal framework to support it, raises several questions. Some of these open questions relate to defining the appropriate set of knowledge holders with whom benefits should be shared. The benefits were to be shared with Kanis of Kerala, but members of the same Kani community occur in the neighboring state of Tamil Nadu. It is also likely that members of other local communities may have shared this knowledge. In the absence of any systematic documentation, these issues cannot be resolved.

Documenting People's Knowledge

Many attempts have been made to systematically document folk knowledge of biodiversity, beginning possibly with the initial codification of Ayurveda at least two thousand years ago, and continuing through *Hortus Malabaricus,* the Wealth of India series, and a national project on ethnobiology in the 1980s, all without explicitly acknowledging the contributions of knowledge providers until the efforts of the Honeybee Network.

An organization called the Foundation for Revitalization of Local Health Traditions (FRLHT) initiated a somewhat different attempt in the 1990s beginning with the preparation of "community biodiversity registers." Through this medium, as well as through many assemblies of folk healers, FRLHT has developed a large database on medicinal uses of herbs in the folk traditions. However, it has made little progress in sharing this knowledge or in disseminating any benefits to knowledge providers. More recently, it has begun to execute prior informed consent (PIC) statements with knowledge providers. While these PIC declarations ensure that the folk knowledge providers are made aware of what is happening, FRLHT does not accept any specific responsibilities over how it will use this knowledge, how the knowledge providers can participate in the use of the knowledge, or how the knowledge providers may benefit from the use of the knowledge.

People's Biodiversity Registers

Documentation of knowledge associated with biodiversity is clearly pertinent in the context of the provisions of CBD for equitable sharing of benefits with knowledge holders. To support this objective, the Indian Institute of Science, Bangalore, broadened the scope of the FRLHT's community biodiversity register, creating "people's biodiversity registers" (PBRs) to include documentation of local biodiversity, relationships between biodiversity resources and people, people's knowledge of biodiversity in the context of medicinal as well as other uses, their ecological knowledge, and their perceptions of ongoing and desired patterns of biodiversity management.

Beginning in 1996, a series of PBRs has been prepared with the help of networks of environment-oriented nongovernmental organizations (NGOs) and high school and undergraduate educational institutions. With experience, and the growing availability and capability of tools of modern information and communication technology, the program has been refined so that much of the information so generated can be pooled together and organized with the help of a relational database management system. In 2002, the Ministry of Environment and Forests, government of India, proposed that these exercises be made a part of the Millennium Ecosystem Assessment (MA). This proposal has been accepted, and the exercises have served as the Indian contribution to the subglobal assessments component of the MA.

People's Knowledge

The PBR exercises have not actively sought to document knowledge of uses of biodiversity, since the legal and institutional framework for the management of such knowledge has yet to be put in place. However, at one PBR site, that of Mala village of Karkala taluk (part of a district with a typical area of about a thousand square kilometers), of the Udupi district in the state of Karnataka, such documentation was undertaken by an associate of the PBR program, Dr. Satyanarayana Bhat, a professor in an Ayurvedic college in Bangalore. Dr. Bhat was prompted to undertake this exercise in 1994–95 because of Mr. Kunjira Moolya, a resident of Mala and a highly respected dispenser of herbal medicines.

Dr. Bhat extensively documented the various medicinal formulations employed by Mr. Moolya, the methods of administration, and the symptoms of maladies for which these remedies were used. However, the group involved in preparing the PBR at the Indian Institute of Science advised Dr. Bhat that he should not make this documentation public until clear measures were in place for protecting Mr. Moolya's IPRs. Some additional material on medicinal uses of plants was also collected later from the following seven practitioners at Mala: (1) Ms. Indira Anantha Marate, (2) Mr. Ganesh Joshy, (3) Mr. Govinda (Menpa) Hegde, (4) Ms. Mutthu Poojarthi, (5) Ms. Muddu Merthi, (6) Mr. C. J. Michael, and (7) Mr. Shrinivasa Prabhu Kadari. All this material was maintained as confidential with the Indian Institute of Science until March 2004.

Biological Diversity Act

India acceded to the CBD in March 1994. Two of the CBD's provisions—(1) sovereign rights of countries of origin over biodiversity resources, and (2) the need to share benefits from commercial utilization of nonformal, often oral knowledge of sustainable use of biodiversity resources of communities or individuals—are of particular interest to India. The process of drafting a

Biological Diversity Act to provide a legal framework for implementing these two provisions was initiated in India in 1996, and a draft act was produced for public discussion in 1999. This process was significant because normally any such legislation to be brought before the Parliament is treated as an official secret until it is tabled in the Parliament. In this case, however, the minister obtained special cabinet approval to place the draft before the public for feedback. The act was tabled before the Parliament in 2000 and was finally approved by the president in 2003.

Institutional Framework

The Biological Diversity Act provides for the establishment of a National Biodiversity Authority (NBA), state biodiversity boards (SBBs) in all states of the Indian Union, and biodiversity management committees (BMCs) at the level of all local bodies, namely village and town councils and city municipalities. Approval by NBA is mandatory for any foreign agency or individuals engaging in research or bioprospecting for commercial use of Indian biodiversity resources and associated knowledge. NBA is also to screen all patent applications in India based on Indian biodiversity resources and associated knowledge and will permit them to be processed only after ensuring that they:

- provide due acknowledgement to the resources over which India has sovereign rights as the country of origin and associated knowledge of Indian origin
- agree to equitable arrangements for sharing of benefits with resource and knowledge providers.

NBA is expected to consult the concerned BMC whenever agreeing to any foreign agency accessing Indian biological resources and associated knowledge as well as when agreeing to any patent application.

A concrete information base needs to be created to permit meaningful consultation by NBA with the tens of thousands of village and town councils and city municipalities that cover India. To this end, the act provides for chronicling of biodiversity resources and associated knowledge by all local BMCs. The rules promulgated under the act further state that preparing this documentation in the form of PBRs constitutes a major function of BMCs. The BMCs are also authorized to regulate access of all outside agents, Indian as well as foreigners, to local biodiversity resources and associated knowledge, and they have the authority to levy collection charges for this purpose.

Safeguarding Intellectual Property

An important issue that arises in this context is the protection of people's intellectual property rights over knowledge with potential commercial application that may be documented during the process of preparing PBRs. If all this documentation were made available to the public, there would be no way to ensure the flow of benefits to people in cases where the products are developed and sold in markets outside India. There is no international agreement in place to permit India's National Biodiversity Authority to persuade foreign enterprises operating outside of India to share benefits in such a contingency. Neither is such an international agreement likely in the near future, especially since the United States has refused to ratify the CBD.

It is therefore vital that details of such knowledge are kept confidential. One possible agency to do this is NIF, established by the government of India in March 2000. NIF has grown out of the Honeybee Network's activity as an agency to promote green grassroots innovations and traditional knowledge. It is presided over by the head of India's Council for Scientific and Industrial Research, with involvement by the Honeybee Network activists. It maintains an information base called the National Register, a repository of all socially and environmentally acceptable information flowing to it from multiple channels, including village-level exploratory trips. A provision exists to maintain the confidentiality of some of the information lodged with the National Register.

The governing body of NIF has decided that NIF will set up an additional database to be named the People's Knowledge Database (PKD) to supplement the existing National Register. The PKD will serve as an electronically searchable, multilingual, and multimedia repository of all people's knowledge recorded through PBRs and other means. It will be maintained either as publicly accessible or as confidential knowledge, as specified by knowledge providers, giving full credit to the individuals or communities concerned. All entries in the PKD will be scrutinized, and those components that meet the criteria evolved by NIF pertaining to environmental and social sustainability will be transferred to the National Register, again maintaining specified restrictions on access and indicating the content to the public in a synoptic form. Entries not accepted for inclusion in the National Register will continue to be maintained in the PKD. The PKD and National Register will form part of a distributed biodiversity information system (BIS), which will also incorporate other relevant scientific, technical, IPR, and market-related information and serve as a knowledge base for NBA, the SBBs, and the BMCs.

Memorandum of Agreement

The Indian Institute of Science has maintained an active dialogue with NIF to explore the use of the National Register as a repository of confidential information pertaining to uses of biodiversity provided by communities or individual knowledge providers in the course of PBR preparation. As a test case, it has employed the information on medicinal uses of plants collected from Mr. Kunjeera Moolya and other knowledge providers of Mala Village. A model of information management for this purpose was evolved during a brainstorming session at the governing body of NIF on March 9, 2004 (figure 13.1).

This model proposes that NIF execute a memorandum of agreement with the knowledge providers, in place of a simple prior informed consent (PIC). The memorandum would acknowledge NIF's acceptance of certain conditions established by knowledge providers under which their knowledge may be shared with third agencies. These third agencies would primarily be research and commercial organizations interested in developing products based on the knowledge. The knowledge providers may specify the kind of agencies that may be allowed access to their knowledge, how these agencies may further manage this knowledge, and the expected benefits from these agencies. NIF may make their knowledge available only after these conditions are met.

Of course, outside agencies need an indication of the nature of the knowledge being held as confidential in the National Register to enter into an agreement with NIF for access to any particular item of knowledge. For this purpose, the National Register would provide a synopsis of the nature of the confidential knowledge. Such a synopsis may, for instance, mention the symptoms of a disease that can be treated with an herbal remedy, while withholding the name of species and other details. If such an arrangement works, it would be an excellent way to bridge the gap between local and global scales and between folk and modern scientific knowledge.

Following the elaboration of this model at the governing body meeting of NIF, a series of discussions was held with the knowledge providers of Mala village as a part of the Indian Institute of Science's activities under the subglobal component of the MA. These discussions, which focused on the form of a memorandum of agreement acceptable to them and to NIF, led to the drafting of a

Figure 13.1
A framework for managing formal and informal knowledge.

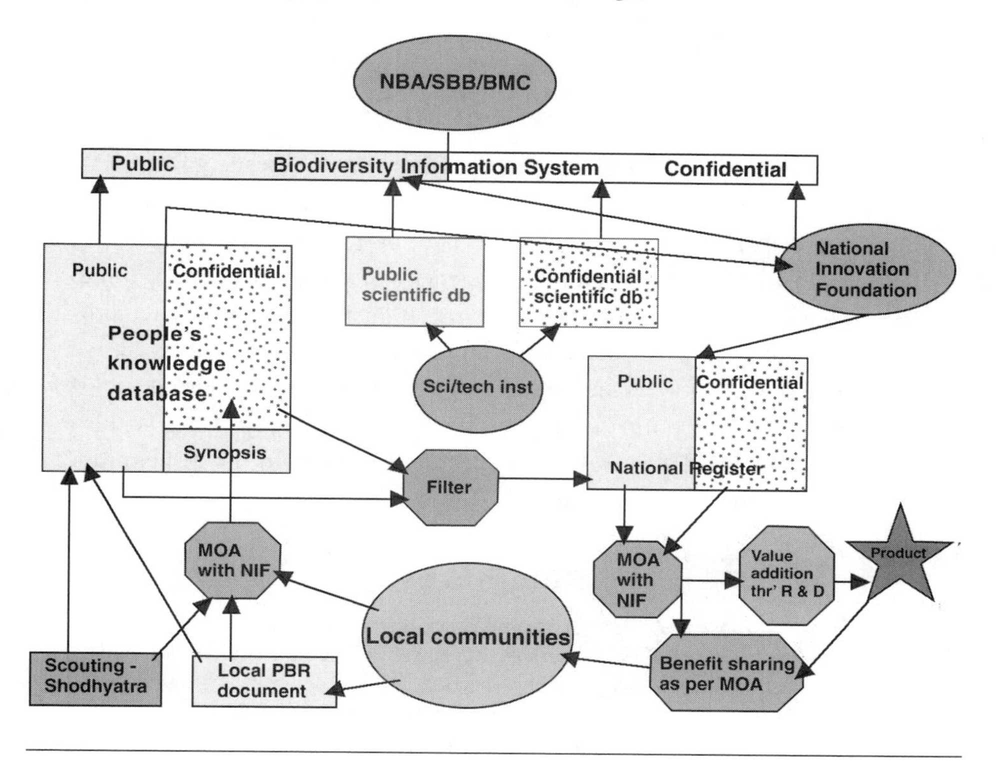

mutually agreeable memorandum. The appendix to this chapter lists the main clauses of this memorandum, which was signed by the knowledge providers and NIF on June 14, 2004, with full concurrence of the village council.

The example of this memorandum of agreement could be a very useful first step in tackling the significant challenge of bridging local and global scale, and folk and modern scientific knowledge. However, a number of issues still need to be addressed. NIF has to develop a good system of links with government, academic, and commercial research and development agencies to help add value to such knowledge. It also has to ensure that the confidentiality of the knowledge in its repository is not violated during the process of collecting and storing the knowledge elements. Moreover, NBA has to decide how it will organize a countrywide BIS, including the mechanisms for maintaining the confidentiality of, while at the same time promoting value addition to, the knowledge flowing from village councils through SBBs to the national level. NBA must also decide on the possible role of NIF in this process. These and many other challenges will need to be addressed in the days ahead.

Acknowledgments

We are thankful to the Ministry of Environment and Forests, government of India, for long-term and flexible financial support as well as for involving the Indian Institute of Science in a pilot project on implementing the Biological Diversity Act. Several other agencies of the government of India, in particular, the Council for Scientific and Industrial Research, have provided vital support to the National Innovation Foundation. Above all, we appreciate the vital inputs and collaboration of traditional knowledge holders from all over the country, and especially from the village of Mala. We also acknowledge financial as well as intellectual inputs from the Millennium Ecosystem Assessment program.

References

Gupta, A. 1999. Conserving biodiversity and rewarding associated knowledge and innovation systems: Honey Bee perspective. Invited paper for the First Commonwealth Science Forum—Access, Bioprospecting, Intellectual Property Rights and Benefit Sharing and the Commonwealth, Goa, India, September 23–25.

Gupta, S. K. 2000. Rustom Jal Vakil (1911–1974)—Father of modern cardiology. A profile. *Indian Academy of Clinical Medicine* 1 (2, July/September).

Manilal, K. S. 1980. *Hortus Malabaricus with annotations and modern botanical nomenclature.* 12 vols. Thiruvananthapuram, India: University of Kerala.

National Innovation Foundation. 2002. Value addition to local Kani tribal knowledge: Patenting, licensing and benefit sharing. A case study based on the data collected from Kani tribe, Kerala, India. W.P.No.2002-08-02 (August).

Appendix

Salient Features of Memorandum of Agreement (MoA)

1. MoA between individual knowledge holder/community with NIF to include information in PKD and also possibly in the National Register being prepared by NIF. This will help knowledge holder/community to retain the claim and confidentiality, if needed, over the knowledge deposited with NIF without changing right of knowledge holder/community over it. This does not mean that this traditional knowledge or innovation or practice may not

have been reported by some third party already or may not be reported directly later or may not already have been put in public domain.

2. MoA highlights need of differentiating between information already in public domain/documented without the consent of the knowledge holder/s and the documentation with mutual agreements such as MoA/PIC.
3. NIF is engaged in scouting, documenting, augmenting, and adding value to the innovations and traditional knowledge of the innovators at the grassroots level. NIF is mandated to develop a National Register of traditional knowledge and contemporary unaided grassroots innovations. NIF is also engaged in strengthening R&D linkages between the scientific institutions and grassroots innovators and traditional knowledge holders so as to promote commercial and non-commercial applications of grassroots innovations and traditional knowledge.
4. NIF also wishes to enter into an agreement with the traditional knowledge holder/community so as to add value, wherever possible, to the people's knowledge, innovations and practices of both contemporary and traditional origin and disseminate the same, protecting *inter-alia* the intellectual property rights of the knowledge holders as applicable in each case and ensuring equitable share of benefits wherever applicable.
5. Now therefore both the parties hereto agree as follows:
   a) That the traditional knowledge holder will provide the complete information/particulars to NIF in order to enter the traditional knowledge in its data base (PKD) and if possible in the National Register. NIF may also consider the traditional knowledge to be included in the list for the award in the next and subsequent biennial competitions.
   b) That the traditional knowledge holder/community has agreed for publishing indicative information of traditional knowledge along with contact address on the internet/Honey Bee magazine or any other media with the precaution so that their detailed traditional knowledge does not become public.
   c) That the traditional knowledge holder/community has agreed to share the traditional knowledge with the third party(s) on exclusive and /or non exclusive basis only if the written consent from traditional knowledge holder(s)/community for sufficient amount of money is received in return as per the milestones of value addition and/or commercialization where applicable. However the traditional knowledge holder(s)/commu-

nity can share the traditional knowledge for individual use and/or for further R&D in order to add value to it.

d) That the traditional knowledge holder/community has agreed to allow NIF to use the information for product research and development purpose so long as the intellectual property rights are intact/protected and traditional knowledge holder(s) is/are going to receive the benefit out of it. NIF will take care that in any circumstances, the confidentiality of the knowledge is maintained by research team involved in the product research and development process.

e) That NIF will add the information/particulars, pertaining to a specific traditional knowledge to People's knowledge database and/or National Register if found suitable. The information can be made available to a third party only with informed written consent of the traditional knowledge holders(s) (or in case he/she has expired, his/her legal heirs) and on the terms and conditions including benefit sharing indicated by the traditional knowledge holder/community.

f) That the benefits, arising from the possible commercialization of the traditional knowledge being improved by NIF on the basis of the basic information provided by traditional knowledge holder/community, will be shared among various stakeholders (including other communities providing same or similar information, third party researchers/business plan developers) as per the terms and conditions agreed upon by the concerned innovator(s)/traditional knowledge holders in consultation with NIF.

g) NIF can facilitate IPR in cases where applicable.

h) That in case of substantial improvement being done by the scientist(s) contracted by NIF, the concerned scientist(s) may be named as the co-inventor and a part of the benefit may be shared with him/her as well as other stakeholders such as the institutions like GIAN, NIF or their sister institutions, for meeting institutional overheads or for conservation of nature or community development or innovation fund for helping other communities or innovators etc. as per the mutual consent of the traditional knowledge holder/community and the concerned person/s and NIF.

i) That in the case of the publication of the outcome of the research and development the prior informed consent will be taken from the traditional knowledge holder(s)/community and in the publications due credit will be shared with the traditional knowledge holder(s)/community.

6. That both parties shall indemnify, defend, protect and hold harmless each other and its respective successors in case any of the party fails to discharge its obligations.
7. The MoA shall remain in force for a period of TWO years, and can be renewed for two additional terms of two years each that is for six years after which it will be reviewed. Review can take place earlier also through mutual consent. It is however assured that the confidentiality of the knowledge deposited with NIF on conditions specified will be respected in perpetuity unless otherwise agreed upon in writing by the knowledge holder(s)/community regardless of this process of review. It is possible that knowledge provided by a particular individual/community may have been communicated by another individual(s)/community(ies) directly to NIF or may already exist in public domain due to prior documentation by third party. In such cases NIF may share such knowledge as per the existing conditions but without sourcing the community which has provided knowledge in PBR unless so authorized.
8. That all disputes arising out of this agreement shall be settled through conciliation by a mutually agreed person, and shall be governed by the provisions of the Arbitration and Conciliation Act 1996. The place of conciliation shall be at Ahmedabad, Gujarat, India.

Signed by Traditional knowledge holder, NIF representative and BMC Chairperson/Panchayat Secretary.

CHAPTER 14

Barriers to Local-level Ecosystem Assessment and Participatory Management in Brazil

CRISTIANA S. SEIXAS

One of the Millennium Ecosystem Assessment (MA) challenges is to "develop procedures that can integrate local knowledge with data collected at the regional or global level and produce information that is salient, credible, and politically legitimate to the decision makers that are a major audience for the results of the MA" (Millennium Ecosystem Assessment 2004). This chapter aims to contribute to such effort by capturing lessons from participatory local-level ecosystem assessment and resource management in Brazil.

Participatory approaches in natural resource management have the potential to bridge knowledge systems (e.g., local ecological knowledge and scientific knowledge), enable knowledge flow across scales, empower local people, speed up technological adaptation, enhance human capital, and increase adherence to resource management goals (Chambers 1991; McAllister 1999). Since the Rio 92 Conference, the discourse of participatory management has been incorporated into several government policy agendas, particularly in high-biodiverse developing countries with a history of centralized (top-down) natural resource management.

In Brazil during the military regime (from the mid-1960s to the mid-1980s), environmental conservation policies were based on a command-and-control approach, with no attention paid to local knowledge and needs, leading more often than not to ecosystem degradation (Fearnside 1979, 1987; Moran 1983; Hecht and Cockburn 1989; Becker 2001)—the pathological resources management (Gunderson, Holling, and Light 1995). Since 1990, when the first extractive reserve was created by the federal government in the Amazon, the Brazilian legislation has contained a legal (formal) mechanism to promote participatory management and, hence, to bridge local and scientific knowledge.[1] The cross-scale institutional arrangement in this case is a formal comanagement—a shared responsibility between government, users, and other stakeholders in resource management.[2]

Interestingly enough, informal comanagement arrangements—in which government approves area-specific regulations based on local demand and local ecological knowledge—already existed before 1990; for instance, Seixas and Berkes (2003) documented informal comanagement occurring at the Lagoa de Ibiraquera since 1981. All these formal and informal arrangements created a space for users and decision makers from different levels to share their needs, concerns, and knowledge about the resource conditions in order to better understand management problems and improve management regulations.

Over the years, these formal and informal comanagement arrangements started to "replicate" throughout the country, without taking into account sociocultural differences among localities.[3] Moreover, most of the arrangements were, in fact, state initiated (sometimes with support from nongovernmental organizations [NGOs] or research units) and state run, with local users having limited real input (Sick 2002). In light of this situation, several questions are raised here: To what extent are policy makers prepared to accept local knowledge as a credible knowledge system that may complement scientific knowledge? To what extent are local resource users (who are used to paternalistic, top-down decision making) prepared to engage in ecosystem assessment and management? To what extent are fieldworkers (government and NGO staff, including science-trained researchers) trained to mediate the flow of knowledge between bureaucrats and resource users or to accept different understandings of ecosystem dynamics? We see a huge gap between theory and praxis in conducting participatory research and management in the field and in combining local and scientific knowledge across political levels for ecosystem assessment.

Thus the objective of this chapter is to identify some of the driving forces that impede local-level ecosystem assessment and participatory management in Brazil. To our knowledge, local-level ecosystem assessment has only recently (since the late 1990s) started in Brazil, and few reports are available about it (e.g., NMD 2004). Hence, we focus our attention on cases of participatory management, which often include participatory research, in order to bring out some important issues that may be broadened to assessment activities, particularly concerning bridging different knowledge systems and information across scale. For this purpose, we analyze four case studies of participatory fisheries management, based on government-, NGO-, or research-driven initiatives in different regions of Brazil.

Methods

This chapter analyzes four cases studies of participatory fisheries management from different regions of Brazil (table 14.1; see also the online appendix at http://www.islandpress.org/bridging_scales), none of which is an MA case. The first three cases were extracted from the literature. The fourth—also a Brazilian initiative of conducting local-level ecosystem assessment[4]—is based on project reports and joint team project experiences in which we were partially involved. All analyses presented in this chapter are based on our own interpretation of the publications and may not totally reflect the authors' opinions.

We chose these cases because they are examples of current or past fisheries comanagement in Brazil, which has been well documented by the scientific literature. Future research could expand this sample to include other also quite interesting fisheries comanagement cases, such as the "fishing accords" at the Lower Amazon River (Castro 2000; Castro and McGrath 2003) and the fisheries management at the Mamirauá Sustainable Development Reserve (Queiroz and Crampton 1999).

Case Studies Background

Historically, three of the study cases occurring in natural ecosystems (Extractive Reserve, Lagoa dos Patos, and Lagoa de Ibiraquera) experienced successful community-based resources management (CBRM; i.e., resource abundance, high catches, and few user group conflicts) until the early 1960s (Seixas and Berkes 2003; Silva 2004; Kalikoski and Vasconcellos 2005). In 1967, the Federal Fisheries Agency (SUDEPE) was created (later replaced by the Brazilian Environmental Agency, or IBAMA, in 1989) as part of the national policy.[5] Between 1967 and the establishment of comanagement arrangements (respectively in 1997, 1996, and 2002), government centralized management was the norm in these localities and locally devised rules were no longer respected. An exception was the advisory comanagement agreements between fishers and

Table 14.1
Case studies in participatory fisheries management in Brazil

Case Study (region)	Management Arrangement (establishment date)	Source of Information
Ceará Reservoir Fisheries Project (northeast region)	Informal comanagement between users and federal government (supported by international development agency) (1989/1990)	Christensen et al. (1995) Barbosa and Hartmann (1997) Hartmann and Campelo (1998)
Maritime Extractive Reserve in Arraial do Cabo (southeast region)	Formal comanagement between users and federal government (1997)	Lobão (2000) Silva (2004)
Forum Lagoa [lagoon] dos Patos (south region)	Multistakeholder body (forum) (1996)	Reis and D'Incao (2000) D'Incao and Reis (2002) Kalikoski, Vasconcellos, and Lavkulich (2002) Kalikoski and Vasconcellos (2005)
Lagoa de Ibiraquera Project (south region)	Multistakeholder body (forum) (2002) Ibiraquera Local Ecosystem Assessment Project (2001)	NMD (2004) Freitas (forthcoming)

SUDEPE/IBAMA at the Lagoa de Ibiraquera during the 1980s and early 1990s (Seixas and Berkes 2003).[6] The fourth case study, the Ceará Reservoir fisheries system, experienced government centralized top-down management from the time the reservoir was built until the Reservoir Fisheries Project was implemented (Barbosa and Hartmann 1997).

All of the initiatives described here aim to promote sustainable fisheries through participatory management. The Ibiraquera project is the only one still in the early stages of participatory management (i.e., ecosystem assessment, capacity building, and community organization through establishment of a local Agenda 21 forum). The other three initiatives have advanced into fisheries management and were able to influence government to pass new fisheries regulations for their localities.

Results and Discussion

Barriers to Participatory Research and Management

Several factors impede the full success of participatory research and management in Brazil. Below we present and discuss those found in at least three of the four cases analyzed here (table 14.2).

Barriers to User Participation

The degree of fisher involvement in the several stages of participatory management—environmental assessment (data gathering, data analysis), planning (decision making), implementation, monitoring (including enforcement), and evaluation—varies significantly from case to case. In the three cases where participatory research was reported (Ceará Reservoir project, Forum Lagoa dos Patos, and Ibiraquera project), fishers were usually a source of information or helpers in collecting data and samples but, according to our understanding, were never involved in data analyses, which were carried out by outside researchers. Nevertheless, results from data analysis were discussed with fishers in fishing meetings at the community level and supported fishers to formulate management recommendations in at least two cases (W. D. Hartmann, personal communication, 2004; P. F. Vieira, personal communication, 2004).

Barriers to user involvement in participatory research and management were, in general, related to a history of socioeconomic and cultural marginalization of artisanal fishers and the culture of patron-client relations established in Brazil. Much prejudice still exists against fishers' knowledge and their perceived "low" cultural and literacy level; fisher participation, although advocated by many, is in fact "undermined and sabotaged at many levels and by many organizations" (Barbosa and Hartmann 1997, 442).

Misrepresentation of fishers within their organizations and in the decision-making process hinders the potential of local knowledge use in decision making. Quite often, long-established fisher organizations are controlled by a local elite that neither represents the interests of most artisanal fishers nor holds their knowledge.[7] This misrepresentation may reflect a lack of organizational skills of fishing communities, which have experienced many social and economic influences during the past four decades (e.g., exposure to new values brought by outsiders, modernization of fishing gear and transportation systems, opening to a market-oriented society, and change from subsistence to

Table 14.2

Barriers to participatory research and management. The barriers to participatory research and management shown here were present in all four cases analyzed (with the exceptions noted).

Barriers to User Participation

- Socioeconomic and cultural marginalization of artisanal fishers
- Culture of patron-client relations and corruption
- Prejudice against user knowledge and literacy level by researchers and decision makers (except Ibiraquera project)
- Misrepresentation of fishers within their associations and in decision making
- Physical and economic threats to those involved in assessment and enforcement (except Forum Lagoa dos Patos)
- Existing conflicts and hierarchies (except Forum Lagoa dos Patos)

Government-related Barriers

- Lack of government support to or recognition of comanagement institutions
- Ambivalent support from government representatives
- User lack of trust of government agencies with a stake in the participatory management
- Ineffective enforcement by government
- Conflicting government policies and agendas (all levels) (except Ceará Reservoir project)

Governance Challenges

- Low-level comanagement: decision making not totally shared; government holding the last word
- Lack of a clear property rights system in the area
- Lack of effective government presence
- Lack of commitment and support from all stakeholders, particularly government agencies
- Lack of capacity (funds, training, and experience) from different partners (except Ceará Reservoir project)

commercial fisheries). These influences have in turn produced a breakdown of their traditional management system.[8]

Previous existing conflicts among stakeholder groups that have not been properly addressed by the initiative also hamper user participation in resource management and, consequently, the full potential of bringing together different understandings of ecosystem and resource conditions. Some community members involved in ecosystem assessment, planning, and rule enforcement have been physically, emotionally, or economically threatened by rule transgressors.

Government-related Barriers to Participatory Management

Theoretically, government at different levels should play an important role in facilitating and enabling cross-scale participatory management, in particular comanagement (Pomeroy and Berkes 1997). Nevertheless, in all four initiatives, some government agencies (different levels and sectors) with a stake in the management process (but not the major ones involved in it) demonstrate little or no support to or recognition of local comanagement institutions. This situation involves two factors in particular. First, a high degree of multiplicity and fragmentation of government exists at all levels. For instance, within the same government agency, two offices may have distinct management agendas (e.g., IBAMA's enforcers and regulation decision makers). Also, power disputes and conflictive agendas between government agencies from different sectors and political levels are quite common. The second factor is that support by government staff to participatory management depends more on staff members' own beliefs about CBRM and not so much on the organization's agenda (ambivalent support) (Barbosa and Hartmann 1997). Hence, conservative representatives of government agencies (who are used to top-down management) tend to hinder the participatory management process.

Another problem that sometimes hampers the participatory process is the involvement of government agencies related to environmental control and enforcement within the new cross-scale management institution. This may transmit an inaccurate image of the new management institution as another enforcement organization, thereby repelling user involvement in the process because of users' lack of trust in such enforcement agencies, which they see as corruptive and inefficient. In fact, in all cases, ineffective enforcement by government agencies (e.g., lack of resources and personnel, and unprepared or corruptive agents) at local and regional levels was a major problem hampering resource management.

Governance Challenges

Many governance-related problems were observed in these four cases of participatory management. Of particular interest are those related to (1) decision-making power, (2) level of decision making and use of local knowledge, (3) share of responsibility, and (4) institutional capacity in conducting participatory management.

Despite the fact that all four cases offer a democratic space through which fishers can express their values and knowledge, the final step of the decision-making process does not remain in the fishers' hands. In all different forms of comanagement (formal and informal arrangements), locally devised rules need to be sanctioned by IBAMA at the federal level to be enforced by government agencies. In other words, these are low-level comanagement arrangements in which decision-making power is not totally shared. The situation was poorer in the case of Forum Lagoa dos Patos, where relatively few fishers either were consulted or participated in the decision-making process at the local level. The fishing community was merely informed of decisions; hence, rule compliance became quite low because the resource users did not perceive the rules as legitimate—a fact well discussed in the common property literature (Ostrom 1990; Ostrom et al. 2002).

Two problems were identified concerning the level of decision making and the use of local knowledge in the cases studies. First, formalizing locally devised rules at a higher political level (national in all cases) increased the rules' legal status but at the same time decreased flexibility for rule changes (i.e., hindering rapid feedback mechanisms for rule adaptation according to local resource dynamics and climatic conditions). On the other hand, frequent (yearly) rule changes (adaptive management) at the Ceará Reservoir project weakened management impact because it generated confusion and insufficient time to evaluate the effects of management rules on fishery (Hartmann and Campelo 1998).

Second, government staff and researchers tend to prefer a few generally applicable and easily controllable rules to reduce transactional costs (i.e., the "one-size-fits-all" syndrome [Berkes 2003]); fishers, in contrast, desire specific rules for each locality within a large ecosystem, resulting in many different rules for this large ecosystem. This is clearly a case of institutional misfit (Folke et al. 1997; Brown 2003) in resource management, especially in a very large heterogeneous ecosystem, as in the case of Lagoa dos Patos (D'Incao and Reis 2002).

The question of responsibility over resource management is key. Comanagement is, theoretically, a way for government to share responsibility with users. Nevertheless, after decades of command-and-control top-down fisheries management, many fishers have shown that they are not used to taking (or willing to take) responsibilities for resource management. Moreover, communities are demanding better, more effective actions and support from government for resource management, not less. For instance, communities want government agencies to enforce locally devised rules sanctioned by IBAMA. They also want recognition of local ecological knowledge and support to comanagement from

Table 14.3

Use of local and scientific knowledge in participatory management

Use of Local and Scientific Knowledge	Ceará Reservoir Project	Extractive Reserve	Forum Lagoa dos Patos	Ibiraquera Project
Influences of scientific knowledge				
Environmental education about local ecological processes	X			X
Research information feedback to fishing communities	X		X	X
Researchers advising users when local knowledge is not sufficient		X		X
Local decisions largely influenced by scientific/technical knowledge		X (likely)	X	
Influences of local knowledge				
Participatory research	X		X	X
User participation in policy making at higher levels (conference, watershed committee)	X			X
Local decisions partially based on previous informal management system		X		

government agencies directly or indirectly involved in the process. This is by no means a return to command-and-control management.

Participatory research and management requires an interdisciplinary approach (i.e., bridging different disciplines), and in Brazil the higher-level education is still very disciplinary. Thus, because participatory approach is relatively new in Brazil, well-trained people (government and NGO staff and researchers) able to conduct the process are in short supply. Another problem is that government funds to carry out participatory management are insufficient, despite the extensive recent advocacy toward such an approach.

Knowledge Flow across Scale

Institutions for Combining Local and Scientific Knowledge

Each initiative's effort toward sharing and combining technical/scientific and local knowledge systems varied largely (table 14.3). The Ibiraquera project's first proposal was to carry out a participatory local-level ecosystem assessment that documents local knowledge for further integration with scientific knowledge. Later, the Ibiraquera project focused on initiating a local Agenda 21 forum for future resources comanagement and ecodevelopment. The other three initiatives focused mainly on improving resource management; sharing and combining knowledge systems were a means toward this end for some participants.

Mechanisms used to share technical/scientific information with local people were found in all cases. These included environmental education about local ecological processes; research information feedback to fishing communities; and researchers advising users for decision making when local knowledge was insufficient (see table 14.3). Environmental education was conducted through courses and seminars given to fishers, local schoolteachers, and the community in general. Pure scientific or participatory research findings were presented to fishers during meetings to discuss resource management, and sometimes to propose management regulations. University researchers speaking at some of these meetings advised local users when local knowledge was not sufficient to make a decision (e.g., at the Extractive Reserve and the Ibiraquera projects). Clear mechanisms that enable use and integration of local knowledge in resource management were more difficult to find.

Despite the existence of formal or informal comanagement arrangements, it is difficult to measure how much local knowledge has been used in decision making in each case, particularly because this analysis uses secondary data. In the Ceará Reservoir project, until 1997, all community-proposed management measures were ratified by IBAMA, becoming fisheries regulations (Barbosa and Hartmann 1997). However, based on the information given by Barbosa and Hartmann (1997) that (1) "there [were] no local traditions of [fisheries] resource use and management" in the area because of the reservoir's recent origin and (2) environmental awareness training about local ecosystem process was provided to fishers, it seems that local decisions were quite influenced by technical knowledge.

Concerning the Extractive Reserve, according to Silva (2004), the reserve management plan was in part based on a long-standing informal arrangement of resource access (codified in 1921 by the old fisher organization). However, local decision has also been influenced by the Scientific Technical Council formed by university researchers, which is linked to the Associação da Reserva Extrativista Marinha de Arraial do Cabo (an association responsible for comanaging the reserve with the government) (Lobão 2000).

At the Forum Lagoa dos Patos, decision making has been largely influenced by scientific and technical knowledge (Reis and D'Incao 2000; Kalikoski, Vasconcellos, and Lavkulich 2002), while user knowledge has been overlooked (Kalikoski and Vasconcellos 2005). Nevertheless, some initial effort toward participatory research has happened (D'Incao and Reis 2002), and the forum has triggered more management-oriented research by university teams to deal with questions raised by the forum (Kalikoski, Vasconcellos, and Lavkulich 2002).

Even when a project's primary objective is to integrate knowledge systems, as for the Ibiraquera project, the distance between objectives and results is large. Until January 2004, three years after the project started, the Ibiraquera project had not been able to create a database integrating data from all research teams involved in the local-level participatory assessment (fisheries, aquatic invertebrates, birds, game and domestic animals, landscape, agriculture, water quality, health, socioeconomic-political-cultural issues); moreover, each research team has collected, analyzed, and documented its data separately. At the time of this writing, some results had been presented at meetings of working groups of the recently established Forum of the Lagoa de Ibiraquera (e.g., fisheries working group), but no overall summary of data had been presented to communities in a systematic method allowing for discussion, validation, and use by the communities—despite being anticipated in the project methodology.

In fact, the Forum Lagoa de Ibiraquera has been quite active in bridging local and scientific knowledge to lobby decision makers and in attempting to improve regulation enforcement and environmental policy (Freitas, forthcoming). The forum members have tried to influence decision making by inviting government agents (municipal, state, and federal) as guests to their meetings. Some forum members and community representatives have also participated in a regional fisheries conference intended to influence policy at state level. The involvement of local resource users in subregional management institutions (e.g., a watershed management committee) was also noted at the Ceará Reservoir Project (Barbosa and Hartmann 1997).

Impediments to Knowledge Flow across Scale

Table 14.4 presents impediments to knowledge flow across scale found in the four cases. In most fishing areas in Brazil except conservation areas, there exists no legal mechanism that compels government organizations to consult resource users for management decision making. Of the four cases analyzed here, only the Extractive Reserve provides such a mechanism—a formal comanagement arrangement in which a management plan has to be developed by a local organization of resources users to be later analyzed and approved by the government (IBAMA). In two other cases of informal comanagement arrangement, the Ceará Reservoir project and the Forum Lagoa dos Patos, government consultation with civil society and the use of local knowledge in resource management depend largely on the government staff's own beliefs about the value of local knowledge and the potentials of community-based management, and not so much on the organization's agenda (Hartmann and Campelo 1998).

In the fourth case, the Ibiraquera project, the situation is poorer compared to the other three cases. This is for three primary reasons: (1) because many government agencies at the municipal and state levels do not support the civil society initiative of establishing a forum to manage local resources (NMD 2004; Freitas, forthcoming); (2) because the municipal government did not accept a representative indicated by the forum on the Municipal Environmental Board (i.e., it created a barrier to knowledge flow) (NMD 2004); and (3) because government agencies claimed that before they take actions to reverse degradation processes, more scientific studies (which usually take a long time to complete) have to be carried out to prove local knowledge and perceptions about ecosystem degradations (NMD 2004).

In fact, many government agencies and even some researchers do not accept and value local knowledge, and some government staff members do not accept user rights for comanaging. For instance, a segment of the Brazilian Navy does not recognize fisher rights to comanage the Extractive Reserve; moreover, the Navy's research institute continues to carry out research within the reserve area without interacting with local fishers (Lobão 2000). Another example, despite some initial effort toward participatory research (D'Incao and Reis 2002), Kalikoski and Vasconcellos (2003) argue that exchange of knowledge between fishers and scientists has not been very intense and that fisher knowledge has not yet received the required attention by this forum despite its role in helping maintaining a productive and resilient fisheries system before the 1970s. Indeed, these authors point out that "illiteracy and socio-economic marginalization create low expectations of the management value of fishers' knowledge among scientists and decision makers" (p. 452)

Another limitation to knowledge flow relates to the lack of institutions to create an integrated coastal zone management plan for the Brazilian coast (Kalikoski, Vasconcellos, and Lavkulich 2002). Integrated coastal zone

Table 14.4

Impediments to knowledge flow across scale

Impediments to Knowledge Flow across Scale	Ceará Reservoir Project	Extractive Reserve	Forum Lagoa dos Patos	Ibiraquera Project
Some government staff not accepting and valuing local knowledge (prejudice)	X	X	X	X
Lack of legal mechanisms that compel government agencies to consult fishers	X		X	X
Some government agencies not accepting user rights for comanaging	X	X		X
Local knowledge use depending on government staff's own beliefs about potentials of community-based resource management	X		X	
Overall management process is still top-down based on conventional scientific approach			X	X
Lack of an integrated coastal zone management plan			X	X
Conflict between users and scientists about resource conditions			X	X
Limited participatory research and exchange of knowledge			X	
Misfit between institutions and ecosystems that hinders use of fisher knowledge in management			X	
Lack of funding for participatory, local-level ecosystem assessment				X

management has the potential to bring together resource users and government agents from different economic sectors, geographical scales, and political levels to exchange knowledge and experiences in order to develop and implement a management plan. Indeed, many problems affecting local resource management are external to the scale or sector being managed. For instance, many of the factors affecting the Lagoa dos Patos fisheries are related to the industrial fisheries on the coast outside the estuarine zone (Kalikoski, Vasconcellos, and Lavkulich 2002). At the Lagoa de Ibiraquera and the Extractive Reserve, local fisheries are also affected by human actions at a larger ecosystem scale (i.e., the coastal zone).

Related also to the issue of scale, there is often a misfit between management institutions governing a large ecosystem and the local characteristics of (and local knowledge about) its parts. For instance, fishing rules that may be appropriate for one area of a large ecosystem may not be so for another area, and the local knowledge held by a group of fishers for the first area may differ from the local knowledge held by another group of fishers for the second area. This fact may hinder fishers' stewardship of resources and the use of their knowledge in managing large ecosystems (Kalikoski and Vasconcellos 2005).

The Ibiraquera project was the only one of the four initiatives that clearly aimed to link (and integrate) information systems related to the conditions of resources and ecosystems at different scales. However, three years after the project started, the team had spent so much time searching for funding (the project was funded only after June 2003) and trying to coordinate team members that almost no effort had been made by then toward elaborating a complete database of the local assessment—not to mention integrating it to other government scientific information systems encompassing larger ecosystems.

Finally, despite the effort of certain people engaged in some comanagement arrangements in combining local and scientific knowledge for resource management, common understanding of the problems and agreement on measures still may not be reached. Conflicts among users, scientists, and decision makers over resource conditions may still occur; hence, stakeholders must craft mechanisms that facilitate conflict resolution and consensus building.

Challenges in Conducting Local-level Ecosystem Assessment and Participatory Research

Several challenges emerged in these initiatives when conducting participatory research. A major challenge is how to congregate and coordinate an interdisciplinary, transdisciplinary research team (i.e., researchers from different disciplines with different understanding and approaches to user participation in research and management) (NMD 2004; Lobão 2000). The task becomes even more difficult when considering the long periods of time required in participatory assessment and management—in many cases, researchers are students, which results in a rapid turnover of team members (NMD 2004).

For the Ibiraquera project—the only one focusing on local-level ecosystem assessment—a reflective analysis by team members shows that other major challenges include lack of research funding for participatory assessment; lack of an internal team assessment of the process of participatory appraisal; and communication problems during meetings (locals have difficulty understanding researchers' objectives and limitations) (Freitas, forthcoming). Other issues related to participatory research noted in some of these cases are fatigue of community members involved in community organization and research projects for long periods of time; the need for researchers and development agents to adapt to users' schedules and time availability and to spend very long periods of time in the field; and the pressures researchers receive from fishers for rapid research feedback (results) in order to change regulations more quickly.

Creating New Arenas for Bridging Knowledge through Cross-scale Institutional Management

All these initiatives have created new arenas for cross-scale institutional management, with the potential to bridge knowledge systems and perhaps compile information from ecosystem assessment at different scales. In particular, they have created a space for political inclusion of a working-class, traditionally socially excluded group—the fishers. They have given an opportunity for fishers to express their needs, knowledge, and concerns. What can be seen in all four cases is much learning-by-doing and exchange of knowledge and experience. Most, if not all, of the initiatives have built on existing experience from elsewhere. The Forum Lagoa dos Patos, for example, was established based on two successful experiences of community-based management in nearby lagoons, which were initiated by the same organization two years earlier (Reis and D'Incao 2000). The Ibiraquera project initially used a research method developed in India (NMD 2004). The Extractive Reserve initiative drew on the available government institutional framework, in which "extractive reserve" is one of the Brazilian categories of protected areas (Lobão 2000).

Within the learning and sharing experiences context, these initiatives also have served (or intend to serve) as a model for other projects in the same region or in another region of the country. For example, the Ceará Reservoir project was initiated as a pilot project in two reservoirs. Later, project activities extended to five reservoirs within the same watershed (Barbosa and Hartmann 1997). The same project has been considered as a model for similar endeavors by various organizations on state and regional levels. For instance, the experience of the Ceará Reservoir project in community empowerment and strengthening of citizenship—and particularly in promoting social learning, participatory democracy, discursive design of management, and comanagement—led the project staff to assist the state government in organizing community members for participation in a commission of reservoir users for integrated water resource management (Barbosa and Hartmann 1997; Hartmann and Campelo 1998).

At the Extractive Reserve, local fishers have shared experiences with fishers from other places intending to create new maritime extractive reserves (Lobão 2000). One goal of the NMD/UFSC research team coordinating the Ibiraquera project is to replicate the methodology at a large scale (that of a larger watershed) (NMD 2004).

Positive learning feedbacks from resource management have helped expand the actual arena of cross-scale institutional linkages. For example, the positive outcome of a legal dispute led by the Forum of Lagoa de Ibiraquera to close a shrimp farm at the lagoon has strengthened the forum's credibility among community members, government employees, and businesspeople as a space to discuss the lagoon's problems and search for solutions (Freitas, forthcoming). At the Forum Lagoa dos Patos, something similar has occurred in which participants "are developing the means to achieve a better internal organization to cope with the external influences" (Kalikoski, Vasconcellos, and Lavkulich 2002).

Conclusions

Policy makers' preparation to accept local knowledge as a credible knowledge system that may complement scientific knowledge varies largely. Acceptance of local knowledge seems to depend more on each policy maker's beliefs about the potential of CBRM than on the agenda of the person's organization. Of course, other cases may exist in which policy makers willing to promote CBRM are constrained by the agenda of their organizations—but no such situation was reported in any of the cases analyzed here.

Concerning the extent to which local resource users (who are used to paternalistic, top-down decision making) are prepared to engage in participatory research and management, this chapter shows that some users seem not yet prepared for such challenge. Much capacity building concerning community organization and empowerment is needed, in particular to overcome decades of socioeconomic marginalization and to find a way out of the patron-client culture in resource management. Capacity building to engage in participatory research and management is needed not only by resource users but also by fieldworkers (government and NGO staff, including science-trained researchers). Most of the initiatives have demonstrated a lack of qualified personnel who are able to accept a different knowledge system (i.e., a different understanding of resource condition, ecosystem dynamics, and management problems) and who are able to mediate conflicts and facilitate the flow of knowledge between bureaucrats and resource users.

The conflicting agendas and power disputes among many government agencies, and within some agencies, is another major constraint in implementing participatory cross-scale management (and thus in bridging different knowledge systems); they have no tradition for such an approach. In fact, all four cases have faced several degrees of management constraints because of lack of support from some government agencies at different political levels and economic sectors.

The role of each initiative in combining local and scientific knowledge to improve policy varied, but our overall impression after reading all the publications is that scientific and technical knowledge still plays a major role in decision making, despite the fact that the first round of decisions is made locally by resource users and civil society (i.e., before regulation proposals are submitted to federal government approval).

In the end, despite the advocacy from government agencies and individual efforts to promote participatory management, decision making is still centralized at the federal level. Moreover, in some other Brazilian experiences, the participatory management "slogan" has been used to engage resource users in management in order to legitimate assessments based on scientific knowledge or a decision-making process, which is in fact manipulated to achieve the goals of government or of more powerful stakeholders (R. R. Freitas, personal communication, 2004).

In theory, both formal and informal comanagement arrangements may enable knowledge flow (both local and scientific) across levels. In practice, a lack of mechanisms exists for integrating the knowledge base and management efforts at the local level with those at larger scales. The challenge is to create more multilevel institutions to help understand ecosystem dynamics at different scales and how ecosystem management at one level affects management at lower and higher levels.

Finally, although all of these experiences have created new arenas for bridging knowledge through cross-scale institutional linkages, much remains to be done to fit management institutions with one another and with the scale of the management problems they are addressing.

Despite of all the challenges highlighted above, however, all four cases have positive aspects that contributed to improving participatory fisheries management in Brazil and to bridging epistemologies and scales in resource management and ecosystem assessment. For instance, all four cases promoted the involvement of resource users in decision making to an extent not seen before in those areas. At least two cases (the Ibiraquera project and the Ceará Reservoir project) have contributed to fishers' empowerment and enhanced local human capital. In addition, at least one initiative (the Ibiraquera project), and probably the other three, has tried to influence decision making by inviting government agents (municipal, state, federal) as guests to its meetings. Moreover, three initiatives (Forum Lagoa dos Patos, Extractive Reserve, and Ceará Reservoir project) were able to influence federal government to pass fisheries regulations specifically for their localities. These regulations very likely resulted from efforts to bridge epistemologies (local and scientific knowledge) and information assessed at different scales.

References

Allegretti, M. H. 1990. Extractive reserves: An alternative for reconciling development and environmental conservation in Amazonia. In *Alternatives to deforestation: Steps towards sustainable use of the Amazon rainforest,* ed. A. B. Anderson, 253–64. New York: Columbia University Press.

Barbosa, F. I., and W. D. Hartmann. 1997. Participatory management of reservoir fisheries in North-Eastern Brazil. In *Inland fishery enhancements,* ed. T. Petr, 427–445. FAO Fisheries Technical Paper 374. Rome.

Becker, B. K. 2001. Amazonian frontiers at the beginning of the 21st century. In *Human dimensions of global environmental change—Brazilian perspectives,* ed. D. J. Hogan and M. T. Tolmasquim, 301–23. Rio de Janeiro: Academia Brasileira de Ciências.

Begossi, A. 1998. Resilience and neo-traditional populations: The caiçaras (Atlantic Forest) and cablocos (Amazon, Brazil). In *Linking social and ecological systems: Management practices and social mechanisms for building resilience,* ed. F. Berkes and C. Folke, 129–57. Cambridge: Cambridge University Press.

Berkes, F. 2003. Alternatives to conventional management: Lessons from small-scale fisheries. *Environments* 31 (1): 5–19.

Brown, K. 2003. Integrating conservation and development: A case of institutional misfit. *Frontiers in Ecology and the Environment* 1 (9): 479–87.

Brown, K., and S. Rosendo. 2000. The institutional architecture of extractive reserves in Rondônia, Brazil. *Geographical Journal* 166 (1): 35–48.

Castro, F. 2000. Fishing accords: The political ecology of fishing intensification in the Amazon. PhD diss., Indiana University, Bloomington.

Castro, F., and D. G. McGrath. 2003. Moving toward sustainability in the local management of floodplain lake fisheries in the Brazilian Amazon. *Human Organization* 62 (2): 123–33.

Chambers, R. 1991. Shortcut and participatory methods for gaining social information for projects. In *Putting people first: Social variables in rural development,* ed. M. M. Cernea, 515–37. 2nd ed. Oxford: Oxford University Press.

Christensen, M., W.J.M. Soares, F.C.B. Silva, and G.M.L. Barros. 1995. Participatory management of a reservoir fishery in Northeastern Brazil. *Naga* 18 (2): 7–9.

Cordell, J., and M. A. McKean. 1992. Sea tenure in Bahia, Brazil. In *Making the commons work: Theory, practice, and policy,* ed. D. W. Bromley, 183–205. San Francisco: ICS (Institute for Contemporary Studies) Press.

Diegues, A. C. 1983. *Pescadores, Camponeses e Trabalhadores do Mar* [Fishers, peasants, and sea workers]. Rio de Janeiro: Editora Ática.

D'Incao, F., and E. G. Reis. 2002. Community-based management and technical advice in Patos Lagoon estuary (Brazil). *Ocean and Coastal Management* 45:531–39.

Fearnside, P. M. 1979. The development of the Amazon rain forest: Priority problem for the formulation of guidelines. *Interciencia* 4:338–43.

———. 1987. Deforestation and international economic development projects in Brazilian Amazonia. *Conservation Biology* 1 (3): 214–21.

———. 1989. Extractive reserves in Brazilian Amazonia: An opportunity to maintain tropical rain forest under sustainable use. *BioScience* 39 (6): 387–93.

Folke, C., L. Pritchard Jr., F. Berkes, C. Colding, and U. Svedin. 1997. *The problem of fit between ecosystems and institutions.* Beijer Discussion Paper Series No. 108. Stockholm: Beijer International Institute of Ecological Economics, The Royal Swedish Academy of Science.

Freitas, R. R. Forthcoming. Manejo costero integrado y participativo: Breve descripción del proyecto de ecodesarrollo en la laguna de Ibiraquera, Santa Catarina (Brasil) [Integrated and participatory coastal management: A brief description of the ecodevelopment project at the Lagoa de Ibiraqurea, Santa Catarina (Brazil)].

Gadgil, M., K. Achar, A. Shetty, et al. 2000. *Participatory local level assessment of life support systems—A methodological manual.* Bangalore: Center for Ecological Science, Indian Institute of Science.

Gunderson, L. H., C. S. Holling, and S. S. Light, eds. 1995. *Barriers and bridges to the renewal of ecosystems and institutions.* New York: Columbia University Press.

Hartmann, W. D., and C.M.F. Campelo. 1998. Ambivalent enforcers: Rules and conflicts in the co-management of Brazilian reservoir fisheries. Paper presented at the Seventh Conference of the International Association for the Study of Common Property—Crossing Boundaries, June 10–14, Vancouver, British Columbia.

Hecht, S. B., and A. Cockburn. 1989. *The fate of the forest: Developers, destroyers and defenders of the Amazon.* New York: Verso.

Kalikoski, D. C., and M. Vasconcellos. 2003. Fishers' knowledge role in the co-management of artisanal fisheries in the estuary of Patos Lagoon, southern Brazil. *Putting fishers' knowledge to work,* ed. N. Haggan, C. Brignall, and L. Wood. Conference proceedings, University of British Columbia, August 27–30, 2001. Fisheries Centre Research Report 11 (1): 445–55.

———. 2005. The role of fishers' knowledge in the co-management of small-scale fisheries in the estuary of Patos Lagoon, southern Brazil. Chap. 14 in *Fishers' knowledge in fisheries science and management,* ed. N. Haggan, B. Neis, and I. G. Baird. Oxford: Blackwell Science / UNESCO (United Nations Educational, Scientific and Cultural Organization).

Kalikoski, D. C., M. Vasconcellos, and L. Lavkulich. 2002. Fitting institutions to ecosystems: The case of artisanal fisheries management in the estuary of Patos Lagoon. *Marine Policy* 26:179–96.

Lobão, R.J.S. 2000. Reservas Extrativistas Marinhas: Uma Reforma Agrária do Mar? [Marine Extractive Reserves: A land reform of the sea?]. Master's thesis, Fluminense Federal University, Niterói, Brazil.

McAllister, K. 1999. *Understanding participation: Monitoring and evaluating process, outputs and outcomes.* IDRC Series. Ottawa, Ontario: International Development Research Centre (IDRC).

Millennium Ecosystem Assessment. 2004. Millennium Ecosystem Assessment Web site. http://www.millenniumassessment.org/ (accessed March 2004).

Moran, E. F., ed. 1983. *The dilemma of Amazonian development.* Boulder, CO: Westview.

NMD (Núcleo de Meio Ambiente e Desenvolvimento—Center for Environment and Development) at the Santa Catarina Federal University (USFC). 2004. Avaliação Local Participativa de Ecossistemas Litorâneos no Sul do Brasil: Projeto Piloto de Criação de uma Agenda 21 Local na Área da Lagoa de Ibiraquera, Municípios de Imbituba e Garopaba, Estado de Santa Catarina [Local participatory assessment of coastal ecosystems in South Brazil: A pilot project to create a local Agenda 21 at the Lagoa de Ibiraquera area]. Partial Project Report to CNPq. Florianópolis, Brazil: USFC.

Ostrom, E. 1990. *Governing the commons: The evolution of institutions for collective action.* Cambridge: Cambridge University Press.

Ostrom, E., T. Dietz, N. Dolšak, P. C. Stern, S. Stonich, and E. U. Weber, eds. 2002. *The drama of the commons.* Washington, DC: National Academy Press.

Pomeroy, R. S., and F. Berkes. 1997. Two to tango: The role of government in fisheries co-management. *Marine Policy* 21 (5): 465–80.

Queiroz, H. L., and W.G.R. Crampton, eds. 1999. *Estratégias para Manejo de Recursos Pesqueiros em Mamirauá.* Série Estudos de Mamirauá, V 5, Coleção Bases Científicas do Plano de Manejo de Mamirauá [Strategies for fisheries management in Mamirauá]. Brasilia: Sociedade Civil Mamirauá, MCT-CNPq, DF.

Reis, E. G., and F. D'Incao. 2000. The present status of artisanal fisheries of extreme Southern Brazil: An effort towards community-based management. *Ocean and Coastal Management* 43:585–95.

Seixas, C. S., and F. Berkes. 2003. Dynamics of social-ecological changes in a Lagoon fishery in Southern Brazil. In *Navigating social-ecological systems: Building resilience for complexity and change,* ed. F. Berkes, J. Colding, and C. Folke, 271–98. Cambridge: Cambridge University Press.

Sick, D. 2002. *Managing environmental process across boundaries: A review of the literature on institutions and resource management.* Minga Program Initiative. Ottawa, Ontario: International Development Research Centre.

Silva, P. P. 2004. From common property to co-management: Lessons from Brazil's first maritime extractive reserve. *Marine Police* 28: 419–28.

CHAPTER 15

Integrating Epistemologies through Scenarios

Elena Bennett and Monika Zurek

As Folke et al. (2002, 437) write: "The goal of sustainable development is to create and maintain prosperous social, economic, and ecological systems." These systems are intimately linked; however, our study of them is often discrete. We might study the ecology of a region with a model that largely ignores human impact on the ecosystem. We might study the people of the same system without recognizing the impact that the ecosystem can have on their interactions. When interlinked systems are studied in a discrete way, important dynamics, driving forces, and interactions that help explain the system may be overlooked. Understanding these complex systems requires combining the knowledge and perspectives from many different ways of knowing (Lubchenco 1998).

Recently, many scientists, policy makers, and others concerned about the state of the world have pointed to the increasing urgency of environmental problems (Ehrlich 1997; Millennium Ecosystem Assessment 2005a) and the poor state of our ability to overcome these challenges with disciplinary research (Kinzig et al. 2000; Lubchenco 1998).

Interdisciplinary research, and research that involves perspectives from inside and outside the academic sciences, can mobilize a wider range of understanding and sources of information (Berkes and Folke 1998; Olsson and Folke 2001). Such broader approaches are less likely to be brittle and therefore are more likely to succeed in the long term (Holling, Gunderson, and Ludwig 2002).

These types of approaches are expected to be a key source of feasible solutions to today's critically intertwined environmental problems.

Conventional science has therefore recently turned its attention to working across disciplinary boundaries to solve tough environmental problems (Kinzig 2001). It has also begun to look at ways of knowing that come from outside the academy to add new vision to resource management. The benefits of using this apparent synergy between traditional knowledge and a knowledge gap in Western understanding may result in better ecosystem management. For example, in western Ecuador, interactions between indigenous and scientific knowledge yielded collective action to preserve ecosystem services and biodiversity in a communally owned watershed (Becker and Ghimire 2003).

Although it is widely recognized that integration of many perspectives is needed to understand social-ecological systems, few practical methods for doing so exist. In this chapter, we consider not only various disciplines within the academy of conventional Western science but also the incorporation of local and traditional knowledge and information gained outside the academy. There are often critical disconnects in language, approach, bounding of the problem, and even paradigm among different epistemologies that make communication across this divide extremely difficult.

Each way of knowing basically amounts to a paradigm through which members understand the world (Mingers 2001). This paradigm includes notions of truth, rules of evidence, and standards of rigor. Knowledge is gathered and stored based on a particular collection of assumptions, theories, and methods for understanding the world. These assumptions, and even the conceptual structure of each paradigm, often remain hidden or unspoken, rarely surfacing to a conscious level. Integrative scenarios compel participants to discuss and challenge their assumptions with others who hold different beliefs, an important first step toward better integration.

Here, we present scenario development as a method for improving decision making about social-ecological systems and for building an understanding of these systems that is open to knowledge from many different ways of knowing. In other words, a set of scenarios can be, and often is, a single product that is the result of several worldviews and information from many different perspectives. We begin by briefly discussing different ways of knowing and by exploring the difficulties of integrating information from different epistemologies. We propose and explain the use of scenario development as a tool for integrating and synthesizing across epistemologies. The bulk of the chapter revolves around four examples of scenario exercises carried out as part of the Millennium Ecosystem Assessment (MA), each of which serves as an illustration of integrating across a different set of epistemologies.

In the first case study, the MA global scenarios provide an example of incorporating qualitative and quantitative information into scenarios. The second case study shows multiple academic disciplines coming together to build a set of scenarios for the Caribbean region. In the third example, academic scientists and local stakeholders work together to build scenarios. Finally, in the last case study, we highlight an example of traditional knowledge and academic science coming together to build scenarios.

The Difficulties of Integrating Epistemologies

Although many experts are talking about the importance of using multiple types of information and incorporating different paradigms in resource management, we struggle for methods to do so. Integrating knowledge from different sources can be hampered by differing methodologies, vocabularies, ways of assigning merit, and even worldviews.

Western scientific traditions have typically dealt with the mind-boggling complexity of systems by reducing the complexities to a manageable number of elements. Doing so necessarily means setting system boundaries. Disciplinary understanding affects how system boundaries are chosen. Differing time horizons of research, organizational structures, and institutional traditions (such as the means of giving credit for research) also complicate interdisciplinary collaboration. Traditional ecological knowledge (TEK) faces further difficulties because it may not be written and because the practitioners of TEK often do not interact with those gathering conventional scientific information about ecosystem management.

Additionally, issues of power may be problematic. In multidisciplinary or transdisciplinary projects, one paradigm may remain dominant and simply absorb bits of information from other paradigms. Using information without understanding the paradigm in which it originated may lead to overlooking important boundary conditions about how that information can be used. Finally, disregarding some paradigms may cause stakeholders to walk away from the table if they are not allowed to participate in defining the question.

Integrating Epistemologies through Scenario Development

Scenario planning was developed as a creative, systematic way to think about the future (Peterson, Cumming, and Carpenter 2003). Scenario planning has been used in the business community for decades (Schwartz 1996) and has recently become an important part of integrated assessment exercises (e.g., the Intergovernmental Panel on Climate Change and the MA). Scenarios are now increasingly explored by natural resource managers for exploring new management approaches (Bennett et al. 2003). The appeal of scenario building lies in the possibility of bringing many stakeholders and different viewpoints into the process. In fact, multiple perspectives are almost a necessity for a compelling scenario-building exercise. Stakeholder groups might include scientists of many disciplines, TEK practitioners, ecosystem managers, local stakeholders, policy makers at various geographical scales, and others. Needing to get all of these groups to agree on a set of scenarios makes the scenario-building process a useful tool for exploring differences in knowledge systems, learning how these differences might influence decision making, and considering ways to bridge them in the decision-making process.

What Are Scenarios?

The MA describes scenarios as "plausible alternative futures, each an example of what might happen under particular assumptions" (Millennium Ecosystem Assessment 2003). This definition highlights the MA's belief in using scenarios to challenge beliefs about the future. Scenarios are stories about the future, told in a set. Scenarios can be developed in many different ways. Based on previous scenario experiences, the MA has developed a process for building scenarios with participants who had little to no previous knowledge of scenario development (figure 15.1). These steps include considering long-term change in the study site, listing key drivers of current change, distinguishing those drivers with known trajectories from those whose future is uncertain, and finally, telling stories that allow uncertain drivers to unfold differently across the scenarios to examine the results.

Scenarios can be qualitative, quantitative, or both (Raskin 2005). Qualitative and quantitative scenario development techniques are often combined to produce a set of comprehensive narratives supported by a quantitative modeling exercise (e.g., the Intergovernmental Panel on Climate Change's Special

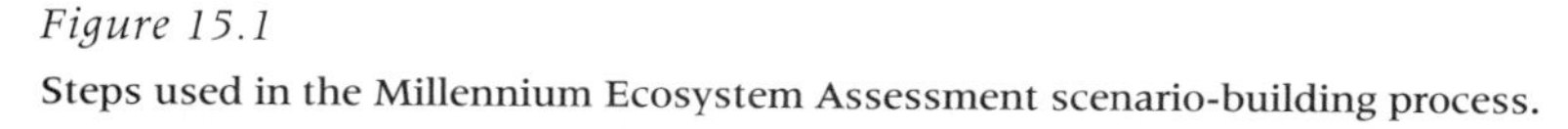

Figure 15.1
Steps used in the Millennium Ecosystem Assessment scenario-building process.

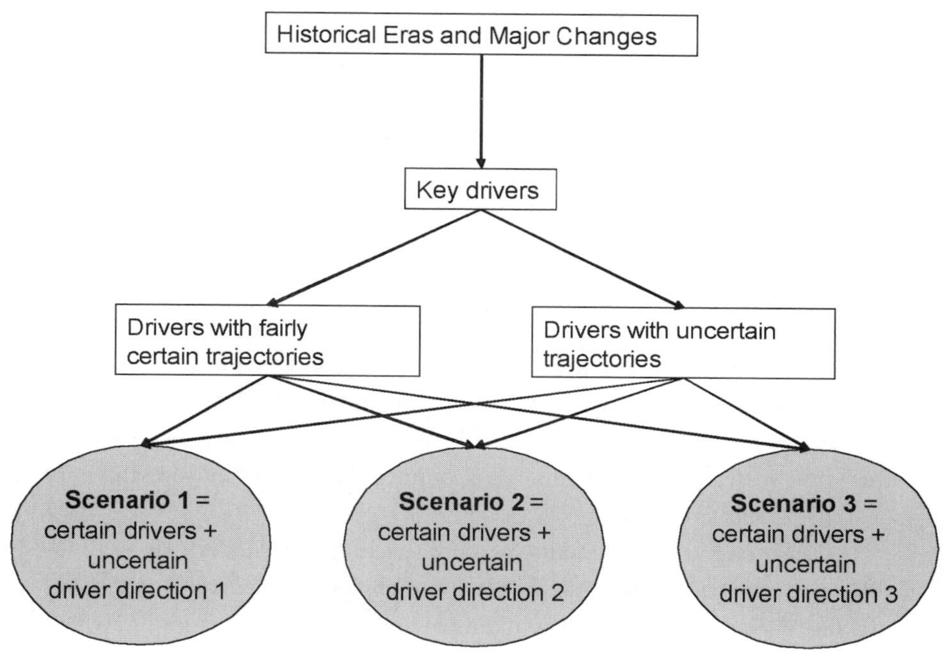

Report on Emissions Scenarios; the MA; and the Global Scenario Group). The qualitative story lines are used to stimulate creative, outside-the-box-thinking. The quantification of driving forces provides a consistency check and can illuminate unanticipated dynamics.

Scenarios are not predictions, forecasts, or projections. In contrast to these methods for describing the future, scenarios do not necessarily assume that the world will remain within today's boundary conditions. Scenarios are, in fact, often based on the assumption that the boundary conditions will change, and each scenario in a set follows the path of a different set of boundary conditions. One of the most useful ways to imagine different boundary conditions is to gather the perspectives of people who come from very different backgrounds and have different concerns about the future.

Scenarios are often part of a decision-making or planning process. They can highlight upcoming choices to be made and potential outcomes of those choices (Rotmans et al. 2000). Scenarios are also useful for thinking about dynamic processes and causal chains that affect the future. In this way, the process of developing scenarios challenges our beliefs and assumptions about how social-ecological systems work. Thus, scenarios are used as "a tool for ordering one's perceptions about alternative future environments in which one's decision might be played out" (Schwartz 1996). Management options can, for example, be tested by exploring how well a given policy works across multiple possible futures.

Using Scenarios as a Method to Integrate Epistemologies

The knowledge that we have, and often the way we have acquired this knowledge, plays a decisive role in shaping our beliefs about the future. Scientists from the same discipline or people representing a particular interest group are likely to use similar concepts and share a common understanding of how the system of interest should be defined; they may find it difficult to communicate with others from different backgrounds. Bridging the unspoken assumptions that come with any given paradigm is a common difficulty of multidisciplinary or transdisciplinary work.

For the scenario development process to succeed, these underlying assumptions must be made explicit so the team can explore their impact on decision making. The process of scenario building often leads to conversations in which individuals from different backgrounds challenge one another about the focal questions of the process, key drivers, and assumptions about how the world works. This type of discussion can lead to a critical examination of assumptions, which helps to assess and deepen the understanding of the system components and their interactions.

The first step in MA scenario building—discussing key uncertainties—is often the scenario development team's first discussions about what is known and unknown from the perspective of their discipline or source of knowledge (table 15.1). Something that one team member believes to be an uncertainty may be clarified by knowledge from another team member. Likewise, something that a team member believes to be certain may be questioned by another team member. These discussions help the scenario-building team explore the knowledge, and the gaps in knowledge, of each member of the team, including themselves.

The second step in scenario building is agreeing on the key drivers of change in the system. Splitting these drivers into those with a fairly certain trajectory versus those for which the trajectory is uncertain helps to make the transition from talking about the social-ecological system in the abstract to deciding which

Table 15.1

How bridging epistemologies occurs at different stages of scenario development

Scenario Development Stage	How Bridging Is Achieved
Discussion of main uncertainties about the future, focal questions	• Voicing different viewpoints about the focal questions • Presenting different pieces of knowledge about the main uncertainties • Discussing assumptions about how uncertainties will play out in the future
Discussion of main driving forces of change	• Voicing viewpoints on the importance of specific drivers • Providing information about how drivers are changing
Putting the scenarios' story lines together	• Developing elements of the story lines • Enriching the story lines by adding particular pieces of knowledge • If models are used, using model results to ground-truth assumptions
Analysis of implications for main stakeholders across the scenarios set	• Analyzing from different viewpoints • Questioning beliefs and assumptions of all knowledge • Understanding the influence of the paradigm in which information is collected and how this affects the way we use information

variables should play a major role in the scenarios. Drivers for which the trajectory is fairly certain are played out similarly across the scenarios. On the other hand, drivers with uncertain trajectories are the ones around which the important stories are told. Involving people with different knowledge bases in the process of scenario development can broaden the perspective of which drivers are important as well as what trajectories those drivers may take in the future. Similar to step one, these discussions bridge the gap between different knowledge systems by helping the scenario team talk through their assumptions, including where these assumptions come from and how they affect beliefs about the future.

In the third step—the development of each story—the opportunities shift from understanding the assumptions and worldviews behind different epistemologies to actually bridging the gaps that have been uncovered. Because scenarios can be qualitative or quantitative, room exists for expressing the same thing many different ways. Because each scenario can follow a different logic, there is room for expressing many different ideas as well. By systematically talking through important uncertainties and "stories" about how they might play out, individual participants can add their perspective and their piece of knowledge to the scenarios process.

Finally, after the scenarios have been developed, they can be interpreted and their implications for different stakeholder groups analyzed. Analyzing the plausible long-term consequences of various decisions through the lenses of many groups or disciplines helps to reinforce the many different visions that people have for the future. It also allows for the combined use of several paradigms by compelling participants to discuss why they believe what they believe. Shedding light on the differences in interpretation helps to understand which information is really used by which people, how it is processed, and how conclusions are drawn. This analysis reveals the influence of the structure and paradigm used by each different epistemological group and how these backgrounds influence people's decisions.

Examples from the Millennium Ecosystem Assessment

The MA aimed to provide scientifically sound information to decision makers and the public to improve ecosystem management and thereby contribute to human well-being. Part of the MA assessment process was a scenarios exercise to describe plausible changes in ecosystem services and their consequences for human well-being at a global scale. The MA also supported a number of subglobal assessment exercises, some of which also built scenarios.

In this section, four different MA-related scenario exercises are described to illustrate how scenario building can be used to integrate information from across epistemologies into a single coherent message. In the first example, the global scenario exercise incorporated qualitative and quantitative information. The second case shows multiple academic disciplines coming together to build a set of scenarios for the Caribbean region. In the third example, academic science and local stakeholders work together to build scenarios. And finally, in the last case study, we highlight an example of traditional knowledge and academic science building scenarios together. In each of the following sections, we describe the process of each scenario-building exercise, including the problems faced, how these problems were solved, and what insight was gained from the process.

Integrating the Qualitative and the Quantitative: The MA Global Scenarios

The MA developed a set of four global scenarios about the future of ecosystem services and human well-being, combining quantitative modeling tools with qualitative approaches. Quantitative results were desired for their scientific credibility, and qualitative results were needed because they could easily incorporate nonlinearities that the models could not address.

Each of the four MA scenarios describes how social-ecological systems might develop between 2000 and 2050 based on different assumptions about how demographic, economic, sociopolitical, cultural, technological, and biophysical factors might develop in the future. The scenarios were developed in a series of eight workshops over three years by a group of about seventy experts from around the world and from many different academic disciplines. The team included ecologists, economists, sociologists, scenarios experts, and global modelers. The process required bringing together knowledge from several different academic disciplines and harmonizing qualitative story lines with quantitative model results. The group was led by two ecologists and two economists. Facilitation and conflict resolution methods were used whenever necessary by involving a professional facilitator in the process.

To achieve integration between quantitative and qualitative information, a "storyline-and-simulation" approach was used (Alcamo 2001). According to this method, a set of qualitative narratives, or story lines, are developed first and are then translated into model variables. The models are used to quantify the results of the stories. Harmonizing the story lines and the models is an iterative process in which both the story lines and the models are compared with each other and adjusted for consistency.

The story lines of the MA global scenarios addressed two main types of uncertainties whose combinations were seen as potentially leading the world into fundamentally different trajectories: the degree of connectedness of countries, markets, and institutions, and the degree to which ecosystem management is reactive or proactive. Each story line was translated into a set of variables that served as inputs to global models that were used to calculate outcomes for various ecosystem services, such as crop production, fish harvest, or water quality. Five different models were used, and each model was run separately for each story line with input values based on the story lines. The results of the model runs were then compared with the narratives to verify the assumptions, to check the story lines for internal consistency, and to add quantitative information. The final product for each scenario was a qualitative narrative that contained quantitative information. Ultimately, the set of scenarios was analyzed for their implications for different stakeholders and for the provisioning of ecosystem services in the future (MA 2005b).

The MA global scenario exercise is an example of how one can harmonize qualitative story lines and quantitative model results to strengthen the final message. In our efforts to quantify the story lines, they had to be simplified in a way that was not always comfortable to those most familiar with the stories. This difficulty was overcome through conversation about what features of the story lines could be simplified and which could not. It was also overcome by allowing the story lines to be told both as narrative and in numbers. Thus, while the narratives were simplified to be useful to the modelers, they could retain their complexity in the final qualitative telling.

A related difficulty was determining whether the scenarios should be driven primarily by the qualitative story lines or by the quantitative modeling results. Discussions resulted in an assessment of the available modeling tools, which determined that the models alone could not adequately answer the focal questions. The decision was taken to use the models primarily as a consistency check of, and quantitative framework for, the story lines. A professional facilitator and strong, balanced leadership were necessary components of these successful decision-making processes.

The integration process between qualitative and quantitative information resulted in a set of detailed stories about the future of ecosystem services and human well-being that were better than scenarios, which were either solely quantitative or solely qualitative. The qualitative aspects of the scenarios added a richness and ability to deal with nonquantifiable nonlinearities in ecosystem services' changes, while the quantitative aspects of the scenarios served as an important consistency check for the story lines. The discussions among qualitative and quantitative scientists and, in particular, the need to defend assumptions made each group's efforts much clearer and stronger than they would have been alone.

Talking across the Disciplines: The CARSEA Scenarios

The Caribbean Sea Ecosystem Assessment (CARSEA) was an MA subglobal assessment designed to evaluate changes in Caribbean ecosystems and ecosystem services and to develop options for responding to these changes. The assessment team chose to develop scenarios as part of their assessment because of the flexibility of the method for thinking broadly about the future without losing scientific rigor and because of its ability to incorporate a wide variety of expert knowledge. The team was very knowledgeable about Caribbean ecosystems, including information from diverse disciplines. The process was led by a multidisciplinary team of five natural and social scientists who had previous experience with scenarios.

The CARSEA group developed four scenarios, which described plausible developments in the Caribbean region and their outcomes for ecosystem services and human well-being over a fifty-year time horizon. In two workshops over the course of 2003, a team of approximately forty participants discussed the key driving forces and critical uncertainties that they expected would determine the future of the region. Prioritizing uncertainties helped to select the set of scenario story lines to be developed. The most critical uncertainties—which turned out to be the level of reliance on income from outside the region, the level of reliance on tourism and its management, global environmental change such as climate change, and the level of regional cooperation—were used to determine the major differences among the story lines.

The next step was the development of the story lines, which was undertaken in small teams of two or three people. These teams developed draft scenarios that aimed to incorporate as many different viewpoints from earlier discussions as possible. Each story line was then presented to and critiqued by the whole group. The discussion was a consistency check for the proposed story lines in which each group member could question the assumptions made in developing each scenario. Input from across disciplines enriched the scenarios by adding additional detail to the story lines. After a few iterations, the majority of differences in viewpoint were settled and the story lines were finalized. Analysis of the scenario implications was undertaken by a small writing team during the final write-up of the scenarios and the assessment report.

Broad discussion helped to form a common language among all participants. Difficulties arose primarily when the main uncertainties for the region were discussed. These difficulties came primarily from differences among the disciplines and had to do with attaching different levels of importance to particular uncertainties. In general, people thought that the uncertainties closest to their own discipline were the most important. For example, some economists thought that the proposed free trade agreement of the region with the United States would be one of the most important determinants for the future of the region, while some natural scientists stressed the negative impacts of new marine diseases and of a rise in sea level on the tourism industry.

Some of these differences were resolved by bringing one key uncertainty to the foreground of each scenario. For example, one scenario focuses on the consequences of a free trade agreement of Caribbean countries with the United States, and another centers on the impact of sea level rise and marine diseases. Knowing that several key uncertainties could be addressed across the set of scenarios meant that we did not have to keep discussing until all participants agreed on a single uncertainty. The flexibility to have several uncertainties eased tension among participants over whose ideas were the most important.

Bringing a multidisciplinary team of experts together to talk about the future of the Caribbean helped to thoroughly discuss the challenges the region is facing as seen from different scientific viewpoints. Each participant enriched the discussion with knowledge from his or her discipline. This helped to improve the story lines by providing their expertise and by questioning some of the propositions of other disciplines and prompting a deeper level of discussion. In this way, not only did the story lines gain in details but their plausibility was also constantly checked and improved. The scenarios methodology provided a platform for two different outcomes: (1) to develop a common language between the disciplines, and (2) to synthesize information and knowledge from different academic disciplines in a consistent, systematic manner.

On the issue of developing a common language, participants were constantly obliged to state concepts used in their discipline in a way that was understandable to people coming from a different discipline. Multidisciplinary experts were a very important factor in the success of these scenarios because they could translate terminology, paradigms, and theories across disciplinary boundaries, leading to better understanding and agreement among participants.

One of the most difficult parts of the process was to make sure that each participant was comfortable with how his or her assumptions about the future were incorporated into the scenarios. In the CARSEA scenario-development process, the experts were familiar with one another from working together on previous projects, which helped them find a common ground for discussion.

In fact, the participants had been carefully selected by the leaders to be people who were known to be good "team players."

The CARSEA scenario-development process provided a platform for weaving together a set of consistent stories about the future that incorporated information and knowledge from across many different academic disciplines. Different disciplinary perspectives, although sometimes controversial, helped the team to develop new insights about important driving forces of change in the Caribbean region. The experience showed how the scenarios process can lead from incorporating different knowledge types to integrating knowledge into pictures of the future that people had not thought of before. The scenarios team believed that these integrated visions of the future brought unforeseen insights that would not have been realized in a single-discipline study.

Combining Scientific and Local Knowledge: The Northern Wisconsin Scenarios

A workshop was held in September 2002 to develop scenarios for the near future of the Northern Highland Lake District (NHLD) in northern Wisconsin, United States. The goal of the scenarios was to explore the ability of the NHLD to maintain its present desirable social and ecological features despite changes driven from within and from outside the region (Carpenter et al. 2003). We chose scenario building in part because we wanted to generate discussion about the future among stakeholders who normally would disagree strongly or not talk at all about managing the NHLD.

The workshop included participants from federal and state resource management agencies, lake associations, out-of-state owners of lakeshore property, realtors, and Native Americans. In addition, academic experts from around the world were present to act as resource people, bringing expertise in such fields as ecology, human demography, economics, and mathematical models of social-ecological systems. In its focus on stakeholders, the exercise differs from the two presented before. Leadership was provided by scientists who had experience with scenario development and facilitation. The goals were integration across many opinions about managing the region and bridging the gap between local stakeholders and academics.

The scenarios were developed following a methodology similar to that of the CARSEA scenarios. Broad discussions of all participants were followed by small groups developing the actual story lines. In this case, the local participants had a very wide range of different desires for the future of the region. In many other scenario-development exercises, such as those developed for CARSEA, the participants generally agreed easily on what would be a "good" outcome for the future of the social-ecological system in question. In the case of the NHLD scenarios, however, no such agreement existed. Instead, there were obvious differences in the interest groups' hopes for the future. Thus, the potential for conflict among stakeholders was high.

Because of these differences, the scenarios were developed such that the scenarios reflected the social-ecological outcomes of different stakeholders' hopes for the future of the NHLD. Because some hoped that the area would become a thriving commercial center, one scenario told the story of rapid development. Since others hoped that the NHLD would remain sparsely populated, another scenario portrayed a plausible path in which increased development did not take place. Following the consequences of each of these stories helped everyone—both those who preferred the particular outcome and those who did not—understand the benefits and drawbacks of each scenario. Other scenarios described the potential for wildlife disease to affect the area or explored the consequences of a massive and rapid influx of residents to this sparsely populated region.

In addition to stakeholders' preferences, we also used the best scientific information about the current state of the social-ecological system and recent trends. For most scenarios, this was fairly easily accomplished. Local interest determined the basic thrust of the story line, and scientific information provided the details, particularly details about the outcomes for provision of ecosystem services. For example, stakeholders told us that one story line should include increased telecommuting leading to a larger population in the NHLD. Scientific literature helped us understand how an increase in the population of young telecommuters would affect ecosystem services in the region. When the best scientific expertise disagreed with stakeholder beliefs or showed a cost of a favored strategy, conflict arose. Some participants were happy to accept the scientific experts' opinions, but others were not.

The integrated results were thought to be more believable than stories developed without scientific expertise. It was also easier to convince nonparticipants of the validity of the scenarios because of the participation of a wide range of stakeholders and scientific experts. Yet, because the scenarios were still based in the interests and concerns of local stakeholders, they were more interesting to other local residents than purely scientifically determined futures would be. Additionally, local stakeholders who participated in the process showed the scenarios to their friends and family in the region, spreading the discussion beyond the boundaries of the workshop participants.

We learned that it is possible, and even relatively easy, to make stories that are based on scientific information about the social-ecological system and at the same time to have the scenarios address the issues that people are really interested in. The difficulties we faced occurred when people's understanding of the system differed from the scientific understanding. For example, it can be difficult to tell a story about the ecological quality of a system if people believe water quality is getting worse but the scientifically collected data indicate that it is not. Usually, these misunderstandings were worked out through discussion. Where this was not possible, we used this potential disadvantage to our advantage by developing two scenarios that explored the split in beliefs.

Integrating TEK and Western Science: Scenarios for Bajo Chirripó, Costa Rica

The Bajo Chirripó assessment was undertaken by a group of Cabécar indigenous people and a Costa Rican nongovernmental organization (NGO) that works on indigenous peoples' issues. The scenario-building team, including NGO members, representatives of the indigenous communities, and two local scientists, carried out a pilot scenario-building exercise. The goal was to help the Cabécar community derive a common vision of their future and to help them cope with ongoing regional developments, some of which threaten the community's territory and culture. Of particular concern was the loss of traditional knowledge and values. Scenarios were chosen as a method because of their ability to incorporate traditional belief systems along with academic or other "outsider" beliefs and information.

For the pilot study, community members came together in a one-day workshop to develop scenarios. The NGO members and the scientists played a double role of representing the view from outside the community and providing some background information on political and societal developments that might affect indigenous communities. The group developed two pilot story lines, which described plausible changes in the region and in their community over a five-year horizon.

As in the other scenario exercises, the discussion of key forces changing the community allowed the participants to bring their knowledge and experience to the table. After identifying the most important sources of uncertainty for the future of the Cabécar, narrowing the focus of the stories forced participants to determine which forces were affecting the community from outside and which were controllable by the community. This identification process was not easy because the views of the Cabécar community often differed from those of participants from outside the community, including the leaders of the exercise. These differences were not totally resolved in this pilot exercise, but the exercise built understanding of the origin of the discrepancies and how they influenced decision making.

The Bajo Chirripó scenarios are an example of how the scenario-development process can bring indigenous people together with others from outside the community to discuss their perceptions of future developments constructively. The process allowed us to combine two very different kinds of knowledge and still develop consistent pictures of the future. The discussion also helped to clarify which processes the Cabécar community can control and which they cannot. In addition, we discussed possible reactions to both controllable and uncontrollable drivers. Incorporating differing views on drivers and possible responses to them enlarged the perspectives and knowledge of all participants.

Conclusions

Ecosystem management can be improved by using multiple types of knowledge to formulate management plans (Berkes and Folke 1998). Yet, bridging the gap between paradigms can be difficult. Although many consider it an important task, few known methods exist that can be used to integrate multiple sources of information into a single coherent product.

We have suggested that scenario building may be an effective method for bridging the gap between epistemologies. Scenarios themselves will be more informative and useful if they can incorporate multiple perspectives (Schwartz 1996). As such, the MA process for scenario building consists of conversations about what is known and what is not known, providing an ideal space for questioning assumptions made by different disciplines or within different paradigms. The discussions that are required to build multidisciplinary scenarios help participants understand, and then question, how their knowledge and

Table 15.2

Comparison of key players in the scenario exercises and their contribution

Scenarios Exercise	Key Players in the Exercise	Contribution of Each Player to the Scenario Development Process
Global Millennium Ecosystem Assessment (MA) scenarios (integrating the qualitative and quantitative)	Story line developers (experts from different disciplines—ecologists, economists, social scientists)	Understanding of specific driving forces and their impacts
	Global modelers (experts from different disciplines)	Modeling capabilities to support assumptions
	Core scenarios team	Leadership; experience with scenario development
Caribbean Sea Ecosystem Assessment (integrating across disciplines)	Experts from different disciplines (ecologists, economists, social scientists)	Scientific knowledge of driving forces and their impacts
	Local stakeholders	Understanding of main problems and political processes; creativity; knowledge about local institutions
	Core scenarios team	Scenarios methodology
Northern Highland Lake District (combining scientific and local knowledge)	Local stakeholders	Understanding of main problems; knowledge about local institutions
	External experts	Scientific knowledge of driving forces and their impacts; scenarios methodology
	Core scenarios team	Scenarios methodology and experience
Bajo Chirripó, Costa Rica (integrating traditional ecological knowledge and Western science)	Indigenous local people	Understanding of main problems; institutional knowledge
	Local NGO	Scenarios methodology
	External experts	Scientific knowledge of driving forces and their impacts; scenarios methodology

paradigms influence the vision of the future. The process of scenario development helps identify blind spots that each type of knowledge has because it obliges each discipline to explain its beliefs and expose the certainties and uncertainties in conversation with other participants. By understanding how the epistemologies influence the vision, we make progress toward integrating epistemologies into a single, consistent set of stories.

Each of the scenario exercises that we highlighted involved many players, each of whom contributed to the process (table 15.2). Core scenario teams contributed scenario expertise. Experts contributed information from their respective disciplines along with information about the boundary conditions of how that information should and should not be used. Local stakeholders contributed creativity, understanding of the key problems, and knowledge of local institutions. Outside experts, when used, added understanding of other systems and could provide information about how the problems faced in the region were similar to problems being faced in other locations. Without multiple perspectives, each of the sets of scenarios, and the scenario-building process itself, would have been less informative. Scenario development also often builds important networks among people who might otherwise not talk about the future together and creates through its process a "safe" discussion forum to express diverging viewpoints.

Scenarios have limitations too. They are not yet well established within the scientific community as a credible method. Although they are useful for synthesizing existing information and for pointing out where further research is needed, they do not generate new information. As with all multidisciplinary projects, power dynamics can play an important role. If not handled carefully, these dynamics may lead some participants to refuse to participate, essentially ending or severely limiting the project. Similarly, because it is easy to simply incorporate each perspective by adding a single scenario that follows that set of beliefs, it may be too easy to gloss over the difficult conversations needed to make the scenarios truly integrative.

Despite its limitations, however, scenario development may be an important step toward bridging the gap between epistemologies and improving ecosystem management. Scenario building is a method for thinking about the future that is made stronger by incorporating multiple perspectives. The discussion required to incorporate these perspectives into a single set of scenarios encourages scenario builders to consider how their assumptions and backgrounds lead to particular beliefs about the way the world works. These discussions not only make better scenarios but enhance our understanding of our epistemological boundaries and, in so doing, improve our ability to work with others from different backgrounds.

References

Alcamo, J. 2001. Scenarios as a tool for international environmental assessments. Environmental Issue Review 24. Copenhagen: European Environmental Agency.

Becker, C. D., and K. Ghimire. 2003. Synergy between traditional ecological knowledge and conservation science supports forest preservation in Ecuador. *Conservation Ecology* 8 (1): 1. http://www.consecol.org/vol8/iss1/art1.

Bennett, E. M., S. R. Carpenter, G. D. Peterson, G. S. Cumming, M. Zurek, and P. Pingali. 2003. Why global scenarios need ecology. *Frontiers in Ecology and the Environment* 1 (6): 322–29.

Berkes, F., and C. Folke. 1998. Linking social and ecological systems for resilience and sustainability. In *Linking social and ecological systems: Management practices and social mechanisms,* ed. F. Berkes and C. Folke, 1–25. New York: Cambridge University Press.

Carpenter, S. R., E. A. Levitt, G. D. Peterson, E. M. Bennett, T. D. Beard, J. A. Cardille, and G. S. Cumming. 2003. *Future of the lakes: Scenarios for the future of Wisconsin's Northern Highland Lake District.* Madison: Center for Limnology, University of Wisconsin Madison. http://www.asb.cgiar.org/ma/scenarios/training/2.1.R.1.%20Future_Of_The_Lakes.pdf (accessed April 9, 2006).

Ehrlich, P. R. 1997. *A world of wounds: Ecologists and the human dilemma.* Oldendof/Luhe, Germany: Ecology Institute.

Folke, C., S. Carpenter, T. Elmqvist, L. Gunderson, C. S. Holling, and B. Walker. 2002. Resilience and sustainable development: Building adaptive capacity in a world of transformations. *Ambio* 31:437–40.

Holling, C. S., L. H. Gunderson, and D. Ludwig. 2002. In quest of a theory of adaptive change. In *Panarchy,* ed. C. S. Holling, L. H. Gunderson, and D. Ludwig, 3–24. Washington, DC: Island Press.

Kinzig, A. P. 2001. Bridging disciplinary divides to address environmental and intellectual challenges. *Ecosystems* 4:709–15.

Kinzig, A. P., J. Antle, W. Ascher, W. Brock, S. Carpenter, F. S. Chapin III, R. Costanza, et al. 2000. Nature and society: An imperative for integrated environmental research. Report for a National Science Foundation (NSF) workshop, November.

Lubchenco, J. 1998. Entering the century of the environment: A new social contract for science. *Science* 279:491–97.

Millennium Ecosystem Assessment (MA). 2003. *Ecosystems and human well-being: A framework for assessment.* Washington, DC: Island Press.

———. 2005a. *Ecosystems and human well-being: Current state and trends.* Vol. 1. Washington, DC: Island Press.

———. 2005b. *Ecosystems and human well-being: Scenarios.* Vol. 2. Washington, DC: Island Press.

Mingers, J. 2001. Combining IS research methods: Towards a pluralist methodology. *Information Systems Research* 12:240–59.

Olsson, P., and C. Folke. 2001. Local ecological knowledge and institutional dynamics for ecosystem management: A study of Lake Racken Watershed, Sweden. *Ecosystems* 4:85–104.

Peterson, G., G. Cumming, and S. R. Carpenter. 2003. Scenario planning: A tool for conservation in an uncertain world. *Conservation Biology* 17:358–66.

Raskin, P. 2005. Global scenarios: Background review for the Millennium Ecosystem Assessment. *Ecosystems* 8:133–42.

Rotmans, J., M. van Asselt, C. Anastasi, S. Greeuw, J. Mellors, S. Peters, D. Rothman, and N. Rijkens. 2000. Visions for a sustainable Europe. *Futures* 32:809–31.

Schwartz, P. 1996. *The art of the long view—Planning for the future in an uncertain world.* New York: Currency Doubleday.

SYNTHESIS

CHAPTER 16

The Politics of Bridging Scales and Epistemologies

Science and Democracy in Global Environmental Governance

CLARK MILLER AND PAUL ERICKSON

Why should global environmental assessments concern themselves with the complex, costly, and sometimes uncomfortable challenge of bridging scales and knowledge systems? After all, conducting a comprehensive, scientific assessment of environmental change at global scales is hard enough in itself. What benefits can be achieved from further complicating the task to integrate knowledge from alternative epistemological paradigms and subglobal scales?

One answer is to more effectively link knowledge to action by promoting accurate, policy-relevant global environmental assessments. Bridging scales and epistemologies may enable assessments to better integrate local knowledges into global models and data sets, potentially strengthening the accuracy of their findings. Likewise, integrating scientific and indigenous knowledges, or global and national styles of reasoning, may contribute to better translation of assessments into effective policy strategies for addressing global environmental change.

These are important pragmatic reasons to bridge scales and epistemologies. In this chapter, however, we approach the question from a more overtly political standpoint. Viewed politically, global environmental assessments are not only attempts to synthesize scientific knowledge but also elements in reworking the constitutional foundations of global order (Miller 2004a; see also Jasanoff 2003 and Litfin 1998). But what kind of global order are assessments forging? Unfortunately, too often, attempts by assessments to portray science in a unified framework contribute to excluding voices from global decision making and exacerbating ideological divisions in global society (Miller 2004b, 2003). In this manner, assessments contribute to the growing "democratic deficit" that permeates international institutions (Held 2004; Verweij and Josling 2003; Keohane 2001). We contend that, if properly designed and managed, efforts to bridge scales and epistemologies in global environmental assessments could contribute to the converse: promoting inclusion, dialogue, deliberation, and democracy in global governance.

Our argument brings together two literatures; comparative policy analysis and deliberative democratic theory. Combined, these literatures suggest the need to recognize and foster epistemic pluralism and deliberation as an important element in democratizing international governance. Building on the idea that this might be accomplished through "reasoning together" (Jasanoff 1998), we suggest a fourfold strategy for bridging scales and epistemologies in global environmental assessments:

- *Building critical capacity for policy reasoning*—strengthening citizen capacity across the globe to formulate and reflect critically on reasoned justifications for global policy choices
- *Promoting epistemic tolerance and pluralism*— recognizing and facilitating the expression of divergent styles of reasoning about global environmental risks in governing forums
- *Enhancing epistemic dialogue and exchange*—encouraging efforts to bring divergent styles of reasoning into dialogue and exchange as well as cross-cutting reflection and evaluation
- *Orchestrating cross-scale epistemic jurisdiction*—strengthening dialogue and exchange, as well as appropriately delegating authority, across scales of assessment and governance.

We then turn to practical strategies that global environmental assessments might adopt to pursue this more politically oriented approach to bridging scales and epistemologies. In particular, we focus on regionalization. Regionalization, per se, is not the point. As a strategy, however, regionalization has immediate consequences—breaking up the assessment into parts, enabling variations in assessment design and practice across parts, and making possible dialogue among the parts—that may benefit a deliberative approach to global environmental assessment. We compare and evaluate several approaches to regionalization, including those adopted by the Millennium Ecosystem Assessment

(MA), the Intergovernmental Panel on Climate Change (IPCC), the Global International Waters Assessment (GIWA), and the Arctic Climate Impact Assessment (ACIA), illustrating how choices in assessment design and management can affect the ability of regionalization strategies to achieve the goals of our model. We conclude with suggestions for going beyond current strategies of regionalization to more effectively bridge scales and epistemologies in the service of democratizing international governance.

Reasoning and Democracy

Democracy can be understood as the insistence that all individuals affected by policy choices have a voice in the policies' making. Many practical approaches have been designed to try to achieve this goal: majority rule combined with protections for minorities, decentralization and differentiation of power, election of representatives to deliberative bodies, federalism, and so forth. Environmental change challenges most if not all of these traditional approaches to democracy. Natural systems and processes cross the boundaries of existing political jurisdictions and affect people who do not have a voice in policy decisions. Additionally, environmental policy making requires experts whose knowledge is essential for assessing and managing environmental problems but whose epistemic frameworks and styles of reasoning may disenfranchise other stakeholders. The need to bridge scales and epistemologies is thus endemic if environmental policy choices are to comport with the core tenet of democratic governance.

The environmental challenge to democracy and the need to bridge scales and epistemologies are particularly acute in global environmental governance. Global environmental change crisscrosses thousands of local and national jurisdictions, and epistemic pluralism is pervasive. Fundamental approaches to reasoning about risk vary across cultural contexts (see, e.g., Thompson and Rayner 1998; Wynne 1995; Krimsky and Plough 1988; Douglas and Wildavsky 1982), including between science and policy, lay, and indigenous communities' social and ecological knowledges (Lachmund 2004; Martello 2004a, 2004b; Iles 2004a, 2004b; Ellis and Waterton 2004) and also across national regulatory sciences (Parthasarathy 2004; Daemmrich and Krucken 2000; Jasanoff 1995, 1986). Even scientific disciplines differ in preferences regarding models, instruments, methods, and styles of reasoning (Hacking 2002). Consequently, bridging scales and epistemologies is not simply a matter of increasing the spatial or temporal resolution but of stitching together multiple knowledge systems that encompass divergent paradigms and operate from distinct assumptions and evidentiary standards, ideological commitments, and frames of meaning (Miller 2000). In this sense, bridging scales becomes a special case of bridging epistemologies, as epistemic frameworks emerge as a key difference across scales.

To address these challenges, Jasanoff (1998) suggests "reasoning together." Theories of deliberative democracy emphasize the importance for democratic legitimacy of government agencies using reasoned analyses to justify their decisions.[1] They also emphasize the opportunity for public deliberation about the reasoning behind policy choices (see, e.g., King 2003).[2] As discussed, however, styles of reasoning used in justifying policy choices vary across countries, suggesting the need, first, for a dialogue about different styles of reasoning about risk before settling on unified global knowledges. Jasanoff characterizes this kind of intentional deliberation, exchange, and comparative evaluation and critique among epistemic frameworks as "reasoning together."

Global environmental assessments already play important roles in the deliberative justification of global environmental policy making; their primary task is to provide a reasoned analysis for making policy choices. To date, however, they have tended to approach this task as one of developing an objective, global rationale for policy action rather than from the perspective of fostering dialogue and exchange among multiple styles of reasoning. How might they do otherwise? Being primarily concerned with setting up the problem, Jasanoff (1998) offers only sparse practical guidance regarding reasoning together. Here, we elaborate the concept of reasoning together to provide more specific guidance for global environmental assessments.

We propose considering reasoning together in two parts: first, strengthening the representation of divergent epistemic frameworks in global environmental assessments; and second, fostering dialogue, exchange, and mutual evaluation and critique among these divergent styles of reasoning. Each can be further differentiated into two subparts. The goal of strengthening epistemic representation in assessments entails, first, building the capacity of divergent groups to articulate persuasive, credible styles of reasoning and, second, creating institutional spaces that help articulate divergent epistemic frameworks. In other words, an absence of epistemic pluralism in global environmental assessments can result either from an absence of multiple powerful voices or from institutional configurations that exclude or marginalize competing voices.

Likewise, fostering epistemic dialogue, exchange, evaluation, and critique involves two additional tasks: first, creating institutional frameworks that encourage such activity within assessments and, second, orchestrating epistemic dialogue and exchange across assessments at multiple scales and in distinct political jurisdictions. Thus, we suggest four challenges.

Building Capacity for Critical Policy Reasoning

A central element of deliberative democracy is the ability for participants in civic life to formulate, articulate, and critically evaluate reasoned justifications for policy choices. At stake is their capacity to reason deliberatively and to make informed judgments about important policy decisions. Yet, few anywhere can claim a high capacity for making reasoned judgments or for evaluating critically the claims made by others about the planet's future. Global environmental assessments reflect one aspect of necessary capacity, but one that meaningfully reaches only a fraction of the Earth's citizenry and that reflects limited epistemic frameworks. Strengthening capacity for critical policy reasoning on global issues will entail, to some extent, public education; but perhaps more important are institutional innovations that enable communities to feel confident, first, in critically evaluating policy rationales and their relevance to local frames of meaning and, second, in formulating and articulating supporting rationales for their judgments about how to protect the global environment. Numerous transnational movements and institutions are responding to this challenge, but global environmental assessments are uniquely situated to contribute to the integration of scientific reasoning into broader processes of social learning.

Promoting Epistemic Tolerance and Pluralism

Global environmental assessments and other global policy-making forums also need restructuring to recognize, tolerate, and facilitate the expression of divergent styles of reasoning. As capacity for critical policy reasoning about global change expands, engagement and participation in global policy exercises seems likely to grow. Global institutions must find ways to respond appropriately to this demand for the expression of ideas from across divergent scales and epistemologies, lest they suffer further loss of legitimacy (Stiglitz 2002). This seems particularly true for global environmental assessments, which have been criticized for failing to include knowledges that differ from those of transnational scientific networks (Thompson 2004; Rayner and Malone 1998; Agarwal and Narain 1991).

Enhancing Reciprocal Dialogue and Exchange

A third important objective is to restructure scientific assessments to serve as deliberative spaces within global governance. The model of reasoning together, as we conceive it, is one in which mutual learning occurs across scales and knowledge systems. Global environmental assessments can facilitate such learning by (1) making differences across styles of reasoning explicit, (2) structuring comparative evaluations of reasoning techniques, (3) promoting dialogue about the appropriate application of methods and frameworks to global contexts, (4) facilitating cross-cutting evaluation, and (5) communicating these deliberations broadly. The last is essential if deliberations prompted by global environmental assessments are to extend their impact to global audiences other than the individuals who participate directly.

Orchestrating Cross-scale, Epistemic Jurisdiction

Efforts to bridge scales and epistemologies should also be understood as part of a broader exercise of effectively linking local, national, and global governance. Deliberative reasoning needs to occur as much across scales as it does among participants at any given scale. Sorting out when reasoning can be left to local or national epistemic frameworks, as opposed to global standards, can be tricky. Likewise, as the structure and authority of global environmental governance expands, citizens, scientists, and businesses can be expected to join states in demanding greater access to global institutions, including those producing knowledge claims used to justify global policies. To help overcome some of the rifts in global environmental policy making, global environmental assessments need to find ways to be responsive to these shifts—for example, by supporting robust notions of epistemic citizenship for individuals around the globe (Jasanoff 2004).

Regionalization: A Strategy for Reasoning Together?

How might global environmental assessments approach reasoning together, as elaborated here? Here we look at one possible strategy, *regionalization,* the practice of breaking up global environmental assessments into parts, each focused on a geographically bounded region. If the point is to promote epistemic pluralism and dialogue in global affairs, regionalization, as a strategy, has immediate consequences—breaking the assessment into parts, enabling variation in assessment design and management across parts, and making possible dialogue among parts—that may help facilitate deliberative approaches to reasoning together. Regions can articulate different epistemic frameworks and rationales for global environmental policies. Not all approaches to regionalizing global environmental assessments are equally conducive to the model of reasoning together, however. Unless regionalization is designed as an exercise and experiment in reasoning together, it will likely fail to address one or more of the four challenges described above.

During the 1980s and 1990s, global environmental assessments focused on the globe, with little systematic attention to regions. For assessments like the IPCC and the Global Biodiversity Assessment, the primary purpose was to communicate the nature and extent of global environmental risks to negotiators of international treaties (Benedick 1991; Bolin 1994). A key feature of these assessments was their emphasis on the universality of such risks—risks that were framed on the scale of the planet itself (Takacs 1996; Jasanoff 2001; Miller 2004a, 2004b).

These first-generation global environmental assessments faced considerable difficulty from multiple styles of reasoning (Jasanoff and Wynne 1998; Thompson and Rayner 1998). In their search to present a consensus view of scientific knowledge, many experienced protracted contests over different approaches to reasoning about risk. One such disagreement took place during the second IPCC assessment report in the mid-1990s. Economists tasked with monetizing the economic impacts of climate change adopted statistical values for lives lost consistent with measures of lifetime earnings and willingness to pay to avoid loss of life. Their results valued lives in wealthy countries an order of magnitude higher than lives in poor countries, generating considerable scientific and diplomatic debate (Meyer and Cooper 1994). Criticism focused on the methods of valuation underpinning global policy decisions and sharply attacked willingness-to-pay approaches. The episode cost the IPCC considerable credibility, especially among developing country audiences (Masood 1995). In mid-1995, the Indian head of delegation to the Framework Convention on Climate Change wrote to his fellow delegates rejecting the IPCC economists' logic.[3] Angry letters, signed by a broad spectrum of scientific and nongovernmental organization (NGO) leaders, denounced the draft chapter in *Nature* and several major British newspapers.[4]

Since 2000, by contrast, international assessments have begun to incorporate substantial regional components. In 2001, the IPCC subdivided the globe into ten geographic regions and carried out chapter-length assessments of climate impacts for each (McCarthy et al. 2001). The MA has developed a bifurcated strategy, including a global assessment and over two dozen "subglobal" assessments that include both regional and thematic, cross-regional studies. The GIWA has adopted a bottom-up perspective, aggregating watershed-scale assessments for each of the world's major river basins into a global narrative. Regions have also pursued their own assessments, such as the ACIA, a stand-alone assessment of the vulnerability of the Arctic region to changes in the Earth's climate system.[5] Many other stand-alone regional assessments of climate change have been carried out, including the U.S. National Climate Impact Assessment and the German Enquête Commissions.

As global environmental assessments have added regional components, they have minimally acknowledged that a single, global assessment fails to address the needs and concerns of people in different cultural and geographic contexts. In some cases, regional assessments have gone further, helping to pluralize styles of reasoning in global environmental governance by allowing regional assessors to adopt divergent methods and approaches. Regional assessments may also build capacity to conduct and critique assessments in multiple centers, and they are positioned, when conducted as part of a global assessment exercise, to bring multiple assessments into dialogue with one another across localities, scales, and epistemes. As the brief discussion of the MA, IPCC, GIWA, and ACIA suggests, however, regionalization has taken a variety of forms. How do these competing approaches to regionalization fare when evaluated according to our model?

Building Regional Assessments

Regional assessments vary according to a range of design and management options. Four are of particular note here: the integration of regional and global assessments; the degree of methodological standardization across regions; whether regional-to-global linkages are bottom-up or top-down; and whether regional assessments seek to bridge epistemologies as well as scales. Table 16.1 offers a brief comparison of the four assessments considered here across these dimensions.

The IPCC follows a common approach to bridging scales. In 2001, the IPCC

Table 16.1

Comparing regionalization strategies of four assessments

	Intergovernmental Panel on Climate Change	Global International Waters Assessment	Arctic Climate Impact Assessment	Millennium Ecosystem Assessment
Stand-alone regional vs. integrated global and regional	Integrated	Integrated	Stand-alone	Integrated
Standardization across regions	Strong: methods and regional definition	Strong: methods and regional definition	None	Weak: orienting principles
Top-down vs. bottom-up data flow and modeling	Top-down	Bottom-up	Both	Both
Epistemologies	Scientific	Scientific	Scientific and indigenous	Scientific and indigenous

subdivided the globe into ten geographic regions and carried out chapter-length assessments of climate impacts and vulnerability for each (McCarthy et al. 2001). These assessments followed a standardized, top-down approach. Each chapter analyzed regional climate impacts using data downscaled from global climate models. These assessments used only published, peer-reviewed scientific studies, and each chapter was written in a standard format, addressing the same topics in the same order.

GIWA also adopted an integrated, standardized approach. Like the IPCC, GIWA assessors divided the globe into nonoverlapping geographic regions that spanned the globe's surface. In contrast to the IPCC, however, GIWA built its global assessment of water resources by aggregating river-basin assessments (Global International Waters Assessment 2002). Like the IPCC, GIWA insisted on strict methodological standards to ease the task of aggregating regional data to derive a global picture. Also like the IPCC, GIWA insisted on using only scientific knowledge.

In contrast to the IPCC and GIWA, the ACIA focuses on a stand-alone assessment of climate change in the Arctic region (International Arctic Science

Committee 2000). The assessment is, nonetheless, designed to bridge scales. Assessors intend to use downscaled data and projections from climate models and satellite data sets to help create robust understanding of climate change in the Arctic. They also argue for the unique, global significance of the Arctic and, therefore, also for the value of insights from the Arctic in global environmental policy. Assessors label the Arctic "a canary in a coal mine"—a place where changes manifest early, warning of potential future dangers. ACIA also differs from the IPCC and GIWA assessments in that it explicitly bridges epistemologies. Scientists have played key roles; so, too, have indigenous communities, who bring knowledge of Arctic change, who learn about the Arctic's role in broader global environmental processes, and who have the potential to become stronger voices in global environmental forums.

While the IPCC and GIWA held regional assessors to tight standards, squeezing out competing styles of reasoning in favor of methodological consistency, the MA adopted a more flexible, plural approach to its "subglobal" assessments. MA subglobal assessments were not planned from above. Instead, the MA initiated these assessments with a call for proposals. Scientists interested in carrying out a subglobal assessment of ecosystem goods and services were invited to submit proposals describing proposed assessment designs. The MA Board then evaluated these proposals and provided seed funding to assessments that met predetermined criteria. The criteria included (1) likelihood of obtaining additional funding for the assessment from non-MA sources, (2) commitment to assessing ecosystem goods and services in an "integrated manner," meaning paying attention to interactions across multiple goods and services and multiple scales, (3) commitment to establishing ties to policy communities, the public, and indigenous groups, and (4) commitment to participating in the MA Sub-Global Working Group.

These criteria constituted a major element in the regulation of MA subglobal assessments, forming basic orienting principles but not specifying the methodology, scope, or institutional organization of a proposed subglobal assessment. This epistemic flexibility was further encouraged during the MA's ongoing work. Although the MA hired a coordinator for the Sub-Global Working Group, who organized frequent meetings among subglobal assessors, these activities were designed to build mutual understanding and dialogue among diverse assessments, not to encourage standardization. Likewise, although the MA strongly encouraged the exchange of data and people between the global assessment and subglobal assessments, MA leaders insisted that these exchanges facilitate bidirectional flows of insights and information.

The MA's bottom-up approach resulted in divergent subglobal assessments, ranging from highly localized to subcontinent in scale. Although most were "regional" in a geographic sense, some, like the Alternatives to Slash and Burn Agriculture assessment, reflected themes that cut across geographic regions. Even among geographically defined assessments, regional boundaries were often defined by widely divergent criteria: geopolitical boundaries ("China" and "Africa"), natural regions ("Milne Bay"), and natural ("Mekong Delta" and "Salar de Atacama" in Chile) and human-managed ("Stockholm city park") ecosystems. Methodologies varied widely, as well, from ethnographic and focus group approaches to remote sensing and sophisticated computer modeling.[6] MA subglobal assessments also sought to link their activities to divergent policy and public audiences. In this way, the MA enabled subglobal assessors to take advantage of cross-national variation in the methods and integration of risk assessment, enhancing their credibility by tying them to regional evidentiary standards, problem framings, and institutional settings. Strong regional ties have also enabled subglobal assessments to work closely with local and indigenous knowledge holders.

It is worth noting that MA leaders also pursued a parallel approach to bridging scales. In over thirty countries, the MA established "user forums" in which policy and economic actors met regularly to discuss the MA. For each, a local coordinator (individual or organizational) was first identified, who was subsequently responsible for identifying both the rest of the participants as well as the precise modalities and activities of the forum. Like the subglobal assessments, user forums have given considerable flexibility to adapting forums to what "emerges organically in each country," and the resulting forums have taken divergent forms across different countries.[7] In some countries, for example, the forums have taken a strongly technical form, with heavy participation from scientists and midlevel managers from government and the private sector; in other countries, the forums have focused on high-level leadership from the government, NGOs, and indigenous groups. As the MA progresses, a careful, comparative analysis of the subglobal assessments and user forums, paying particular attention to their methodological flexibility and its impacts on issues of communication and engagement with global environmental change, will prove invaluable.

A Practical Approach to Reasoning Together in World Affairs?

What impact do these alternative designs for bridging scales and epistemologies have on the potential for global environmental assessments to promote reasoning together and democratization in global environmental diplomacy? In many ways, it is still too early to offer a full analysis. A large fraction of second-generation global environmental assessments, including the MA, GIWA, and ACIA, are still in progress. That said, the four challenges described above can serve as a starting point.

Capacity Building

Nearly all approaches to regionalization build capacity of some sort—but capacity for whom, to do what? Our model understands capacity very specifically: capacity of individuals and communities around the globe to reason critically about global environmental risks and their implications for day-to-day livelihoods. From this perspective, top-down approaches, such as the IPCC's regional chapters, provide less capacity than approaches that involve regional groups in assessments. Giving regional assessors greater flexibility in design and management (following the MA and ACIA) may also build greater capacity to develop, evaluate, and deliberate methodologies, scope, and meaning derivation than does requiring standardized global approaches (following GIWA).

Epistemic Pluralization

Like capacity building, the multiplication of voices and epistemic perspectives in global environmental governance is stronger in bottom-up approaches to regionalization. Independently organized assessments like ACIA allow regional assessors to diverge sharply from global standards and to choose their own problem framings, evidentiary standards, methodological approaches, institutional models, regional identities, and communication strategies. By contrast, top-down assessments like the IPCC and GIWA frequently generate little in the way of diversity of viewpoint or engagement in their regional assessments.[8] Although they may identify differences in the ways in which global environmental risks play out in regional contexts, they are less likely to fully explore such differences or to connect them effectively to local meanings and policies.

Epistemic Dialogue and Exchange

While both stand-alone and bottom-up approaches to subglobal assessment design provide advantages in terms of pluralizing voices in global policy making, stand-alone assessments offer less potential for creating new deliberative spaces in which multiple styles of reasoning can be brought into mutual dialogue and exchange. The ACIA, for example, is clearly intended as a device not only to help local communities in the Arctic region learn about climate change but also as an effort to communicate the region's vulnerability to climate change to a global audience. The problem with independent assessments like ACIA, however, is that they tend toward "place-based" approaches that are geared solely toward local knowledge and action. For reasoning together to occur, in our model, cultural styles of reasoning must be brought into regular dialogue that promotes mutual understanding and exchange of approaches and ideas. The ACIA accomplishes this to some degree, by bringing global environmental scientists into dialogue with local communities in the Arctic. Other communities are not involved, however.

In many ways, the MA faces the same problem of becoming too place based in its approaches. However, a key facet of the MA subglobal assessments is the collective participation of regional assessors in the MA Sub-Global Working Group. This group meets regularly, is facilitated by a central coordinator at the MA headquarters, and is tasked with producing a subglobal report as part of the MA's publication strategy. Both the subglobal meetings and the report emphasize dialogue and exchange among competing methodologies, approaches, and institutional arrangements as a key element of the Sub-Global Working Group's structure. A preliminary outline indicates that a variety of comparative analyses and efforts to identify best practices from among competing methodologies is a key goal of the subglobal assessment report. Facilitating stronger dialogue between the subglobal and global components of the MA has also occupied an important place in the discourse of the Sub-Global Working Group, and multiple efforts have been made to facilitate exchanges between the MA's subglobal and global participants.[9]

Jurisdictional Orchestration

Have efforts at regionalizing global environmental assessments helped promote appropriate integration and differentiation of multiple styles of reasoning and epistemic frameworks across local, regional, and global scales of decision making? Not much evidence is in yet. It seems clear that the IPCC has not yet contributed to a full integration of national and global climate policies, as the regulatory frameworks and reasoning espoused by the framers of the Kyoto Protocol and the governments of the United States and many developing countries remain far apart from one another. Most of the other assessments, which have taken more flexible approaches to bridging scales and epistemologies, are not yet complete and have not yet been in a position to significantly influence decision making at any scale. One design objective of the MA, however, is to use regional assessments to allow for adaptation to culturally appropriate styles of reasoning that may help promote regional learning, as communities deliberate and exchange views on global issues and rethink their perspectives in forums that are not as politically fraught as global governing institutions. Whether it will achieve this goal or not remains an open question.

Regional assessments may also offer better opportunities than global assessments to link up global environmental governance processes to regional and local policy institutions, enhancing the potential for long-term uptake and implementation of ideas and policies. The key here is that more flexible assessments can be attuned not just to the information needs of regional and local decision makers but also to their frameworks of reasoning. Certainly the MA's subglobal assessments and user forums have developed stronger, more formal, and more long-term connections to policy and business communities at scales other than the globe itself than the IPCC regional assessments have offered. Time will tell whether the global MA is capable of capitalizing on these relationships to better integrate ecosystem governance across scales.

Future Challenges

Regionalizing global environmental assessments is hardly likely to serve as a panacea for overcoming the geopolitical and geographic divides that haunt global environmental governance at the start of the twenty-first century. The push toward conceptual and methodological pluralism is likely to spark resistance among those who see the current impasse on climate change and biodiversity loss primarily in terms of either a failure by scientists to communicate the true extent and consequences of global environmental risks effectively or the unwillingness of political leaders and public to undertake necessary economic, social, and policy reforms. The added cost and organizational complexity of conducting multiple regional assessments is also likely to deter many assessments from investing in regionalization. The MA, for example, has devoted only a fraction of its budget to conducting regional assessments, and the proportion looks only somewhat better when one includes resources used to support coordination of the Sub-Global Working Group.

The explicit mixing of global and subglobal assessments nonetheless offers an interesting line of thought and analysis. Global environmental assessments, however organized, form a central element in the emerging civic epistemology of global civil society. This epistemology can perhaps better capture and communicate the heterogeneities of global environmental change and its meanings for the peoples of Earth if, rather than adopting a single, top-down perspective, it permits expression of a diversity of voices. Such an approach would also better reflect the uncertain state of global epistemologies in international diplomacy. Methods and approaches for producing policy-relevant knowledge on behalf of the entire planet are deeply contested at the moment. Allowing methodological pluralism, reflection, and dialogue within global environmental assessments seems an appropriate response.

We should not forget that science can significantly shape the character of democratic institutions and of democratic civil societies. The design and organization of international scientific assessments may factor strongly in shaping the emergence and success of democracy in global governance. Fostering the capacity of many parts of the globe to reason critically, to express their voices in pluralist forums, to deliberate and exchange ideas, and to coordinate across distinct governance regimes would be a valuable contribution to strengthening global civil society and global democracy. Achieving these goals will require global environmental assessments to go further even than the MA in explicitly bridging scales and epistemologies. To conclude, we offer four thoughts.

First, subglobal assessments must not fall back into the easy comfort of "place-based" assessments: local assessments of local concerns. Subglobal assessments can speak to regional perspectives on global risks as well as assess their regional manifestations. As assessors identify subglobal variations in the causes and impacts of global environmental change, they should also elicit subglobal variations in frames of meaning and styles of reasoning for producing knowledge about global risks.

Second, subglobal assessments should abandon their fixation on geography as the defining organizational characteristic. The point of bridging scales and epistemologies is to find alternative ways to slice up global problems for analytical purposes. Many subglobal processes are not confined geographically. Consider the floral industry. Today, airfreight enables growers scattered across the globe to transport flowers to consumers overnight, creating a global market. Assessing changes in ecological services associated with this market might provide valuable insights into ecosystem dynamics across the globe but might not be captured in a standard regional or global environmental assessment. Consider, too, as another example, the ecological consequences of diasporas, which displace cultural ideas, expectations, and practices across multiple regions.

Third, assessments must reach out in their deliberative mechanisms beyond the experts who participate in the assessment itself. Suppose assessors do manage to find effective means of reasoning together. Will the communities that they represent be able to follow their new logics without themselves being engaged in deliberative activities? If global environmental assessments are to help reduce ideological fissures in global society, they must cease being isolated exercises of expert analysis and start becoming focal points by which whole communities can begin to learn to reason together.

Finally, much more needs to be done to fully evaluate the implications of both reasoning together as an approach to democratizing international governance and of using regionalization as a strategy for achieving this democratization. How do we move beyond the bimodal regionalization strategies (i.e., global and regional) currently used in global environmental assessments to more nuanced, multiscale approaches? What implications would this have for the challenge of orchestrating appropriate jurisdictional relationships among competing epistemic frameworks? Other than regionalization, how might global environmental assessments be reconfigured to promote reasoning together? These questions go beyond the scope of this chapter but will be extremely important in future analyses.

We believe efforts to bridge scales and epistemologies in global environmental assessments must be understood in political as well as epistemic terms, as core elements in the process of creating constitutional foundations for international governance. Acknowledging this fact will inevitably increase the complexity and politicization of efforts to bridge scales and epistemologies. Ignoring it will guarantee that global environmental assessments both fail to live up to their potential as experiments in global democracy and also risk perpetuating deep-seated political inequalities and further exacerbating ideological divides in world affairs.

References

Agarwal, A., and S. Narain. 1991. *Global warming in an unequal world.* New Delhi: Center for Science and the Environment.

Benedick, R. 1991. *Ozone diplomacy: New directions in safeguarding the planet.* Cambridge, MA: Harvard University Press.

Bolin, B. 1994. Science and policy making. *Ambio* 23 (1): 28.

Daemmrich, A., and G. Krucken. 2000. Risk versus risk: Decision-making dilemmas of drug regulation in the United States and Germany. *Science as Culture* 9 (4): 505–33.

Douglas, M., and A. Wildavsky. 1982. *Risk and culture.* Berkeley: University of California Press.

Ellis, R., and C. Waterton. 2004. Environmental citizenship in the making: The participation of volunteer naturalists in UK biological recording and biodiversity policy. *Science and Public Policy* 31 (2): 95–107.

Global International Waters Assessment. 2002. Methodology: Detailed assessment, causal chain analysis, policy option analysis. Report. http://www.giwa.net/methodology/GIWA_Methodology_DA-CCA-POA_English.pdf.

Hacking, I. 2002. *Historical ontology.* Cambridge, MA: Harvard University Press.

Held, D. 2004. *Global covenant: The social democratic alternative to the Washington consensus.* Cambridge, England: Polity.

Iles, A. 2004a. Making seafood sustainable: Merging consumption and citizenship in the United States. *Science and Public Policy* 31 (2): 127–39.

———. 2004b. Patching local and global knowledge together: Citizens inside the U.S. chemical industry. In *Earthly politics: Local and global in environmental governance,* ed. S. Jasanoff and M. Martello, 285–308. Cambridge, MA: MIT Press.

International Arctic Science Committee. 2000. *Arctic climate impact assessment: Report on the 3rd meeting of the Assessment Steering Committee and a scoping workshop.* Stockholm: International Arctic Science Committee.

Jasanoff, S. 1986. *Risk management and political culture.* New York: Russell Sage Foundation.

———. 1995. Product, process, or programme: Three cultures and the regulation of biotechnology. In *Resistance to new technology,* ed. M. Bauer, 311–31. Cambridge: Cambridge University Press.

———. 1998. Harmonization—The politics of reasoning together. In *The politics of chemical risk,* ed. R. Bal and W. Halffman, 173–94. Dordrecht: Kluwer.

———. 2001. Image and imagination: The emergence of global environmental consciousness. In *Changing the atmosphere: Expert knowledge and environmental governance,* ed. C. Miller and P. Edwards, 309–38. Cambridge, MA: MIT Press.

———. 2003. In a constitutional moment: Science and global order at the millennium. *Sociology of science yearbook.* Dordrecht, The Netherlands: Kluwer.

———. 2004. Science and citizenship: A special issue. *Science and Public Policy* 31:90–171.

Jasanoff, S., and B. Wynne. 1998. Science and decisionmaking. In *Human choice and climate change: The societal framework,* ed. S. Rayner and E. Malone, 1–87. Columbus, OH: Battelle.

Keohane, R. 2001. Governance in a partially globalized world: Presidential address, American Political Science Association, 2000. *American Political Science Review* 95 (1): 1–13.

King, L. 2003. Deliberation, legitimacy, and multilateral democracy. *Governance* 16 (1): 23–50.

Krimsky, S., and A. Plough. 1988. *Environmental hazards: Communicating risks as a social process.* Dover, MA: Auburn House.

Lachmund, J. 2004. Knowing the urban wasteland: Ecological expertise as local process. In *Earthly politics: Local and global in environmental governance,* ed. S. Jasanoff and M. Martello, 241–62. Cambridge, MA: MIT Press.

Litfin, K. 1998. *The greening of sovereignty in world politics.* Cambridge, MA: MIT Press.

Martello, M. L. 2004a. Global change science and the Arctic citizen. *Science and Public Policy* 31 (2): 107–17.

———. 2004b. Negotiating global nature and local culture: The case of Makah whaling. In *Earthly politics: Local and global in environmental governance,* ed. S. Jasanoff and M. Martello, 263–84. Cambridge, MA: MIT Press.

Masood, E. 1995. Temperature rises in dispute over costing climate change. *Nature* 378:429.

McCarthy, J., O. F. Canziani, N. A. Leary, D. J. Dokken, and K. S. White, eds. 2001. *Climate change 2001: Impacts, adaptation, and vulnerability.* Oxford: Oxford University Press.

Meyer, A., and T. Cooper. 1994. Ten-to-one against: Costing people's lives for climate change. *Ecologist* 24 (6): 204–6.

Miller, C. 2000. The dynamics of framing environmental values and policy: Four models of societal processes. *Environmental Values* 9:211–33.

———. 2003. Knowledge and accountability in global governance. In *Partial truths and the politics of community,* ed. M. Tetreault and R. Teske, 315–41. Charleston: University of South Carolina.

———. 2004a. Climate science and the making of a global political order. In *States of knowledge: The coproduction of science and social order,* ed. S. Jasanoff, 46–66. New York: Routledge.

———. 2004b. Resisting Empire: Globalization, relocalization, and the politics of knowledge. In *Earthly politics: Local and global in environmental governance,* ed. S. Jasanoff and M. Martello, 81–102. Cambridge, MA: MIT Press.

Parthasarathy, S. 2004. Regulating risk: Defining genetic privacy in the US and Britain. *Science, Technology and Human Values* 29:332–52.

Rayner, S., and E. Malone, eds. 1998. *Human choice and climate change: The societal framework.* Columbus, OH: Battelle.

Stiglitz, J. 2002. *Globalization and its discontents.* New York: Norton.

Takacs, D. 1996. *The idea of biodiversity: Philosophies of paradise.* Baltimore, MD: Johns Hopkins University Press.

Thompson, C. C. 2004. CITES and the African elephant. In *States of knowledge: The coproduction of science and social order,* ed. S. Jasanoff, 67–86. London: Routledge.

Thompson, M., and S. Rayner. 1998. Cultural discourses. In *Human choice and climate change: The societal framework,* ed. S. Rayner and E. Malone, 265–343. Columbus, OH: Battelle.

Verweij, M., and T. Josling, eds. 2003. Deliberately democratizing multilateral organization. Special issue, *Governance* 16 (1): 1–21.

Wynne, B. 1995. Misunderstood misunderstandings. In *Misunderstood misunderstandings,* ed. A. Irwin and B. Wynne, 19–46. Cambridge: Cambridge University Press.

CHAPTER 17

Conclusions

Bridging Scales and Knowledge Systems

Fikret Berkes, Walter V. Reid,
Thomas J. Wilbanks, and Doris Capistrano

Bridging Scales and Knowledge Systems is not an assessment of available knowledge—like its parent, the Millennium Ecosystem Assessment (MA)—nor is it a scientific review. Rather, it is a set of papers exploring issues related to bridging scales and knowledge systems, in particular those concerning the intersection of the two in scientific assessments. The idea of building bridges across scales and knowledge systems is not novel. Geographers have been dealing with scale issues for decades, and a sophisticated literature exists on scale and environmental management (e.g., Cash and Moser 2000). Similarly, the idea of seeking bridges across knowledge systems goes back at least to the 1950s, to C. P. Snow's famous analysis of the divide between the sciences and the humanities (Snow 1993).

Although much experience with global and large regional assessments exists, understanding the processes that affect ecosystem services and human well-being also requires attention to subglobal levels and the plurality of scales and epistemologies. What happens at the global level cannot simply be scaled down to provide an understanding on the ground, and what happens at the local level cannot simply be scaled up to interpret global phenomena (Young 2002). Scale does truly matter (see chapter 2 of this volume). Understanding a complex system, such as a global ecosystem, requires an understanding of all the levels in a hierarchy and the relations among them.

In terms of epistemology, relevant questions include the following:

- What is the appropriate kind of knowledge to deal with ecosystem services and human well-being?
- How should assessments deal with diverse kinds of knowledge, including knowledge held by those who live in a particular place?
- How and to what extent is bridging these knowledge systems possible, desirable, and doable?

A strong argument has been made for searching and accessing the full range of available knowledge (see chapters 9, 10, and 11). However, bringing different epistemologies to the same table is not without its transaction costs. What constitutes legitimate knowledge? How can one mediate between kinds of knowledge in a way that helps the decision maker use the most relevant information and interpretation regarding a particular issue? There is a "politics of knowledge" (chapter 7), just as there is a "politics of scale" (chapter 3).

The twin problems of scale and epistemology are coming under scrutiny in several efforts tackling the broader context of environmental issues (e.g., Walker et al. 2004; Kates et al. 2001). The 1992 United Nations Conference on Environment and Development in Rio de Janeiro drew attention to the significance of traditional knowledge. The Implementation Plan of the 2002 World Summit on Sustainable Development drew attention to scale by "encourag[ing] relevant authorities at all levels to take sustainable development considerations into account in decision-making" (United Nations 2002, sec. 3.18). The phrase "at all levels" appears eighty-one times in the fifty-page document.

Thus, the issues of both scale and epistemology have been on international agendas related to environmental management. Nevertheless, a systematic approach to investigate issues of scale and knowledge systems *together* is relatively novel. It is this area that the chapters in this book explore in connecting environmental sustainability to human needs. The chapters further the development of assessments by asking the questions of how to address issues of scale, how to embrace different knowledge systems in assessments, and how these two kinds of questions may be related.

Elements of Bridging

Addressing issues of scale and knowledge systems in assessments and dealing with other interlinked aspects of ecosystem management and human well-being require pluralism in ideas and approaches, as argued in postnormal science (Funtowicz and Ravetz 1993) and sustainability science (Turner et al. 2003). All of the chapters in this book make this point either explicitly or implicitly. The scope, complexity, and uncertainties around issues of ecosystem and human well-being interactions make it impossible for any one perspective, discipline, or approach to monopolize the answers and solutions. Thus, pluralism that recognizes differences in people's values, interests, institutions, or practices as legitimate and autonomous while also helping people work together in a coherent, mutually beneficial way is a practical necessity.

The growing realization that conventional science based on Western paradigms and systems of knowledge is no longer adequate to deal with complexities of environmental management (Ludwig 2001) and that knowledge is contextual has opened the space for considering other systems of knowledge in scientific assessments. However, as several authors here stress, scientific assessments are social and inherently political processes in which competing interests, values, worldviews, and options for action are negotiated. The definition of boundaries, the selection of scale, and the explicit or implicit framing of hierarchy of values and systems of knowledge are all part of this negotiation (chapters 3, 7, 8, 11, and 16).

Scientific assessments require sharing of information, deliberative exchanges, or "reasoning together" among key stakeholders—policy makers, resource managers, the private sector, the civil society, and the public at large—which presents opportunities for mutual learning (chapter 16). This mutual learning can constitute one form of bridging scales and knowledge systems. The process of "reasoning together" enhances legitimacy of policies and promotes more democratic environmental governance when the process is designed and managed well, provides for broad representation of stakeholder views, and involves different stakeholders from different levels (chapters 3, 7, and 16).

The cases included in this collection provide examples of attempts to cross the many dividing lines that hinder the communication, mutual learning, participation, and collaboration needed for assessments to successfully address interlinked issues of scale and knowledge. Several strategies have proven critically important for positive outcomes.

First is the recognition and judicious use of a mix of perspectives, methodological approaches, tools, and techniques that allow for broad stakeholder participation and the accommodation of nonformal, undocumented, or localized knowledge (chapters 9, 10, 11, 12, and 16). This includes, for example, exchanging and cross-validating paradigms (chapter 11); complementary use of qualitative and quantitative measures as well as participatory and conventional methods from a range of disciplines (chapters 9 and 14); the combination and innovation of indicators, means of measurement, and monitoring (chapters 10, 12, and 14); and development of shared visions and narratives of a common future (chapter 15).

Second is the use of methodologies and analytical approaches that allow a more complete description and understanding of the relationships across scales and of the similarities and differences of processes and phenomena at different scales. This includes the analysis of scale-dependent and scale-independent factors, the use of upscaling and downscaling techniques, the identification of characteristic scales of different processes or phenomena, and the design of monitoring systems to detect relevant changes at different scales (chapters 2, 4, and 5).

Third is the creation of forums and platforms for negotiation, conflict resolution, decision making, trust building, and joint action. This sometimes requires new mechanisms, such as multistakeholder consultations, and redefinition of roles and patterns of interaction among key actors (chapters 6, 9, and 10). It can also involve creating different types of institutions or finding new ways of responding to threats and opportunities (chapters 12 and 13) and providing for flexibility to allow for additional interested stakeholders to participate (chapter 10).

Fourth is capacity building and development of new skills for cross-scale analysis (chapter 2) and new skills among stakeholders, particularly those who have traditionally been excluded or marginalized (chapter 12). Training, exposure, and other modes of experiential learning can help level asymmetries in information, skills, and levels of confidence among stakeholders and can facilitate communication and more mutually beneficial interactions across the various divides (chapters 10 and 14).

Fifth is facilitation, mediation, and translation of information and meanings between and among stakeholders. Individuals, groups, or organizations can play these roles that have proven essential to bridging across scales and systems of knowledge (chapters 6, 7, 10, 11, 12, and 14). This includes reporting back on assessment results and research findings to informants and participating stakeholders, which unfortunately is rarely practiced (chapters 10 and 14).

Bridging Knowledge Systems

If one point of agreement exists among the authors, it is that bridging knowledge systems is not easy. However, most agree that it is nevertheless important and necessary. The barriers to bridging include power differences (chapters 7 and 8), centralization and domination of decision making by government (chapters 8, 14, and 16), and scientists' common lack of respect for local and traditional knowledge (chapters 8 and 14). This list of barriers is not meant to be comprehensive. Other barriers also exist, including the following two that emerge from the analysis by Ericksen et al. (2005) of subglobal MA cases: (1) the lack of a common "language" and of an agreed set of assumptions about how the world works, and (2) the absence of a common means of verifying the veracity of knowledge.

The issue of power as a barrier to bridging is a systemic difference from which a number of other barriers emerge. In fact, the power issue is so fundamental that an entire school of thought argues that indigenous knowledge and science should not be bridged. According to this argument, "bridging" results only in taking indigenous knowledge out of its cultural context and inserting it into the very structures that disempower indigenous people in the first place. This not only perpetuates but also exacerbates existing power imbalances (Nadasdy 1999). Some argue that the politics of power may mean that an attempt at bridging could result only in co-option (box 17.1).

Power differences are a problem not only with indigenous communities but also with other minorities and perhaps with resource-dependent rural groups in general. The issue is recognized in the development literature in Sen's treatment (1999) of the idea of development as being all about human empowerment. It follows, therefore, that mechanisms for bridging need to address the issue of power, among the various other barriers.

Joint problem solving, which appears in several chapters, is one mechanism to help indigenous knowledge holders operate as equals with scientists and technical people. In the Arctic Borderlands Ecological Knowledge Co-op (discussed in chapter 10), aboriginal parties are engaged with scientists in long-term joint management of the environment. When local experts from the Gwich'in community of Old Crow reported that wetland lakes were drying up,

Box 17.1

Indigenous People's Views on Risks Associated with Bridging Knowledge Systems

Participants from indigenous groups in the workshop "Bridging Epistemologies—Indigenous Views," held during the Bridging Scales and Epistemologies Conference in Alexandria, Egypt (March 17–20, 2004), prepared a summary of the workshop in the form of comments made during the workshop. The comments included the following.

- A bridge between epistemologies is not possible or not desirable, because it produces invasion and domination. We can only—consciously—sit down at a table of negotiation and dialogue in a world where many worlds (or epistemologies) are welcome, where we can talk between us, and also talk with modern science.
- There is an ethical responsibility for scientists to be clear about the values, world-views and cosmovisions that are embedded in their approaches to ecosystem assessment and whose purposes are being served by that assessment. Scientists and development agents need to be critical and clear about the risks and benefits for indigenous people from assessing ecosystem goods and services, and of course they need to engage indigenous people in this risk assessment from the outset and develop mutually agreed positions.
- Local people can easily cross the bridge to modern science. As a matter of fact, they have been trying to adjust to the modern world dominated by modern science for generations. Because of the assimilationist attitude of modern science, local people have started to realize the losses of their identity, culture and self.
- To build bridges, indigenous communities need to be empowered to translate their own science in a culturally appropriate way for all people to understand and move forward and thus control how and where traditional knowledge is used, without outsiders being the "expert."

scientists followed up on these observations and confirmed the findings with remote sensing studies. In chapter 11, the task is the in situ conservation of native plants of the Andes. The indigenous people there are in the lead, and outside technical experts come to work with campesinos "with eyes, ears and heart wide open." In chapter 12, although technical experts produce the weather forecasts for communities in semiarid southeastern India, their communication to the local level strongly depends on understanding and valuing villagers' perceptions of rainfall prediction.

The development of working relationships among holders of different kinds of knowledge takes time, typically on the order of ten years based on the comanagement literature (Berkes 2002). Hence, the ten-year-old case in chapter 10 is not an exception. Mutual trust and respect, both of which are slow to build, are preconditions for bridging epistemologies, because trust lubricates collaboration (Pretty and Ward 2001). The building of both trust and mutual respect can be assisted through appropriate institutional arrangements. Both are important for the social learning that can arise from collaborative problem solving, consistent with Wenger's emphasis (1998) on learning as participation.

Several of the cases in this book involve what Cash and Moser (2000) have called "boundary organizations." Originally, the term applied to organizations at the scientist–decision maker boundary. But more broadly, the term may apply to organizations that mediate the relationship of science to local and traditional knowledge and that stimulate collaboration. In chapter 11, the Andean Project for Peasant Technologies plays this role. In chapter 12, it is the community-managed village knowledge centers. In chapter 10, it is the Arctic Borderlands Co-op itself.

A unique mechanism for bridging involves the use of scenarios (chapter 15). The authors report on four MA experiences that used scenario development as a method for incorporating multiple epistemologies. The results seem mixed but promising; "storytelling" as the basic idea behind scenario development works well with indigenous thinking. In chapter 15, Bennett and Zurek consider the experiences successful in generating and integrating both qualitative and quantitative information into the scenarios. However, one needs to be cautious about the issue of "what counts as knowledge," given that there could be a major gap between "local locals" who often speak in metaphors and indigenous advocates who claim to speak for them (chapter 7).

Some of the various ways of bridging knowledge systems are summarized in table 17.1. In the first three cases, knowledge production is local and knowledge integration is generally guided by the local partner. In the other four cases, local knowledge and views supplement the scientific and technical approach or are integrated into it (or both).

Bridging Scales

An objective of "bridging scales" can mean a variety of things: understanding how processes and phenomena differ according to scale (geographic, temporal, or institutional), understanding how processes and phenomena interact across different scales, and focusing on a single scale of interest but ensuring

Table 17.1

Characterizing the cases in this volume by style of knowledge bridging and the degree to which the bridging process is dominated by one kind of knowledge or another

Case	How Knowledge Systems Are Bridged
Conservation of Andean cultivated plants (chapter 11)	Technical experts contribute to plant biodiversity assessment, within a framework guided by an indigenous approach to in situ conservation and indigenous worldview (cosmovision).
People's biodiversity registers (chapter 13)	Biodiversity management committees, organized at the level of municipalities and village councils, document local biodiversity and associated knowledge at the grass roots.
Arctic Borderlands Ecological Knowledge Co-op (chapter 10)	Ecological monitoring uses both kinds of knowledge in a program designed to involve the two kinds of knowledge and knowledge holders as equal partners.
Rainfall prediction in Tamil Nadu, southeast India (chapter 12)	Weather forecasts produced by technical experts respond to community needs and are "translated" by village knowledge centers into practical information that can be used alongside traditional weather forecasting.
Use of scenarios to integrate different kinds of knowledge into four Millennium Ecosystem Assessment cases (chapter 15)	Various stages of scenario development seek to incorporate information and views from more than one body of knowledge through the participation of a diversity of stakeholders and perspectives.
Southern African Millennium Ecosystem Assessment (chapter 9)	Informal and tacit assessments of the local people are brought into the process to improve the robustness and coverage of the assessment.
Portugal Millennium Ecosystem Assessment (chapter 4)	Scenario development is used as a mechanism to fully involve stakeholders in the assessment process. In addition, a qualitative approach to ranking the condition of ecosystem services provides a mechanism to integrate qualitative and quantitative information.

awareness of the possible importance of a multiscale context (Millennium Ecosystem Assessment 2003).

Approaches to bridging scales have generally involved three kinds of strategies (Wilbanks 2003): (a) integrating scale-related information at a single scale of interest, often either an intermediate ("regional") or a local scale, (b) seeking a metascale synthesis, or (c) concentrating on cross-scale interactions and mechanisms, such as boundary organizations.

Several of the chapters in this volume are concerned primarily with bridging scales (chapters 2, 4, 5, 9, and 16). Others are multiscale in perspective but not cross-scale in focus (Chapters 6, 10, and 14). Still others are local in perspective but within a larger structural context (chapters 11, 12, and 13), emphasizing potentials to learn from local knowledge. Regardless of the focus, however, all show that environmental assessment is rooted in a definition of the scale of attention, that scale matters (chapters 2 and 3), and that a particular scale cannot be totally divorced from other scales.

Barriers to effective cross-scale analysis are legion. Data are rarely available for processes at all relevant scales; even where comparable data may be available, rarely have studies explored the relevant causal mechanisms for different processes at different scales (chapters 2, 4, and 9). In some cases, relevant information concerning processes at particular scales may be held by local people or practitioners, but the array of barriers to bridging knowledge systems effectively makes it difficult to fully incorporate that knowledge in an assessment (chapters 3 and 10). Where data are available only for certain scales, progress has been made in developing techniques for upscaling and downscaling information; but questions remain about the challenge of understanding what types of information are scale dependent or scale independent (Wilbanks 2003).

Methodologically, the most serious challenges in bridging scales are in tracing out and understanding cross-scale interactions, for two principal empirical reasons. First, most databases are scale specific rather than scale crossing. For example, the regional climate and weather forecast information described in chapter 12 does not include locale-specific information that would ultimately describe the local weather patterns. Second, most environmental analyses and assessments focus on a particular scale of interest rather than on cross-scale linkages and transfers. Partly as a result, conceptual frameworks are also incompletely developed, although some basic dimensions have been identified (chapter 2; see also Association of American Geographers 2003).

As a whole, the chapters suggest a number of directions for further investigation in bridging between scales, although these few studies can hardly be considered the final word (table 17.2). One issue cutting across any discussion of conventional models for bridging scales is the intent of the bridging, especially when the objectives are related to governance and decision making rather than knowledge enhancement (chapter 3; see also MA 2003).

Perhaps most significant of all, taken together the chapters demonstrate

Table 17.2
Approaches for bridging scales

Approach	Examples	How Scales Are Bridged in Assessments	Issues
Integration at a single scale	Chapters 10 and 11; Schellnhuber and Wenzel 1998; Mark 2000.	Use of scenarios; participatory deliberation involving parties knowledgeable about different scales; use of graphics; upscaling and downscaling of data; local experts to assist in data integration and interpretation	Combining quantitative/qualitative data; combining analytical and deliberative processes of reasoning; local capacity; tunnel vision in losing sight of processes at other scales that are of significance
Multiscale synthesis	Chapters 2, 4, 5, 6, and 9; Kasperson, Kasperson, and Turner 1995; Wilbanks 2003	Aligning "issue sets" in comparative attention to different scales; use of scenarios and narratives; acceptance of plurality of views and perspectives; use of graphics; comanagement of assessment processes; use of assessment products	Data availability; lack of consistency in form and quality; lack of comparability in assumptions behind different data sources; frequent shortage of conceptual structures for synthesis; demands for thoughtful and creative deliberation
Single scale but with analysis of cross-scale linkages and flows	Chapters 6 and 14; Association of American Geographers 2003; Cash and Moser 2000	See the cell directly above, especially comanagement	Data limitations; conceptual limitations

that bridging scales is not only desirable but also possible in many cases. This insight can help increase the sensitivity of assessments and the effectiveness of actions based on them. It also shows that the available tools and perspectives are sufficient to support multiscale bridging, even where cross-scale linkage and flow data are limited.

The Intersection of Scale and Knowledge

The interrelationships between issues of scale and knowledge illuminated by the chapters of this book are complex, but several patterns are apparent. First, incorporating multiple knowledge systems can benefit the information content and use of assessments undertaken at any scale, but the "scope" over which different knowledge systems have the potential to contribute differs across scales. For example, local knowledge of weather indicators adds value to forecasts over periods of days or weeks but is less important for forecasts of seasonal or annual weather variation (chapter 12). Conversely, global scientific information can characterize global patterns of climate change effectively, but it has serious shortcomings in providing solutions given the site-specific context and constraints in which any solution must be implemented (chapter 6).

Because climate change and other complex systems phenomena occur at multiple scales, no single level is the "correct" one for analysis. Climate change cannot be understood at the global level alone, just as it cannot be understood at the local level alone. Since coupling occurs between different levels, the system must be analyzed simultaneously across scale. Hence, the overwhelming emphasis on global circulation models in climate change research has created a mismatch between global science and the knowledge that is needed to act locally (Wilbanks and Kates 1999). Although important elements of the needed local information can be generated from indigenous knowledge, there are limits as to the kind of information that can be accessed or used (Berkes and Jolly 2001).

The scope over which different knowledge systems can contribute is bounded both by scale and by issue, although these boundaries tend to be much more encompassing than is commonly assumed (as illustrated by chapters 10 and 13). But there are limits regarding the kind of useful information. For example, scientific knowledge can add little value to traditional understanding of the local cosmovision, and traditional knowledge will add little value to understanding the paleorecord of Earth's climate history.

Second, no simple scale-dependent hierarchy related to knowledge systems exists. This is not to say that there are no scale-dependent features at all. For example, local and traditional knowledge tends to be more context dependent than scientific knowledge, and thus some aspects of this knowledge may be more relevant or meaningful at local scales. But at the same time, many aspects of this knowledge are highly relevant at other scales. Indeed, as Brosius notes in chapter 7, the tendency has been for scientists to turn to local knowledge holders for their understanding of the natural world (an aspect of knowledge that may be very scale dependent) yet to ignore their knowledge of the political world (an aspect that may be highly relevant at other scales). What emerges is a view of highly overlapping features concerning the value, relevance, and utility of different knowledge systems at different scales.

Depending on the issue and the scale being addressed, the utility of different forms of knowledge may vary. Nevertheless, it would be extremely rare to encounter an issue related to environment and development where multiple knowledge systems (local, traditional, natural science, social science, practitioner knowledge, and so forth) did not add value to the information or the influence of the assessment or both. The design choice is not simply one between a centralized versus decentralized assessment system. Rather, it is one that integrates the unique capacities at the top, bottom, and middle of the scale (Cash and Moser 2000). Since different kinds of knowledge correspond to different scales, bridging both scales and knowledge helps to bring complementary knowledge, skills, and capacity to bear on the assessment challenge.

While the chapters document the potential value of multiple knowledge systems across scales, they also provide clear evidence that society does not take full advantage of this potential. A number of barriers exist. At any given scale, there are few mechanisms to enable the incorporation of different systems of knowledge into an assessment or planning process, and the appropriate mechanism may well differ at different scales. Several chapters in this book examine early attempts to establish such mechanisms (for example, the case studies of participatory fisheries management in Brazil described in chapter 14, the weather forecasting mechanism in India described in chapter 12, or the use of scenarios discussed in chapter 15).

But in many cases these mechanisms are limited by an unsupportive institutional context or a lack of respect or recognition by other stakeholders (chapters 8, 11, and 14). Even where the institutional context is supportive, significant challenges remain. These include the difficulty of developing mechanisms that validate knowledge effectively; difficulties in communicating concepts and ideas; and fundamental gaps in the capacity of people holding different types of knowledge to represent that knowledge effectively in novel processes or arrangements. Box 17.2 provides a practical checklist for environmental assessment practitioners to help address issues of multiple scales and epistemologies.

Finally, the chapters document the fundamental political dimension of this intersection of scale and knowledge. The choice of scale (and the linked choice of what systems of knowledge will contribute most significantly) or the choice of knowledge systems (and the linked choice of what scale will dominate) influences, and is often influenced by, the agenda for decision making; it also influences which interests are most strongly reflected in the findings (chapters 3

Box 17.2

Strategies for Bridging Scales and Knowledge Systems: A Checklist for Assessment Practitioners

- Does the assessment allow for pluralism by recognizing a mix of perspectives?
- Does the assessment recognize a mix of methodological approaches, tools, and techniques that allow for broad stakeholder participation?
- Does the assessment accommodate nonformal, undocumented, or localized knowledge?
- Does the assessment use methodologies and analytical approaches that allow a more complete description and understanding of relationships across scales?
- Does the assessment use methodologies and analytical approaches that allow an understanding of similarities and differences of processes and phenomena at different scales?
- Does the assessment use forums and platforms for negotiation, conflict resolution, decision making, trust building, and joint action?
- Does the assessment undertake capacity building and development of new skills for cross-scale analysis?
- Does the assessment foster the development of new skills among stakeholders, particularly for those who have been usually excluded or marginalized?
- Does the assessment undertake facilitation, mediation, and translation of information and meanings between and among stakeholders?
- Does the assessment facilitate the building of mutual trust and respect between holders of different kinds of knowledge?
- Does the assessment allow for enough time for building mutual trust and respect?
- Are there individuals, groups, or organizations (boundary organizations) involved in the assessment that can play bridging roles?
- Does the assessment report back on assessment results and research findings to informants and participating stakeholders?
- Does the assessment use a variety of modes of communication, including those with heuristic value, such as scenarios and graphics, and processes of group deliberation?
- Are there opportunities for mutual learning?

and 7). Chapter 16 provides a positive vision for how this political dimension, if incorporated into assessment design, could in fact help to democratize environmental governance. But this political dimension also leaves the decision-making process open to strategic interventions by particular stakeholders to shape outcomes in their own interests through the choice of scale. There is some understanding of this phenomenon in the area of indigenous knowledge (Nadasdy 1999) and comanagement (Agarwal 2001; Berkes 2002), but an explicit recognition of this political dimension of scale and knowledge in the assessment literature is overdue.

Conclusions

The chapters in this book demonstrate that both the information contained in assessments and the influence of assessments can be enhanced by incorporating multiple knowledge systems and multiple scales. No one scale, time frame, or approach to creating knowledge is fundamentally privileged over others. All offer insights, and each has contributions to make. For instance, the book demonstrates the value of global scales in capturing broad understandings from science, technology, and global trends, but also the value of local scales in capturing local knowledge and better understanding of certain processes. It demonstrates the rich texture of realities rooted in local-scale, fine-grained interactions, as sources of learning and essential elements in ensuring that action agendas are effective and equitable, without detracting from the importance of knowledge and resources for action that also exist at more general scales.

Yet, the selection of scale and knowledge systems to incorporate in an assessment is not politically neutral. The choice of scales and sources of knowledge in an assessment may be primarily driven by the desire to enhance the quality of information in the assessment or its use by decision makers, or it may be driven by the desire to empower (or disempower) specific groups or to serve an advocacy role. This political dimension is an inherent feature of assessment design that deserves to be more explicitly recognized by practitioners. While no assessment could be entirely politically neutral, it is clear that assessments that strive to incorporate information and perspectives from multiple scales, and that do not create artificial barriers to legitimate sources of knowledge, are likely to be more credible, balanced, and accurate from the vantage point of all stakeholders.

Addressing scale and knowledge issues *together* brings further potential benefits. Since different kinds of knowledge correspond to different levels, bridging both scales and knowledge helps do a better job than bridging scales or bridging knowledge alone. Different social actors at different levels of organization will possess complementary knowledge, skills, or capacities. The potential efficiency in partnerships can be captured by bringing together these comparative advantages. Doing so also serves the political problem referred to above, since a single scale and a single knowledge system will be more likely to favor particular stakeholders than will a broader process. In the long run, this approach can help to democratize environmental governance.

Bridging scales and knowledge systems is realistically possible in many cases; it is not just an academic ideal. But significant barriers do exist, and both research and further experience will be needed to reduce those barriers. There is no one formula for bridging knowledge systems or bridging scales; bridging may take one of many forms, as appropriate to the situation. But there are at least three institutional and procedural characteristics shared by the effective experiences described in this volume.

First, boundary organizations often play an important role in helping to bridge scales and epistemologies. To be effective, most institutions must focus on particular scales; we cannot expect all institutions to deal with all scales and all systems of knowledge. But an important niche exists for individuals and institutions that can establish expertise and experience in helping to promote information flow and analysis across scales and across knowledge systems.

Second, processes designed to bridge scales and knowledge systems require considerable time and effort. Time is needed to address many logistical and procedural issues, such as agreeing on a conceptual framework and harmonizing the data. Most important, time is necessary for building trust and developing mutual respect, the two preconditions for effective bridging processes.

Third, bridging usually calls for using a variety of communication modes rather than choosing a single "optimal" mode. It is very rare that a single mode of communication will in fact be optimal at all scales and for all different knowledge systems. In the experiences examined in this volume, the most effective mechanisms for communication were typically those with strong heuristic value, such as scenarios and graphics, and processes of group deliberation, such as scenario building and visioning.

The costs in both time and expenses associated with assessment processes that embrace multiple scales and multiple knowledge systems can be high, and depending on the goal or purpose of an assessment, these costs may not be easy to justify. Historically, it has been the exceptional assessment that has used multiple scales or multiple knowledge systems. But in our view, we are now at a stage where it should be assumed that an assessment process would address multiple scales and incorporate multiple relevant systems of knowledge, unless a more limited assessment could be justified. The benefits of bridging are clear, and while many obstacles remain, a wide array of methods, tools, and examples now exists that can inform future assessments.

References

Agarwal, B. 2001. Participatory exclusions, community forestry, and gender: An analysis for South Asia and a conceptual framework. *World Development* 29:1623–48.

Association of American Geographers. 2003. *Global change and local places: Estimating, understanding, and reducing greenhouse gases*. Cambridge: Cambridge University Press.

Berkes, F. 2002. Cross-scale institutional linkages for commons management: Perspectives from the bottom up. In *The drama of the commons,* ed. E. Ostrom, T. Dietz, N. Dolšak, P. C. Stern, S. Stonich, and E. U. Weber, 293–321. Washington, DC: National Academy Press.

Berkes, F., and D. Jolly. 2001. Adapting to climate change: Social-ecological resilience in a Canadian western Arctic community. *Conservation Ecology* 5 (2): 18. http://www.consecol.org/vol5/iss2/art18 (accessed May 3, 2006).

Board on Sustainable Development, National Research Council. 1999. *Our common journey: A transition toward sustainability.* Washington, DC: National Academy Press.

Cash, D. W., and S. C. Moser. 2000. Linking global and local scales: Designing dynamic assessment and management processes. *Global Environmental Change* 10:109–20.

Ericksen, P., E. Woodley, G. Cundill, W. Reid, L. Vicente, C. Raudsepp-Hearne, J. Mogina, and P. Olsson. 2005. Using multiple knowledge systems in sub-global assessments: Benefits and challenges. Chap. 5 in Millennium Ecosystem Assessment, *Ecosystems and human well-being,* vol. 4 of *Multiscale assessments: Findings of the Sub-global Assessments Working Group.* Washington, DC: Island Press.

Funtowicz, S., and R. Ravetz. 1993. Science for the post-normal age. *Futures* 25:739–55.

Kasperson, J., R. Kasperson, and B. Turner, eds. 1995. *Regions at risk.* Tokyo: United Nations University Press.

Kates, R. W., W. C. Clark, R. Corell, J. M. Hall, C. C. Jaeger, I. Lowe, J. J. McCarthy, et al. 2001. Sustainability science. *Science* 292:641–42.

Ludwig, D. 2001. The era of management is over. *Ecosystems* 4:758–64.

Mark, D. M. 2000. Geographic information science: Critical issues in an emerging cross-disciplinary research domain. *URISA Journal* 10:45–54.

Millennium Ecosystem Assessment (MA). 2003. Dealing with scale. In *Ecosystems and human well-being: A framework for assessment,* 107–26. Washington, DC: Island Press.

Nadasdy, P. 1999. The politics of TEK: Power and the "integration" of knowledge. *Arctic Anthropology* 36:1–18.

Pretty, J., and H. Ward. 2001. Social capital and the environment. *World Development* 29:209–27.

Schellnhuber, H.-J., and V. Wenzel, eds. 1998. *Earth system science: Integrating science for sustainability.* Berlin: Springer-Verlag.

Sen, A. K. 1999. *Development as freedom.* Oxford: Oxford University Press.

Snow, C. P. 1993 [1959]. *The two cultures.* Cambridge: Cambridge University Press.

Turner, B. L., II, R. E. Kasperson, P. A. Matson, J. J. McCarthy, R. W. Corell, L. Christensen, N. Eckley, et al. 2003. A framework for vulnerability analysis in sustainability science. *Proceedings of the National Academy of Sciences of the United States* 100:8074–79.

United Nations. 2002. *World Summit on Sustainable Development plan of implementation.* Johannesburg, South Africa: United Nations.

Walker, B., C. S. Holling, S. R. Carpenter, and A. Kinzig. 2004. Resilience, adaptability and transformability in social-ecological systems. *Ecology and Society* 9 (2): 5. http://www.ecologyandsociety.org/vol9/iss2/art5/ (accessed May 3, 2006).

Wenger, E. 1998. *Communities of practice: Learning, meaning and identity.* Cambridge: Cambridge University Press.

Wilbanks, T. J. 2003. Geographic scaling issues in integrated assessments of climate change. In *Scaling issues in integrated assessment,* ed. J. Rotmans and D. Rothman, 5–34. Linne, The Netherlands: Swets and Zeitlinger.

Wilbanks, T. J., and R. W. Kates. 1999. Global change in local places: How scale matters. *Climatic Change* 43:601–28.

Young, O. 2002. *The institutional dimensions of environmental change: Fit, interplay and scale.* Cambridge, MA: MIT Press.

NOTES

CHAPTER 4. **Assessing Ecosystem Services at Different Scales in the Portugal Millennium Ecosystem Assessment**

1. We define *intensification* as the increase in the level of production per unit of land; *extensification,* as the decrease in the same.

CHAPTER 7. **What Counts as Local Knowledge in Global Environmental Assessments and Conventions?**

1. In making a distinction between these two kinds of actors, I should note that a significant amount of work is being done on the issue of representation by scholars interested in theories of deliberative democracy. See Benhabib 1996, Dryzek 1990, and O'Neill 2001.
2. Particularly influential in this respect have been Eric Wolf (1982), Sidney Mintz (1985), William Roseberry (1989), and Immanuel Wallerstein (1974). See also Schneider and Rapp 1995 and Dirks, Eley, and Ortner 1993.

CHAPTER 8. **Bridging the Gap or Crossing a Bridge?**

1. For a discussion on the contingent nature of scientific knowledge, also see Turnbull (2000).

CHAPTER 11. **Cosmovisions and Environmental Governance**

1. These CBOs are small institutional setups founded formally as nongovernmental organizations by graduates of the Course on Andean Campesino Agriculture that PRATEC offered from 1990 to 1999 in agreement with the state universities of Ayacucho and Cajamarca. They are autonomous from the

administrative and financial points of view.

2. GEF introduced a new project category in the development vocabulary: incremental-costs projects. The idea is that GEF provides additional funds to make projects that have been chosen for implementation by national governments environment-friendly. However, the understanding of incrementality adopted here is much more encompassing.
3. The *ayllu* is the extended family inhabiting a *pacha,* which comprises not only humans but the deities and natural entities as well.
4. Applying the term *organicity* to Andean communities refers to an attribute pertaining to a living organism. This term contrasts with the "organization" brought by external, state institutions for implementing development projects.
5. One such ritual pilgrimage takes place in June a few days before the winter solstice in the region of Apu Ausangate, a sacred mountain in the Cusco area, under whose protection large numbers of pilgrims congregate in a regional festival with the name of Qoyllor Riti.
6. However, "sameness" has to be appreciated as belonging to the particular Andean cosmovision while taking into account that translation may distort the meaning.
7. According to its Latin etymology, "to accompany" means to share bread together.
8. An activity now abandoned is the promotion of seed festivals because they were found to be alien to seed regeneration as practiced by the campesinos.
9. PRATEC conducted an annual course on Andean peasant agriculture from 1990 to 1999 and currently offers a master's program on biodiversity and Andean Amazonian campesino agriculture in agreement with a national university.
10. Bruno Latour (1999b) proposes building good common worlds.

CHAPTER 14. **Barriers to Local-level Ecosystem Assessment and Participatory Management in Brazil**

1. State extractive reserves existed since 1988 in the state of Acre. Both state and federal extractive reserves are a consequence of the rubber-tappers' grassroots movement, supported by environmental groups, which lobbied the government to create a new form of conservation unit in Brazil (Fearnside 1989; Allegretti 1990). This movement started during the 1970s in response to rubber-tappers' displacement from the forest areas due to unregulated increase of the agricultural frontier (Brown and Rosendo 2000).
2. In formal arrangements local people have legal right to participate in decision-making (i.e., extractive reserves), whereas in informal arrangements,

local peoples' participation in decision making depends on a government agent's willingness to accept such participation.

3. In September 2005, there were in Brazil forty-two federal extractive reserves (thirty-one inland and eleven marine) and many others were being created; one federal sustainable development reserve (another category of formal co-management arrangement); and several less formal participatory initiatives of resource management (e.g., the "fishing accords" in the lower Amazon river [Castro 2000, Castro and McGrath 2003]).
4. This project follows the MA methodological approach and started as a potential pilot project for the Millennium Local-level Ecosystem Assessment after the workshop "Linking Local and Regional Assessments to International Ecosystems Assessments," World Resource Institute, Winnipeg, Manitoba, Canada, September 20–21, 1999.
5. SUDEPE: Superintência para o Desenvolvimento da Pesca; IBAMA: Instituto Brasileiro do Meio Ambiente e dos Recursos Naturais Renováveis.
6. This has been a result of key government agents' willingness to listen to fishers' concerns, demands, and ecological knowledge—an exception during the 1980s.
7. The local elite often encompasses fishers who became intermediaries and who now fish rarely, fishers who are more capitalized and usually hire others to fish for them, and fishers with higher education (and other sources of income) who are able to take advantage of their formal knowledge.
8. The breakdown of traditional fisheries management systems because of outside socioeconomic influences seems a trend in many coastal fishing communities in Brazil (Diegues 1983; Cordell and McKean 1992; Begossi 1998; Seixas and Berkes 2003; Kalikoski and Vasconcellos 2005).

CHAPTER 16. **The Politics of Bridging Scales and Epistemologies**

1. For example, the U.S. Environmental Protection Agency must publish the scientific reasoning it uses in setting regulatory standards in the *Federal Register*.
2. Regulatory standard setting in the United States provides numerous opportunities, for example, for public comment periods and adversarial administrative hearings in which diverse views can be expressed.
3. Letter from Kamal Nath, Indian environment minister and head of Indian delegation to the UN Framework Convention on Climate Change Conference of Parties, to Heads of Delegation, March 24, 1995.
4. See, e.g., T. Wakeford et al., letter to the editor, *Nature,* December 12, 1995.
5. Considerably greater detail about the regional components of recent global environmental assessments can be found on the Internet. For example,

IPCC: http://www.ipcc.ch; GIWA: http://www.giwa.net; MA: http://www.millenniumassessment.org; and ACIA: http://www.acia.uaf.edu.

6. Information per author interviews and e-mail surveys of participants in MA subglobal assessments. Additional discussions of the MA subglobal assessments can be found on the MA Web site at http://www.millenniumassessment.org/en/subglobal.overview.aspx.
7. Author interview with Nicholas Lucas, 2003; information about the MA's user engagement strategies can be found on the MA Web site at http://www.millenniumassessment.org/en/partners.aspx.
8. Some readers may object that epistemic pluralism is achieved at the expense of methodological consistency across regions that can inform our understanding of global environmental change in different parts of the world. Part of the point of this chapter, however, is that epistemic differences are not just in the biophysical aspects of environmental risk but also in how environmental risks are understood and valued, and that standardization, however justified in the name of methodological consistency, operates to exclude voices and perspectives from global debate unless it is achieved through open, deliberative processes. In other words, if the objective of global environmental assessments is to build a legitimate empirical basis on which to make global policies, it is important that the epistemic frameworks used to formulate these claims not be prematurely closed.
9. Author interviews with and e-mail surveys of MA Sub-Global Working Group leaders and participants.

LIST OF AUTHORS

The e-mail address of the corresponding author for each chapter is listed.

K. P. (Prabha) Achar, a zoologist based in India, has worked extensively, with the help of college students, to document people's knowledge.

Elena Bennett, an ecosystem ecologist, is an assistant professor in the Department of Natural Resource Sciences and the McGill School of the Environment at McGill University in Montreal, Quebec, Canada. (elena.bennett@mcgill.ca)

Fikret Berkes works at the interface of social and ecological systems and is a professor of natural resources and Canada Research Chair in community-based resource management at the University of Manitoba in Winnipeg, Canada. (berkes@cc.umanitoba.ca)

Emily Boyd, a social scientist, is a postdoctoral researcher at the Centre for Transdisciplinary Environmental Research, Stockholm University, Sweden, and a visiting fellow at the Tyndall Centre for Climate Change Research, University of East Anglia, in Norwich, United Kingdom. (emily@ctm.su.se/ e.boyd@uea.ac.uk)

J. Peter Brosius, an anthropologist, is an associate professor in the Department of Anthropology at the University of Georgia, Athens, in Georgia, United States. (pbrosius@uga.edu)

Doris Capistrano, a resource economist, is the director of the Forests and Governance Programme at the Center for International Forestry Research in Bogor, Indonesia.

Georgina Cundill is a social ecologist at Rhodes University in Grahamstown, Eastern Cape, South Africa.

Chris Davis is a geographer in Seattle, Washington, United States. He directs the Community and Environment Spatial Analysis Center (CommEn Space), a nonprofit organization with offices in the Pacific Northwest and the Northern Rockies that supports nongovernmental organizations through conservation planning, remote sensing, and geographic information systems. (chris@commenspace.org)

Michael Davis, a historian and policy analyst, is an independent specialist and consultant in Woden, Australian Capital Territory, Australia. (mdavis@pcug.org.au)

Tiago Domingos, an ecological economist, is an assistant professor in the Section of Environment and Energy, Department of Mechanical Engineering of the Instituto Superior Técnico in Lisboa, Portugal.

Joan Eamer, an ecologist, was the head of biodiversity and ecosystem science with the Northern Conservation Division of Environment Canada in Whitehorse, Yukon, Canada, and is now the manager of the Polar Programme at UNEP/GRID–Arendal in Arendal, Norway. (joan.eamer@grida.no)

Paul Erickson studies cold war science in the Department of History of Science, University of Wisconsin, Madison, in Wisconsin, United States.

Christo Fabricius, a systems ecologist, is a professor of environmental science at Rhodes University in Grahamstown, Eastern Cape, South Africa. (C.Fabricius@ru.ac.za)

Madhav Gadgil, an ecologist, is a retired professor attached to the Centre for Ecological Sciences, Indian Institute of Science, in Bangalore, India.

Yogesh Gokhale, an ecologist, is an associate fellow with the Energy and Resources Institute in New Delhi, India, and was the coordinator of the Millennium Ecosystem Assessment's subglobal ecosystem assessment program in southern and central India. (yogeshg@teri.res.in)

Anil Gupta is the executive vice chairperson of National Innovation Foundation in India and has worked to promote the recognition of grassroots innovations and skills.

Jorge Ishizawa, a systems engineer, is the coordinator of PRATEC (Andean Project for Peasant Technologies), based in Lima, Peru. (j-ishizawa@speedy.com.pe)

Louis Lebel, an interdisciplinary researcher, is the director of the Unit for

Social and Environmental Research at Chiang Mai University in Chiang Mai, Thailand. (louis@sea-user.org)

Clark Miller researches international science policy and is an assistant professor in the La Follette School of Public Affairs and a senior fellow in the Center for World Affairs and the Global Economy at the University of Wisconsin, Madison, in Wisconsin, United States. (miller@lafollette.wisc.edu)

Henrique M. Pereira, an ecologist, is an assistant professor of environmental science in the Department of Civil Engineering and Architecture of the Instituto Superior Técnico in Lisboa, Portugal. (hpereira@stanfordalumni.org)

Rengalakshmi Raj is an agronomist working with the M. S. Swaminathan Research Foundation in Chennai, India. (rengalakshmi@mssrf.res.in)

Walter V. Reid, an ecologist and policy analyst, is a consulting professor with the Stanford Institute for the Environment in Stanford, California, United States, and was the director of the Millennium Ecosystem Assessment. He is currently the director of the Conservation and Science Program at the David and Lucile Packard Foundation. (wreid@packard.org)

Robert Scholes, a systems ecologist, is a fellow of the Council for Scientific and Industrial Research in Pretoria, South Africa.

Cristiana S. Seixas received her PhD in natural resources and environmental management and is now a postdoctoral fellow at the Natural History Museum, State University at Campinas, in Campinas, Sao Paulo, Brazil. (csseixas@hotmail.com)

Riya Sinha is the national coordinator of the National Innovation Foundation in India and is involved in scouting and documenting local knowledge.

Luís Vicente, a behavioral ecologist, is a professor of animal behavior in the Department of Animal Biology of the Faculty of Sciences (University of Lisbon) and a researcher at the Center of Environmental Biology of the Ministry of Science (Portugal) in Lisboa, Portugal.

Thomas J. Wilbanks, a geographer, is a corporate research fellow and leader of Global Change and Developing Country Programs at the Oak Ridge National Laboratory in Oak Ridge, Tennessee, United States. (wilbankstj@ornl.gov)

Monika Zurek, an environmental economist, is an economist with the Agricultural and Development Economics Division of the Food and Agriculture Organization of the United Nations in Rome, Italy.

MILLENNIUM ECOSYSTEM ASSESSMENT BOARD OF DIRECTORS

INDEX

A

B

D

E

ISLAND PRESS BOARD OF DIRECTORS